RADAR OBSERVER'S HANDBOOK

RADAR
OBSERVER'S HANDBOOK
for Merchant Navy Officers

BY

W. BURGER, M.Sc. (WALES)

Extra Master, F.N.I.

GLASGOW

BROWN, SON & FERGUSON, LTD., NAUTICAL PUBLISHERS

4-10 DARNLEY STREET

First Edition - *1957*
Fifth Edition - *1975*
Sixth Edition - *1978*
Reprinted - *1980*
Reprinted - *1981*
Seventh Edition - *1983*

ISBN 0 85174 314 5 (Sixth Edition)
ISBN 0 85174 443 5

© 1983 BROWN, SON & FERGUSON, LTD., GLASGOW, G41 2SD

Printed and Made in Great Britain

PREFACE

Owing to the introduction of micro-electronics, computer and digital techniques, many changes had to be made in the seventh edition of this book. The section on the Automatic Radar Plotting Aids (ARPA) has been completely revised. Instead of concentrating on particular makes of ARPAs, I have tried to make a general approach although particular and unique features of certain types of ARPAs are mentioned and emphasised. As a result of the development of ARPAs, several automatic plotting devices treated in previous editions have been left out. Descriptions of some radar systems previously classified as ARPAs, for example, the Marconi Predictor and the Kelvin Hughes Situation Display, have been left in. These, according to the IMO Specification, can no longer be designated as ARPAs, but, of course, they continue to be used on board ships.

In connection with these new developments I am much indebted to the staff of Liverpool Polytechnic, Department of Maritime Studies, the staff of South Glamorgan Institute of Higher Education, School of Maritime Studies, Mr. J. H. Beattie of Racal-Decca Marine Radar Limited, Captain J. P. O'Sullivan of Sperry Marine Systems and Mr. K. W. Collier, Radar Research Officer of UWIST. Other acknowledgements and a list of references are shown at the end of the book.

Captain A. G. Corbet of UWIST checked the complete script and made numerous suggestions for which I am extremely grateful.

Much weeding out has been done. For example, the section about Radar Modified Charts is no longer required.

Originally this book was written for the purpose of being a school text book for students on the Radar Observers Course. At present it is used a lot as a practical guide on board ships. I hope, in this edition I have combined the two requirements.

Cardiff
January, 1983

W. B.

v

CONTENTS

CONTENTS

CONTENTS

x   CONTENTS

INTRODUCTION

THE object of this book is to give the Radar Observer a broad, non-technical, but sufficient understanding of his equipment to enable him to make the most efficient use of its potentialities, yet to be aware of its limitations.

To the uninitiated, the picture on the screen of a harbour entrance, for example, presents a series of apparently confused echoes. He cannot compare it with the visual view, as the picture is displayed in two dimensions and no perspective is introduced. It also does not correspond with the chart or the map where projection methods are employed. Many features on the chart are not depicted on the radar screen and often echoes coming from objects which are not charted are seen on the radar display. Interpretation and identification of echoes on a radar screen is difficult, and training and experience in these is essential.

In this country the marine radar set has been designed primarily as an anti-collision aid and in this connection many pitfalls beset the operator.

The ordinary marine radar—there are some exceptions—does not reveal the shape of a ship unless she is very near. A vessel at a long or medium range just yields an intensified spot on the screen and the spot alone does not give an indication of the vessel's course. It is true that there are radar display presentations which give information about the *recent* heading of the target by showing past or predicting future tracks but the present heading is not shown.

Secondly, there is the difficulty connected with the relative motion display where the centre of the screen always represents own ship irrespective of whether she is moving or not. The radar eye—or scanner—moves with the ship, and distances and bearings are recorded with respect to this moving observation post. A stationary object ahead will possess an apparent speed towards own vessel, equal and opposite to the speed of the observer's ship and its echo on the screen will show a movement opposite to the direction represented by own ship. Echoes on the screen of moving objects are therefore subject to two movements, first the motion of own vessel reversed, and secondly, the true motion of the target. The direction and speed of such echoes on the display represent the resultant of the two velocities. It follows then that the motion of the echo does not indicate the true motion of the target and in case of any danger of collision with a ship whose echo is observed on the screen, no effective avoiding action can be taken unless the true heading and the speed of the other vessel have been determined without doubt. To do this, a simple plot has been devised, and it is essential that a Radar Observer is fully conversant with this procedure. No seaman, however practical he may be, can be fully aware of the movements around his own vessel if the process of plotting (manual or automatic) is omitted.

Thirdly, the all important question arises how collision dangers can be recognized and how the Collision Rules should be applied for vessels which are not sighted but detected on the radar display.

Lastly, and this is most important, the observer must be trained to keep the many limitations of radar for anti-collision purposes always in mind. To meet this requirement and if one wants to obtain the full benefit of radar, i.e. extra safety and perhaps saving of time (with which the owner is concerned), radar needs one or more trained persons to attend it. The attendant(s) cannot be employed for other duties. Collection of information must be efficient (trained observer), interpretation rapid (quick plotting) and the Master must be able to understand the interpretation and act correctly upon it.

The Master and owner should understand that if no extra or no trained personnel is available, a marine radar set is useless for saving time, as no information can be extracted which can warrant a course alteration without reduction of speed. Under these circumstances the radar set, however, can still act as an extra safety aid. Speed can be reduced in abundant time and the engines stopped even before the fog signal of an approaching vessel is heard.

The second function of a marine radar set is as a navigational aid. Here, also, limitations and pitfalls are plentiful. The set itself has characteristics which must be known and reckoned with. Under certain weather conditions, as for example heavy seas, rain or snow or under certain atmospheric conditions when radar visibility is below normal, response of targets can be greatly reduced. Shadow effects may confuse target interpretation. Then the target itself must be considered. It may not re-radiate strongly, nor scatter an appreciable amount of the energy back in the direction of the scanner. This makes the picture, especially at medium and long ranges, difficult to understand and the chance of errors can be high if a position by radar is obtained. In this respect, radar falls behind the Decca Navigator, which acts solely as a navigational aid.

It was with all these problems of radar usage in mind that, in 1948, Nautical Colleges in the United Kingdom were requested to provide facilities for the training of Radar Observers. Until recently, courses of three weeks' duration were held in all the main ports in Great Britain. Many seafaring countries followed this practice, but nowadays the training for radar usage is often integrated within the nautical syllabus for deck officers.

In 1960 Radar Simulator Courses were started, enabling Shipmasters and deck officers to practise ship manoeuvring and collision avoidance on radar information. These five day courses offer *realistic* experience in dealing with the various problems mentioned in this introduction.

In 1982 Radar Simulator training became an integrated part of the certificate course. At about the same time short courses in ARPA (Automatic Radar Plotting Aids) training were instituted.

CHAPTER 1

FUNDAMENTAL PRINCIPLES OF RADAR

Pulse Transmission and Reception

Choice of Wavelength

The word RADAR has been derived from the phrase " Radio Detection and Ranging ". Its principle is simple. A short burst of electro-magnetic energy is thrown out, bounces off a target and then returns. If we are able to time the interval between the transmission of the pulse and the return of the echo, and if the velocity of propagation is known, then, by multiplying half the time interval by the velocity, the distance to the object causing the echo can be found. The same principle is, in fact, employed in the Echo Sounder. Here, use is made of sound waves but in radar, electromagnetic waves are used. The electrostatic field is usually situated in the horizontal plane and moves outwards in cycles. The length of one cycle is called the *wavelength.*

Marine radar sets generally use a wavelength of 3 cm. (X band) but sometimes a wavelength of 10 cm. (S band) is employed. See also Appendix VI.

The transmitted burst must be short. If a Master of a vessel wants to determine the distance of a cliff by sounding a blast on the whistle and noting how long a time it takes for the echo to come back, he must give a short blast. If the blast is too long, the echo will return while he is still transmitting and he will be unable to hear it. The same is true of radar. In most radar sets, the receiver is blocked during the transmission period and the sooner the transmission is over, i.e. the shorter the pulse, the earlier the set is ready for reception. Hence there is a better minimum range for a radar set with a short pulse than with a long pulse.

Pulse-length

The *pulse* is produced by the transmitter and its duration which is known as the *pulse-length, is expressed in micro-seconds* (millionths of a second). The *radiated pulse* or *radio impulse* has a length depending on the duration of the pulse and *this length can be given*

1

in metres as it is known that a radio wave travels 300 metres in 1 micro-second. The reflected wave is called the *echo* or *reflected signal*; after the reflected energy has been collected by the aerial or scanner, the word *signal* is used to denote it.

Pulse-repetition Frequency

The number of pulses transmitted per second is called the *pulse-repetition frequency* or p.r.f. The p.r.f. is generally between 500 and 4000 pulses per second. Most radar sets have a short pulse and comparatively large p.r.f. for the short ranges and a longer pulse combined with a smaller p.r.f. for the longer ranges. To obtain a good minimum range, it is essential to have a short pulse for the shortest range scale. In order to collect enough echo strength, a great number of these pulses must be sent out per second. In other words, *the output power is a function of the p.r.f. and the pulse-length.*

For longer ranges the minimum distance is not important. The pulse can be made more powerful by extending its length and then the number of pulses per second can be reduced.

Transmission and Reception

The burst of electromagnetic energy is formed by the transmitter and directional intensity is afforded by means of the aerial, sometimes called the scanner.

The earliest form of aerial for cm. waves, and still frequently used, is the metal parabolic reflector which has a horn or flare situated at its focal point. The idea here is that if an energy source is brought to the focus of a parabolic reflector all rays emerging from the reflector will be parallel to the axis. This method works very well for light but does not hold for cm. waves because the wavelength is not negligible compared with the dimensions of the mirror, and diffraction (the property of the wave to bend round obstacles) plays an important role. Nevertheless, as will be seen *Appendix* 1.1), the parabolic type of reflector is of great use.

At present aerials are in use consisting of a straight piece of waveguide with slots cut in the broad or narrow face. The broad or narrow side, respectively, is placed inside a horn which runs along the length of the waveguide. The vertical cross-section of the horn determines the extent of the vertical beam-width. The complete slotted waveguide aerial is placed inside a casing of perspex or fibreglass in order to protect it against weather, soot or dust. See Fig. 1.1.

In shipborne radar sets, the aerial is driven round at a uniform speed by the aerial motor. Motor and aerial together form the scanner unit. The wave pulse is radiated from the horn onto the

reflector and is then re-radiated outwards as a beam of electromagnetic energy. The aeriel not only directs the electromagnetic
energy outwards, it also collects the returning energy in its focal
point—the horn—in a similar way as a magnifying glass is able to
bundle sunrays and bring them together at one spot.

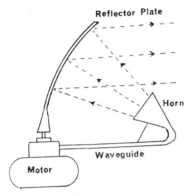

TILTED PARABOLIC SCANNER (side elevation)

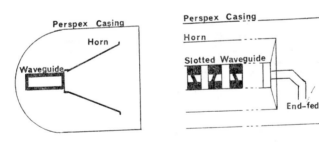

SLOTTED WAVEGUIDE AERIAL (side & front elevation)

FIG. 1.1   TYPES OF SCANNER

The velocity of the pulses of the electromagnetic waves is so great
with respect to the rate of rotation of the aerial that we can assume
that the aerial direction for transmission and reception of each pulse
is the same.

From the aerial the echo is despatched to the receiver where it is
amplified and thence fed to the display unit.

B

Beam-width and Radiation Diagrams

The radar beam can be compared with the beam of a searchlight which rotates and illuminates the vicinity for a short moment while it sweeps over an object. An image of the radar beam, having the same beam-width can be conceived on the display, rotating in unison with the scanner. It would act as a paintbrush, painting the echoes on the screen when it receives the information from the scanner. To obtain an accurate and detailed picture of the surroundings, i.e. good bearing accuracy and discrimination, a fine paintbrush is required, in other words a small angular horizontal beam-width for a radar set is essential. Another point is that undue spreading of the beam will mean a great loss of energy.

A more detailed description of how the horizontal beam is produced is given in *Appendix* 1.1, page 13.

The vertical beam-width should be wide in order to allow for rolling and in order to help pick up low-lying objects at a minimum distance of 50 yards (45·7 metres).

Polar Diagram

In Fig. 1.2 the power values of the radiation from a scanner in a particular plane for the same *range* are plotted against the direction. The aerial is taken as the centre and from this point or pole the power values are laid off in their respective directions. Such a diagram is known as a *polar diagram* and it is seen that the greatest energy is concentrated along the centre of the beam while at angles either side of the centre the power decreases rapidly. The energy along the bearings of *AB* and *AC* (equal to half *AO*) is exactly half the energy along the centre of the beam. In practice, the *angular beam-width is defined as the angle subtended at the pole, A, by the half-power points or angle BAC.*

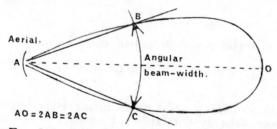

FIG. 1.2   POLAR DIAGRAM AND BEAM-WIDTH

The various polar diagrams show, that in addition to the *main beam* or *lobe*, there are smaller beams, known as *side lobes*. See Figs.

1.3 and 1.7. It is impossible to make a scanner aerial without side lobes (see *Appendix* 1.1) although the design of an aerial affects the magnitude of the side lobes.

The horizontal beam-width for marine radar sets working in the X-band varies between ·6 and 2 degrees, the vertical beam-width between half-power points varies between 15 and 30 degrees.

What is particularly important is to understand that a polar diagram is similar for transmission *as well as for reception*. When the signal returns its electromagnetic field is widely distributed. The scanner, because of its directional properties, will, however, only pick up the strongest re-radiation along its centre-line and between approximately the half-power points, say, for example, for about half a degree on either side of the centre-line in the horizontal plane.

Radiation Coverage Diagram

The diagrams in Fig. 1.3 do not represent polar diagrams, which are purely graphical pictures showing the distribution of energy against azimuth in the horizontal plane and against altitude in the vertical plane for a particular aerial. Fig. 1.3, in fact, shows diagrams which give a more factual presentation of the distribution of radiation strength *as the radio impulse travels along*. Sometimes these diagrams are referred to as *radiation coverage diagrams* and they show the limits of useful radiation. The range varies and the contours are drawn for *a given field strength*.

The radiation pattern in the vertical plane is not so simple as in the horizontal plane. It would be something like the pattern in the horizontal plane if the aerial was tilted upwards so the lobe did not touch the sea surface. This, however, is not the case with a marine radar set and due to reflection against the sea surface, the lobe pattern is modified and distorted. The energy component is now the vector sum of the directly transmitted energy *and* the reflected energy. If the direct wave and the indirect wave are in phase or in step then the resultant energy will be strong and there is a maximum. If, however, the two waves are in anti-phase or out of step then the resultant energy will be weak and there is a minimum. Hence the radiation pattern is broken up. Note that along the lines of maximum radiation strength the lobes have extended in range. They have separated in height and this separation will increase with range; targets some distance away may fall between the lobes and cannot be picked up until they move into a lobe. Rolling has no effect on this particular lobe pattern as it is a function of the earth's surface.

It should be understood that the diagrams in Fig. 1.3 are out of proportion. The lobes may extend many miles depending on the power of the transmitter pulse. The lobe structure in the vertical plane is much finer than depicted in Fig. 1.3; this will be discussed in a moment.

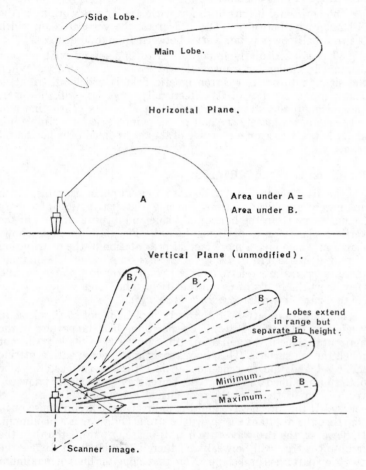

Horizontal Plane.

Vertical Plane (unmodified).

Vertical Plane (modified by surface reflection).

FIG. 1.3   RADIATION DIAGRAMS IN THE HORIZONTAL AND VERTICAL PLANE

Though a coverage diagram for *reception* has the same shape as a coverage diagram for transmission (directional properties are the same), its contours show the limits of detection and the size of the area within the contours depends on the echoing strength of a particular target. Reduction of receiver gain, for example, will result in a shrinkage of the pattern.

Choice of Wavelength

The fineness or coarseness of the lobe structure of the radiation pattern depends on the wavelength. It can be shown that the distance between the minima or maxima along the base-line (scanner to scanner image) equals half a wavelength. For an aerial 15 metres high, for example, using the 3 cm. wavelength the number of lobes developed between the scanner and the sea will amount to 1000; the corresponding number for a 10 cm. installation will be 300. It can also be seen from the diagram that the distance between the lobes will increase with the range from the vessel. It is apparent now from this consideration that the shorter wavelength yields the better sea coverage.

The vertical radiation diagrams in Fig. 1.3 also explain why a small reflecting surface, for example a radar reflector, may give rise to fluctuating echo strength when the ship travels away or towards it. The object will pass through a series of maximum and minimum radiation.

Note especially in Fig. 1.3 that the lower lobe appears to rise only if the earth is considered to be flat. If the diagram were redrawn for a curved earth, it would be seen that the lobes turn gently towards the earth's surface.

Another reason why a short wavelength is used in a shipborne radar set is because there is a relationship between the size of the scanner reflector and the beam-width (study *Appendix* 1.1 and the related formula). The larger the width of the scanner the smaller is the angular beam-width for the same wavelength. A fine radar beam with little spread can only be obtained if the scanner is large in relation to the wavelength of the electromagnetic wave. A narrow horizontal beam-width is required but the horizontal dimensions of a scanner cannot be made too large as this would involve a big motor and high power to drive the scanner round. Hence for practical reasons the scanner is restricted in size and this means that the wavelength adopted should be small.

Other advantages of the cm. wave are, that, because its wavelength is relatively small, this involves a dense electromagnetic field with corresponding power and a *short* pulse-length can be employed. Also, its bending effect around the earth (diffraction) is less than that of waves which have a longer wavelength. The latter property keeps the energy close to the surface and enables it to pick up objects at a good minimum range.

There also exists a relationship between the re-radiation factor and absorption characteristics of a wave and its wavelength. The energy of the radar wave is dissipated in three ways:

(i) Absorption by atmospheric gases (oxygen and water vapour);

(ii) Absorption by water droplets or ice particles;

(iii) Scattering, i.e. reflection.

The reduction of intensity of the wave experienced along its path is known as *attenuation*.

In general, gases act only as absorbers. At wavelengths greater than a few centimetres, attenuation by oxygen and water vapour is very small and can be neglected except in cases where long ranges are concerned. However, at wavelengths below 3 cm. the attenuation due to water vapour begins to be significant and increases steadily with decrease of wavelengths (except for a few " humps " at about 1·25 and 0·15 cm. when the electromagnetic wave reacts with the molecules, which have a permanent electric dipole moment, causing them to oscillate and rotate).

On the other hand, cloud and rain attenuation at wavelengths *below* 10 *cm.* has to be taken into account and can have serious effects for waves in the X band. See Fig. 7 . 10.

The mm. wave (Q band) combined with an ordinary sized scanner would, of course, yield an extremely fine horizontal beam-width (·3 degree) but due to absorption the range is limited to only five miles.

All these arguments narrow down a suitable choice for the wavelength of radar. The radar beam must not weaken too much due to absorption, yet still provide enough re-radiation from any object important for marine navigation. The choice for marine radar sets fell either on the 3 cm. (X band) or 10 cm. (S band) installation.

One of the ideal qualities for a merchant ship radar system would be to have three radar sets, each employing a different wavelength, that is: one using the 10 cm. wavelength for long range detection and for the detection of targets in rain areas, one using the 3 cm. wavelength for intermediate range work and one utilizing the 8 mm. wavelength for close range work where a highly defined picture is desirable.

All this is linked up with the fundamental principles of pulse transmission. The returning echo signal, of course, is very weak. Only a very small part of the energy ejected from the scanner will bounce off an object and a very tiny fraction of the returning energy will find the scanner again. The greater loss occurs on the return path, the outward path being unidirectional, and the return path " hemispherical ". Hence the need for boosting up the echo signal is unavoidable. This is done in the receiver before the echo pulse is sent to the display unit.

Principles of Range Measurement

How is the time interval between the transmission of the pulse and the reception of the signal measured? Well, it is done much in the same way as we would measure the time interval between the blast on the whistle and the returning echo from a cliff, for example. At the commencement of the blast we press a knob and set the hand of a stopwatch into motion from zero. When we hear the echo we press again and take a reading. If, at a later stage, we want to repeat the procedure, the hand of the stopwatch is re-set to zero.

Owing to the fact that radio waves travel so rapidly, the time interval between transmission and reception is extremely short and so an electronic timing device must be used. The echo pulse, when it returns, is fed to a *Cathode Ray Tube* or CRT whose screen can be regarded as the face of the electronic stopwatch.

The CRT is an electronic valve and consists of:

(*a*) The gun, comprising the cathode, grid and one or more pierced anodes, which fires electrons at

(*b*) the screen, which is transparent and acts as the final anode. The echoes are displayed on the screen (Fig. 1.4).

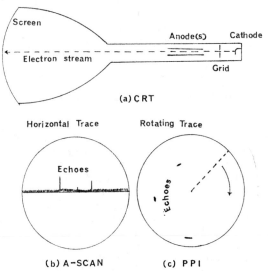

FIG. 1.4 CATHODE-RAY TUBE AND SCREEN

Negatively charged particles, called electrons, are emitted by the cathode. The electron stream bombards the screen at a certain point. A special chemical coated on the screen shows up this point which is called the *spot*. The intensity of the electron stream (and of the spot) is called the *brilliance* or *brightness*. The brilliance can be adjusted and controlled by applying an electrical pulse to either the grid or the cathode. The size of the spot can be regulated by a focus arrangement. Acceleration of the electrons takes place under the influence of the cylindrical anode(s).

The spot performs the same function as the hand of a clockwork stopwatch. It can move over the screen with a uniform velocity. thereby describing a *trace*, sometimes known as *sweep* or *time-base*.

Two types of CRT can be utilized:

(*a*) *A-Scan* or " Short Persistence Tube ". The trace is fixed and horizontal. Echoes are only recorded on one particular bearing, i.e. the direction in which the scanner is pointing at *that* moment. The rotation of the scanner has to be stopped if a range and a bearing is taken. Echoes appear as blips and die away quickly (Fig. 1.4 (*b*)).

(*b*) *Plan Position Indicator* or PPI. This tube is also known as a " Long Persistence Tube ". The trace is moved round in unison with the rotation of the scanner and echoes previously recorded are retained during a period of at least one scanner revolution. The observer is presented with a picture showing a plan representation of many of the objects surrounding the ship which he would be able to see in daylight in clear weather from the scanner position. The echoes appear as intensified spots (Fig. 1.4 (*c*)).

When facing the screen, the trace on the A-Scan is diametric from left to right, the trace on the PPI is from the centre to the circumference and rotating about the point representing the scanner position.

At the *same* instant as the radio impulse, which is generated by the transmitter, leaves the scanner, the spot on the CRT is deflected across the screen in a straight line at a uniform speed. When the signal (echo) returns and is detected by the receiver a blip (on A-Scan) or an intensified spot (on PPI) is formed on the screen (this compares with the moment when the reading of the hand of stop-watch is taken when the echo is heard). The distance the spot has travelled over the face of the tube meanwhile is an indication of the total distance the pulse has travelled. The latter distance, of course, is twice the distance from the ship to the object which has reflected the pulse and caused the echo. If, for example, an object at 10 miles distance gives rise to an echo appearing at the edge of the screen, then an object at 5 miles distance will yield an echo halfway between the starting point of the spot and the echo of the object at 10 miles. Hence, if we know the distance of a target causing an echo at the edge of the screen, then the distance of other objects giving echoes can be deduced and the screen can be calibrated in nautical miles.

The longer it takes for the spot to travel across the face of the tube, the greater is the distance the pulse can travel to the object and back again, in time to be recorded.

Suppose that we want the diameter of the A-Scan or the radius of the PPI to represent 10 miles. This means that the echo of an object at a distance of 10 miles must appear at the edge of the screen. Hence in the time taken by the pulse to travel a total distance of 2×10, i.e. 20 miles, the spot must have covered the distance from its starting point to the circumference of the screen. The velocity of radar waves is 3×10^{10} cm. per second or 186,000 statute miles per second (162,000 nautical miles per second) which is equivalent to 300 metres per millionth of a second. Generally in radar theory, we take the velocity of the radar waves to be 300 metres per microsecond (one millionth of a second). There are 1852 metres in a nautical mile, and therefore the time taken for the pulse to travel 20 miles is $20 \times \dfrac{1852}{300}$ micro-seconds, i.e. 123 micro-seconds. The same time is required for the spot to complete the trace. It follows from this consideration that the spot has to travel much faster over the face of the tube for short range scales than for long range scales.

When the spot has reached the edge of the display, it has to be re-set to its starting point, in the same way as our stopwatch needs re-setting if we want to make another range determination. After the spot has completed the trace, it returns to its starting point at an extremely high speed. During this " fly-back " the brilliance is suppressed and the tube blacked-out so that the return journey is not made visible on the screen. Back at its starting point, the spot will remain at rest until the next transmission takes place when it will start moving again. At the same time the brilliance is increased and the spot will become visible once again. During the rest period of the spot no echoes, of course, are recorded.

Principles of Bearing Measurement

If the radar set is equipped with an A-Scan only, as was the case in the earlier radar sets, it is best to rotate the scanner slowly by hand. When an echo is recorded, stop the scanner. The true bearing of the target can be established by noting the direction of the scanner. A graduated dial should be provided so that the relative bearing can be read off directly. The procedure resembles the taking of a D.F. bearing with a rotating loop aerial.

The PPI yields continuous bearing information as the echoes are retained for a short while. When the scanner points directly ahead, the brilliance is increased for a short moment and the trace is painted on the screen as a bright datum line representing the fore and aft line of the ship (heading flash or heading marker). The rotation of the trace is synchronized with the scanner rotation. Hence the direction of the trace when an echo is recorded, while comparing it with the direction of the heading flash, yields information about the direction of the scanner with respect to the fore and aft line of the ship (relative bearing). The direction of the scanner may be taken to be the same for transmission as for reception, because it turns a negligible amount between transmission and reception, and so represents the direction of the target. This is so because the velocity of electromagnetic energy is so extremely high.

Wavelength and Frequency

We have mentioned already that the electromagnetic field thrown out by the scanner moves outwards in cycles, in the same way as water waves move outwards from a centre of disturbance. The characteristic of the wave is the *wavelength* which is the length of one cycle. In many radar manuals the word " wavelength " is not mentioned however, instead the wave is characterised by its frequency.

The *frequency* of a wave is defined as the number of cycles which pass a stationary observer per second.

If f cycles are passing the observer in one second, then the total length of the wave train is $f \times \lambda$, where λ denotes the wavelength, or the length of one cycle, and f is the frequency. But the distance covered per second is the velocity of the wave and this is for an electromagnetic wave 3×10^{10} cm. per second. Hence we obtain the formula:

$$Velocity = Frequency \times Wavelength$$

For the 3 cm. wavelength, for example, the frequency is $\dfrac{3 \times 10^{10}}{3}$

i.e. 10^{10} cycles per second or 10,000 megahertz where a megahertz is equivalent to one million cycles per second.

Generally the transmission in this wave band (X Band) is done between 9300 and 9500 MHz corresponding to 3·2 cm. wavelength.

In the same way it can be calculated that for 10 cm. transmission a frequency of 3000 MHz is required.

Generally the transmission in this wave band (S band) is done between 3000 and 3246 MHz, the latter frequency corresponding to 9·2 cm. wavelength.

Appendix 1.1

To understand how the radar beam is produced, a study of Figs. 1.5 and 1.6 is required.

Fig. 1.5 shows the horizontal cross-section through a parabolic type of scanner. The parabola is the locus of all points for which the shortest distance to a certain fixed line (directrix) equals the distance to a fixed point (focus) so that for point A $AD = AF$ and for point B $BC = BF$. The radar waves emerge from F as a series of expanding spheres and due to electromagnetic induction the points on the reflector plate act as an infinite number of secondary sources and

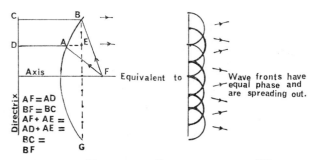

FIG. 1.5   EQUIPHASE DISTRIBUTION OF WAVES

re-radiate the beam (with a 180° change in phase). All the points on the reflector plate become the source of a wavelet that travels out in an expanding sphere and the envelope of these spheres form the new wavefront. Fig. 1.5 shows that $AF + AE = BF$, in other words, in the vertical plane through BG, perpendicular to the axis, all the re-radiated waves have equal phase. This means then that the parabolic reflector can be used to obtain an equiphase distribution across a plane aperture.

The total effect in the field is the algebraic sum of the amplitudes which each wave would cause separately and this is illustrated in Fig. 1.6. The re-radiated waves have the same frequency and have to travel different distances to reach a point in the field, in other words, they have different phases at that point. However, for a certain distance away from the scanner, the points lying on the bisector of the plane AB (Fig. 1.6) are for all practical purposes equidistant to the points across the aperture AB and it is along this line where all the waves come into phase, in fact this line forms the centre-line and indicates the direction of the strongest radiation or *main lobe*.

In order to calculate the extent of the main lobe, consider a point D in the field (Fig. 1.6) so that $BD - AD$ equals a wavelength λ. Then the path lengths of A to D and C to D will differ by $\frac{1}{2}\lambda$. Hence

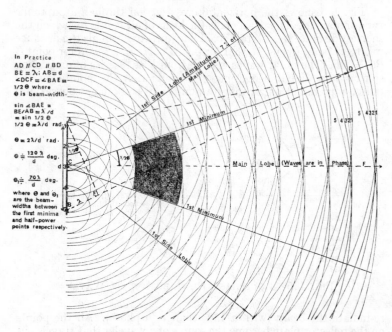

FIG. 1.6   FORMATION OF MAIN LOBE AND SIDE LOBES

the waves re-radiated from A and from C will differ 180° in phase at D and their intensities will cancel out at D. The point adjacent to A (below A on the diagram) will emit a secondary wave which at D will cancel out the effect of the wave generated by the point adjacent to C (just below C on the diagram). Hence we can group together elements of the aperture AB, separated by half the scanner width d, which contribute equal but opposite amounts of intensity at D.

On the assumption that AD, CD and BD are parallel, it will be seen from triangle ABE that $\sin \frac{1}{2}\theta = \lambda/d$.

Therefore the *horizontal beam-width* $\theta \simeq \dfrac{120\lambda}{d}$ *degrees* between the first two minima.

This formula gives the beam-width to zero power points, but in practice the beam-width stated refers to the width between the points where the power has fallen to one-half of the power along the centre-line of the main lobe. It can be proved that $\theta_1 = \dfrac{70\lambda}{d}$ *degrees*, where θ_1 represents the beam-width between half-power points.

For example, using a six-foot (1·83 metres) scanner in the X band,

$$\theta_1 = \frac{70 \times 3\cdot2}{183} = 1\cdot2°.$$

A second minimum, each side of the main beam, takes place where $\sin \tfrac{1}{2}\theta_2 = \dfrac{2\lambda}{d}$, and between the first and second minima the resulting field has a certain intensity, but far below the intensity of the main beam. This part of the field is known as a *side lobe*.

The overall radiation of a parabolic scanner therefore consists of a *strong main lobe and a number of side lobes of decreasing amplitude*. In Fig. 1.7 a typical radiation pattern is plotted showing the intensity in decibels (see page 357, Appendix) against the beam-width in degrees. The conventional beam-width is the beam-width between half-power points or — 3 db points.

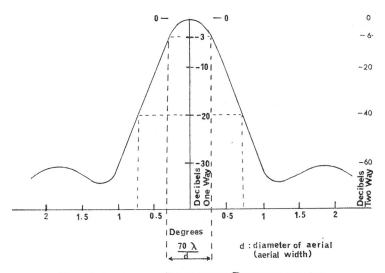

FIG. 1.7   AERIAL RADIATION PATTERN IN THE
HORIZONTAL PLANE

One great advantage of the parabolic type of scanner is that the strongest re-radiation takes place from points near the axis. The re-radiation is net uniform and decreases towards the edges of the reflector plate. This reduces the intensity of the side lobes. In fact, comparing it with a uniformly radiating source, the intensity of the first side lobe is reduced from 22% to 7% of that of the main lobe.

It can also be seen from Fig. 1.6 that the wavefront near the scanner is nearly straight and it is only at a distance away from the scanner where the actual main lobe begins to take shape. This means that within a certain distance (the *Raleigh distance*) the radiated field and the returning signal (the azimuthal width of the echo) is a function of the scanner width and not of the horizontal beam-width, as will be seen at a later stage. See Fig. 1.8. This distance varies directly as the square of the scanner width and inversely as the wavelength used. For example, the Raleigh distance for a six-foot (1·83 m.) scanner radiating a 3·2 cm. wavelength, amounts to just over one cable. For a 12-foot (3·66 m.) scanner this distance would be 4½ cable, if the same wavelength is employed.

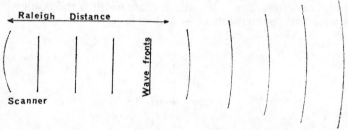

FIG. 1.8 THE RALEIGH DISTANCE

CHAPTER 2

COMPONENTS AND THEIR SITING

Function of Components

The main units of a marine radar set are:
>Transmitter;
>Pedestal Unit, including the scanner;
>Mixer and Receiver;
>Display Unit;
>Alternator or Inverter.

The *transmitter* comprises a Trigger Circuit and Modulator (which determine the pulse-repetition frequency or p.r.f.) plus the Magnetron where the cm. pulse is produced. From there the electro-magnetic energy is guided along the inside of a hollow tube, called the "waveguide" to a horn or to a series of slots positioned in the *scanner unit*. The horn is placed at the focus of a parabolic plate and the radiation is released after reflection by this plate. Alternatively the scanner may be of the slotted waveguide type. In this type an end section of the guide about 1·8 m. (or 3·6 m.) long, with slots cut in the wall of the guide, is turned in a horizontal plane. The slots are sometimes cut obliquely, but vertically polarized aerials are in use with horizontal slots cut in the broad face of the waveguide at both 3 and 10 cm. Also, polarizing is sometimes done in the window of slotted waveguide aerials or sometimes by blocks alongside the slots. The side of the waveguide with the slots is placed inside a horn which runs along the length of the waveguide. It is found that the slots radiate the energy. The horizontal length of the scanner and the vertical cross-section of the horn determine respectively the extent of the horizontal and vertical beam-width. The complete slotted waveguide aerial is placed in a casing of perspex or fibreglass in order toprotect it against the weather, soot or dust (Fig. 1.1).

In the majority of radar sets the same aerial unit emits the pulse and collects the reflected signal. This means that transmitter *and* receiver are connected to a common waveguide and during transmission the receiver is automatically paralysed. The unit responsible for this action is the T/R (Transmit/Receive) Cell which acts like a shutter when transmission takes place, but otherwise is open.

The waveguide must not be too long, otherwise the pulse and signal attenuate too much. Dirt and moisture also decrease the signal strength, and in some sets, drying units are supplied.

The echo pulse, collected by the scanner, enters the *receiver* via a *crystal mixer* where another frequency is injected. The latter frequency is generated by a local oscillator (a klystron or gun diode, see Appendix III) and differs only slightly, relatively, from the frequency of the incoming signal. The resulting output is the difference between the two frequencies, sometimes called the "beat frequency". The purpose of this mixing or "heterodyning" is to lower the frequency in order to simplify the amplification of the signal in the receiver section. The wave guide technique is used in that part of the set where the energy of the very high frequency (r.f. or radio frequency) cm. wave pulse is conveyed. After the signal has gone through the mixer, other types of feeders are used instead of the waveguide.

After the mixing the resulting frequency becomes known as the Intermediate Frequency or I.F. It has retained its pulse character, and is now amplified and smoothed out (demodulation). The gain and anti-clutter control circuits form part of this section.

In the *display unit* further amplification takes place and range ring and range marker circuits are added. The final output signal is then fed into the Cathode-Ray Tube or CRT.

Other connections to the CRT are :—

   (i) Time-base circuit so that the range-scale of the display can be changed.
  (ii) Scanner rotation signal so that the bearing of an object can be determined on the Plan Position Indicator (PPI).
 (iii) Brightening circuit producing the spot sweeping the PPI so that echoes can only become visible during the time-base period.
 (iv) Heading marker pulse showing the fore and aft direction of the vessel, thus giving a reference line for picture alignment and bearing measurement.

A block diagram is shown in Fig. 2.1. Note that the range marker circuits, the time-base and brightening circuits are all synchronized in their action by the trigger circuit; they all start their operation at the same instant as transmission begins.

A more detailed description of the function of the main units, for interested readers, is given in Appendix III.

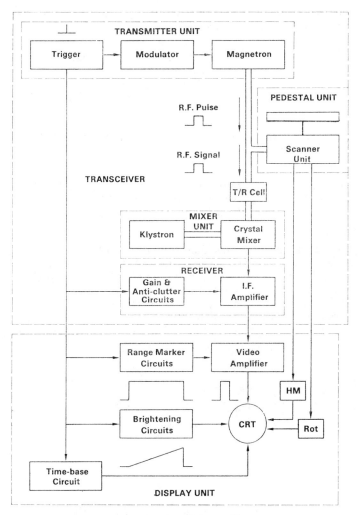

FIG. 2.1   RADAR BLOCK DIAGRAM

Siting of Components

Transmitter

The transmitter is sometimes found in the chartroom, sometimes on the monkey island in a protective container. As in the case of all radar units, it should be marked with the minimum distance at which it should be mounted from the standard and the steering magnetic compasses. Electric interference to other radio equipment should be suppressed as much as possible.

In some sets, part of the transmitter (and the mixing crystal) are contained in the pedestal unit. Such sets are economic to run and little loss of power occurs in the waveguide because it can be kept short; also positioning the scanner is made simpler as no long lengths of waveguide are needed from the pedestal unit. However, the installation is not looked upon with favour by the maintenance engineer when the mentioned part of the transmitter fails or the crystal burns out.

Receiver

Generally the transmitter and the receiver are housed in the same unit, which is then known as the *transceiver*, although some sections of the receiver are contained in the display unit.

Scanner

The main factors in deciding the position of the scanner is the avoidance of blind or shadow sectors and the avoidance of indirect echoes. Indirect echoes are discussed in Chapter 8. The positioning of the scanner in the horizontal and vertical planes is necessarily a compromise on merchant ships which generally have many obstructions in the form of masts, funnels and samson posts (see Fig. 2.2). Generally a centre-line siting is preferred because a clear

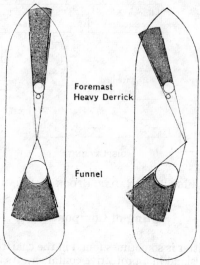

Foremast
Heavy Derrick

Funnel

FIG. 2.2 SITING OF AERIAL AND SHADOW SECTORS

radar view to *starboard and to port* is highly desirable in relation to meeting targets end-on, important for overtaking a vessel, and for vessels passing each other on opposite courses in narrow channels.

Another consideration is that for navigation in narrow waters the heading marker on the display should represent the fore and aft *centre*-line so that no parallax is introduced. This means that the scanner should take a view from that line.

On the other hand, in the event of ship obstructions forward of the scanner, the master might put greater emphasis on a clear view right ahead in narrow waters (where zig-zagging is not always possible) and thus might prefer port side siting.

It is advisable to have the scanner in a horizontal plane below the cross-trees so as not to create a wide shadow sector dead ahead or *downwards*. If this cannot be arranged, its position should be well above the cross-trees to allow for changes in trim.

In nearly every case it is now possible to site the aerial so as not to have shadow sectors forward and foremast installations are becoming more important. In the latter cases it should be remembered that, when observing objects at short ranges, compass bearings on the bridge might differ from radar bearings taken of the same object.

A high scanner increases the distance to the radar horizon, but on merchant ships the navigator is generally more concerned with medium and short ranges than with long ranges. High scanners also afford more opportunity for reflection against the crests of the waves and more sea clutter will be recorded. Sometimes, however, a high scanner is needed and on ice-breakers, for example, a scanner forward of the mast is to be recommended. In such cases, part of the transmitter and the mixing crystal should be included in the pedestal unit so that no long waveguides with resulting loss in power are required.

On ferry-boats equipped with bow rudders, particular attention should be paid to shadow sectors and often they are equipped with two scanners, one forward of and one abaft the funnel, with a change-over switch in the chartroom.

If the scanner unit is too low, then a shadow sector can be caused by the bow and the minimum detection range forward is increased.

Loose halliards and aerials should not be too close to the scanner unit, otherwise they may get entangled with the rotating aerial. If this happens the fuse may blow but in many cases the armature windings of the aerial motor will burn out. It is obvious that safety of crew working nearby, must also be considered.

In some cases, the aerial is contained in a weather-tight container made of perspex or fibreglass. The power of the driving motor and the cost of the components can, in those cases, be reduced.

Display

The siting is generally in the wheelhouse and it is an advantage for the navigator that the PPI of the radar is so located, that on raising his eyes, he can immediately get a visual lookout forward of the beam. For anti-collision purposes in poor visibility on the other hand, it is strongly advisable *not* to combine visual and radar lookout if there are a sufficient number of officers on the bridge. The reason for this will be discussed later in this book.

The wheelhouse siting is, unless one plots on the screen itself, not so suitable for plotting carried out during the night. Several ships nowadays have a second display in a special plotting room or in the chartroom. The latter position also makes it possible for easy comparison with the chart. Modern ships are equipped with combined wheelhouse and chartroom which makes radar usage much simpler.

If a second display is decided upon, one should order each display with its own time-base unit. Not only do displays working from the same time-base unit interfere with each other, but there is the great advantage that different range scales can be selected for each display. One display can have the short range scale in use while the other one can give warning of targets at medium range.

Many ships are equipped with a " True Motion " unit. One display can then be used to show the relative motion of targets and give warning of impending collision danger, while the other display can be adjusted to indicate the true motion.

Radar displays of the true and relative motion types, speed and distance indicator, clock etc. are nowadays arranged and grouped in console mountings.

A viewing hood or curtains round the display should be provided in order to make daylight observation possible. Some modern sets produce a bright picture or employ photography which enables daylight viewing to be carried out without the aid of screens or hoods.

Alternator or Inverter

Either an alternator or an inverter is used for converting the ship's main electrical supply into a form suitable for feeding the radar apparatus. They are generally very noisy and they should be placed so that they neither disturb men sleeping on watch-below nor prevent fog signals being audible on the bridge.

Interswitching

Interswitching of two radar installations is shown in Fig. 2.3. Its purpose is :—

(*a*) to have the radar picture available at all times, even if one or more components break down or are being serviced,

(*b*) to have the use of two radar displays at one time, one display acting as slave to the other, and

(*c*) to have flexibility in display presentation. For example, one radar could work in the X-band and the other in the S-band, or one display on Relative Motion and the other on True Motion (see section Radar Display Presentation, p. 145).

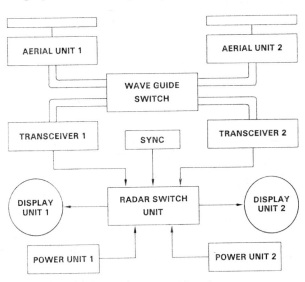

FIG. 2.3   INTERSWITCHING OF RADARS

Three remarks must be made. Firstly a synchronizing unit is required thus making both radars transmit at the same instant. If the p.r.f.s are not the same, special interference suppressor units (see later in this book) have to be fitted. Secondly, with S and X-band systems, it is of course impossible to interswitch r.f. sections (aerial and part of the transceiving units). Thirdly, if the master display is operating on a short range scale with a short transmitter pulse then the slave display will also receive signals from the same short pulse. If it is switched to a medium or longer range scale, the signals will therefore be weaker than when a long pulse is used. Conversely,

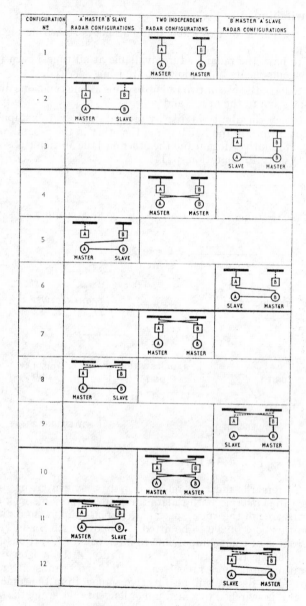

FIG. 2.4 INTERSWITCHING COMBINATIONS

if the master operates on the longer or medium range scale with long pulse, the slave when switched to a shorter range scale, will suffer to some extent through lack of range discrimination (see Chapter 3). In the latter case, however, some degree of compensation can be provided by operating the differentiator (Chapter 4). With the two radar installations and the interswitching systems shown in Fig. 2.4, either of the two displays can be used as the master, which means that if long range results are required at the slave position, this could be switched to master display and thereby use the long pulse.

The flexibility of two X-band systems being interswitched, is illustrated in Fig. 2.4 where 12 combinations are portrayed, thus providing a very high reliability factor.

Note.—More details about siting precautions are given in Notice No. M.983 : *The Use of, and Training in, Radar and Electronic Aids to Navigation.* Section 2.2. Published by the Department of Trade.

CHAPTER 3

CHARACTERISTICS OF SET

The Department of Trade published a Marine Radar Performance Specification and extracts from the latest one (1968) are provided with this book as an Appendix. Since 31st July, 1968, new types of Shipborne Navigational Radar Sets have been tested for compliance with this specification. If the equipment complies with the specification, and if an undertaking is received from the manufacturers that subsequent models will be identical to the specimen tested, the Department of Trade will issue a Certificate of Type-Testing. The validity of certificates issued in relation to the previous specification (1957) will not, however, be affected and the same holds true for certificates issued against the present specification in the event of the present specification being revised.

Echoing area is the projected area of an equivalent sphere which possesses the same echoing strength as one unit of area of the specific type of target. Its magnitude depends on the aspect. If, for example, the echoing area of a target is 10 sq. metres, then for a given direction, one sq. metre of the target will yield the same echoing strength as a sphere, 10 sq. metres in cross-section (about 1·8 metres radius).

Power ratios are given in decibels. An explanatory note is given at the bottom of the Appendix, showing the relationship between power ratio and decibel (page 357).

It is interesting to read through the specification and obtain information about the standards required for a Marine Radar Set. For the observer, however, it is much more important to get fully acquainted with the characteristics and the limitations of the radar set he is using. These characteristics will now be discussed.

Maximum Range

The maximum range of a radar set depends on:

(a) The number of pulses transmitted per second or p.r.f.

There is, however, an upper limit to the p.r.f. for longer range scales as the trace on the PPI has to be completed between two successive pulse transmissions. In other words, the length of the time-base for these longer range scales must be such that echo signals from good ranges can be recorded.

(b) The peak R.F. power.

(c) The shape and size of the aerial. These determine the aerial gain.

(d) The wavelength used. Both (c) and (d) determine the beam-width. The narrower the beam-widths in both the vertical and horizontal planes the greater will be the maximum range.

(e) The receiver sensitivity. This is limited by " noise " arising from thermal agitation in conductors and by the irregular nature of electronic emission in vacuum tubes. In general an echo can only be distinguished if its power is of the same order as that due to the noise. The noise power output varies linearly with the bandwidth. On the other hand, the bandwidth cannot be made too narrow, otherwise severe pulse distortion would take place. Hence a compromise must be made, but for certain long range warning sets it is possible to work with a smaller bandwidth—and less noise—thereby obtaining a longer pulse-length and sacrificing the range discrimination.

(f) The height of the scanner.

Minimum Range

The minimum range is governed by:

(a) The pulse-length. Generally the set is not ready for reception before the total pulse has left the scanner opening. The shorter the pulse, the earlier the set is ready to receive and the shorter the ranges which can be recorded.

(b) The position of the scanner. The fo'c'slehead may cause an unfavourable shadow sector if the scanner is placed too low and the cross-trees may cause a shadow downwards if the scanner is placed too high.

(c) The vertical beam-width and the wavelength used. Sea surface reflection breaks up the vertical coverage pattern (consult Chapter 1). It must be noted, however, that the minimum range is *not* governed solely by the number of degrees of the vertical beam-width stated in the Manual. Strongly reflecting objects nearby can be detected well outside the half-power points (see section: Vertical Beam-width, this Chapter).

(d) The " change-over " time of the T/R cell (see Appendix III). This extends the minimum range and gives a ragged edge to the dark transmission circle which can be seen on the short range scale. The radius of the dark spot can be varied internally so it does not accurately indicate the extent of the minimum range.

Range Accuracy

The range accuracy depends on:

(a) The uniformity and rectilinearity of the time-base or sweep.

(b) The size of the spot, especially on long range scales.

(c) The curvature of the screen (especially near the tube edge).

(d) The height of the scanner which can introduce parallax when the scanner is high and objects are near and low-lying.

The range accuracy of the fixed range rings is generally such that the maximum error does not exceed $1\frac{1}{2}\%$ (1968 Specification) of the maximum range of the scale in use. The range accuracy of the rings should be determined in port or when anchored in a roadstead by comparing the true range with the recorded radar range of a radar conspicuous object. It can be done sometimes on a moving vessel by measuring the difference in radar ranges between two radar conspicuous objects which lie on the same or opposite bearings and comparing them with the chart distances. Alternatively parallel index lines engraved on a rotating bearing mask at intervals of the fixed range rings can be used to measure the radar range between two objects. Even a pair of rubber-tipped dividers could be employed to measure the radar range between two fixed objects. After measurement take the radar range off along the radius of the rings and compare with the chart.

For interpolation between the fixed range rings it may be advisable to use the variable range marker or strobe. Generally the accuracy of the variable range marker is slightly less than that of the range rings. The error may vary for different range scales and also at different ranges on the same range scale.

The variable range marker must be *checked and calibrated against the fixed range rings*. It should be done fairly frequently because it sometimes happens that small drifts are introduced in the error. The method of calibration is as follows:

Bring range rings and range marker simultaneously on the screen. Let the range marker coincide with the range ring concerned. Then take the reading of the variable range marker and compare it with the range indicated by the particular fixed range ring.

Always measure the range of the leading edge of the echo. A radar pulse cannot " feel " the off-side of a target. When taking a range with the range marker, the outer edge of the range marker should touch the inner edge of the echo.

Bearing Accuracy

Bearing accuracy is governed by:

(a) The horizontal beam-width (which depends on the wavelength used in relation to the scanner width). It was explained in

Chapter 1 that a mm. waveband radar set used in connection with an ordinary sized scanner yields a very fine horizontal beam-width.

(*b*) Rectilinearity of the time-base.

(*c*) Correct synchronisation between the revolving scanner and the rotation of the trace on the PPI.

(*d*) The thickness and the correct alignment of the heading marker.

(*e*) The type of bearing marker employed.

It should be pointed out here that bearing accuracy can be affected by other factors such as faulty centring, parallax of the curzor and yaw of own ship but generally these can be reduced and perhaps eliminated by a careful operator. They do not fall under the characteristics of the set and these errors will be discussed fully in the next chapter. It may happen, however, that even after careful centring, the rest position of the spot wanders slightly round the centre of the PPI. In such a case, as will be seen later, *slight errors* may be introduced when a bearing is taken by means of the mechanical curzor of an object giving an echo at the *edge* of the display, but *far greater errors may occur in the bearing of a target which has its echo situated near the centre of the screen.* It is therefore always advisable to take a bearing of an object whose echo appears in the outer half of the display. On many radar sets, the range scales are multiples of the shortest range scale, for example, the scales represent $\frac{3}{4}$, $1\frac{1}{2}$, 3, 6, 12 etc. nautical miles, so that with the exception of the shortest range scale it will always be possible to keep the echo of a target in the outer half of the display.

Let us now look closer at factors (*a*) and (*d*).

Suppose that the horizontal beam width is 2θ degrees and the radar beam sweeps over a thin vertical pole (Fig. 3.1 (*a*)). At any time when the pole is in the beam it will return radiation. During that time the centre of the beam sweeps through an angle of 2θ degrees and an arc 2θ degrees in width will be painted on the display. The middle of the arc is the correct bearing. The same thing happens when the beam strikes the corner of a coastline which is at right angles to the beam (Fig. 3.1 (*b*)). The picture on the screen shows the coastline echoes being extended to an amount of θ degrees. This effect is known as *beam-width distortion*.

In practice, the radiation intensity is neither uniform within the beam, nor sharply limited to exact boundaries, as shown in Fig. 3.1 (this was explained in Chapter 1). Strongly reflecting targets and objects close-by may well be detected outside the half-power points and such targets may give rise to a greater angular extent of the echo. Hence *beam-width distortion depends on the reflecting property and the distance of the target.* It is therefore more accurate to say that for a given distance beam-width distortion is *proportional* to the stated beam-width. For example, a radar set which has a beam-width of two degrees might produce a beam-width distortion with a particular target of three degrees, whereas a radar with a beam-width of one degree will produce a distortion of only 1½ degrees with the same target at the same range.

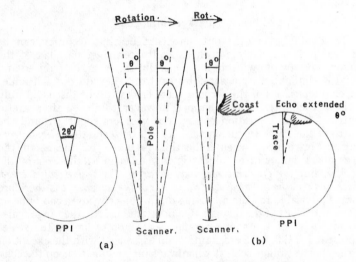

FIG. 3.1 BEARING ACCURACY AND BEAM-WIDTH DISTORTION

Beam-width distortion should not be confused with "*tube-edge*" *distortion*. The latter distortion also increases the size of the echo at the *circumference* of the screen but this is due to a certain amount of de-focusing which takes place at the tube edge.

It is not difficult to eliminate an error due to beam-width distortion for isolated targets. The correct bearing corresponds to the middle of the arc. The corner of a coastline, however, or targets on the coast itself, have, on the PPI, their arcs merged with arcs of other echoes and exact determination of the middle of the arc is difficult. In such a case one can try to reduce the sensitivity or gain and a sharp echo, showing the correct bearing of the object, may come out on the screen. The outline, for example, of the echo of a corner of a coast seems to contract on the display if we reduce the sensitivity. This is because most of the returning radiation comes back when the scanner points *exactly* at the target. The re-radiation which *enters* the aerial just before and just after that moment is weaker in power (due to the directional properties of the aerial) and their echoes on the screen become fainter when we reduce the sensitivity below saturation brilliance of the echo. The divergent parts of the returning signal are, as it were, " cut away ".

Another way of looking at this is by means of the horizontal coverage diagram for reception. Consider a target *well inside* detection range. Reduction of gain will result in contraction of the pattern (Fig. 1.3). The contours each side of the target will close up towards it. These contours show the limit of detection; hence bearing accuracy (and discrimination) will be increased. Targets near the detection range in such a case, of course, will be lost because they move outside the area bounded by the contours.

We have seen in Chapter 1 that during each revolution of the scanner, a line named *heading marker* is painted on the display which should represent the fore and aft line of the vessel and hence the course, not allowing for current. Owing to incorrect positioning of a contact inside the scanner housing or pedestal (see Appendix III) when the radar was installed, the heading marker will differ from the correct position, sometimes by several degrees. When this is the case, errors in bearings will occur. With a " Head-Up " picture, i.e., the heading marker indicates zero degrees, all relative bearings will be out. But this will also be the case with " true " bearings taken on a stabilized display—connected to a gyro-compass transmitter and the heading marker acting as a gyro repeater—when the heading marker, painting an incorrect position of the fore and aft line, indicates the true course reading on the bearing scale. Hence the position of the heading marker is related to bearing accuracy.

It might sometimes happen with end-fed slotted waveguide aerials, that, although the heading marker contact originally was aligned correctly, owing to a slight change in the frequency of the

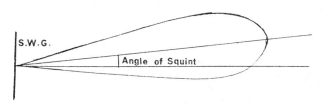

FIG. 3.2   ANGLE OF SQUINT

cm. wave, the lobe pattern of the aerial is slewed round through a small angle—*the angle of squint*—and the axis of the main lobe is no longer perpendicular to the longitudinal axis of the aerial. This is so because the distance between the slots is based on *one* particular frequency and when this frequency is changed, a re-distribution of radiation from the slots will take place. As a result of this the picture on the display will slew round, with respect to the heading marker, through an angle equal to the angle of squint and the heading marker should correspondingly be re-aligned. See Fig. 3.2.

The error caused due to the incorrect position of the heading marker contact in the scanner housing, is constant. All bearings will be wrong by an equal amount. It can be removed mechanically by re-adjustment of the said contact. The maintenance manual can be consulted or an expert called in. Still, it is the operator's duty to find out *if there is an error*. This can be done in port or at anchor or anywhere when there is a target at a distance.

Use the " Head-Up " presentation and set the heading marker to zero degrees. Check that the centre of the rotation of the trace coincides with the centre of the PPI. If there is a discrepancy between the two centres, make an adjustment as described in the next chapter. Select an object which yields a distinct echo near the edge of the display. Check the radar bearing with respect to the heading marker against the visual bearing—taken from near the scanner position—with respect to the fore and aft line of the ship. Record the difference between the bearings. It is good practice to repeat the procedure for different bearings of different objects, if possible, in succession about 90 degrees in azimuth apart. If there is an error and it is due to the placing of the heading marker only, then the results for the different bearings should be practically constant. If a variable error is found then it could be because there is centring error or incorrect synchronism, or a combination of errors.

For example, if the visual bearing and the radar bearing are respectively 280° and 282° relative, then the heading marker is displaced two degrees to port or Red 2°.

At sea, one could bring a small object or a ship, at a distance of not less than two miles, dead ahead and at the same time take a radar bearing. Suppose that the relative bearing is 357° on the bearing scale, then the fore and aft line is three degrees to the left of the heading marker, or in other words, the heading marker in fact represents 3° Green.

A rough idea about the position of the heading marker, when the scanner is sited in the centre-line, can often be obtained by observing the heading flash against the dark sector (caused by the shadow sector of the foremast) projected on rain or sea clutter.

If a heading marker error is found to exist and is not corrected by re-adjustment of the contacts, a notice indicating the error should be displayed prominently near the radar display. Record also in the radar log.

Several ferry-boats, by the way, have stern markers on their equipment.

With a good radar set and an experienced operator the bearing accuracy can be within one degree.

Range Discrimination

Range discrimination is the ability of the set to discriminate between two small objects close together on the same bearing, i.e. to produce separate echoes. The further object must not be in the shadow of the nearer one (Fig. 3.3 (a)).

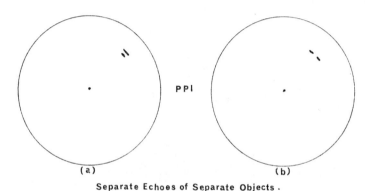

Separate Echoes of Separate Objects .

FIG. 3.3   RANGE AND BEARING DISCRIMINATION

The range discrimination depends on:

(a) The pulse-length. If, for example, the duration of transmission is ·25 micro-second, then the length of the pulse leaving the scanner is ·25 × 300, i.e. 75 metres long.   The theoretical range discrimination is a minimum of half the pulse-length, i.e. $37\frac{1}{2}$ metres. A shorter separation between the two targets will make the echo pulse from the further target join or overlap the echo pulse from the nearer one.   The two targets would be shown as one echo on the display.   This is illustrated in Fig. 3.4.   Here the distance between *two* small objects 'A' and 'B' is 32 metres while the range discrimination amounts to 75/2 i.e. $37\frac{1}{2}$ metres.   The combined reflected pulse-length equals 75+32+32 i.e. 139 metres shown on the radar display as *one* echo 139/2 or $69\frac{1}{2}$ metres in length.

In practice the range discrimination amounts to more than half the pulse-length due to:

(b) The size of the spot.   On a short range scale (i.e. a large scale picture) the influence of the spot-size will be less than when a longer range scale is employed.

Range discrimination can be improved by correct focusing and brightness, by selecting a short pulse-length and by differentiating the echoes.

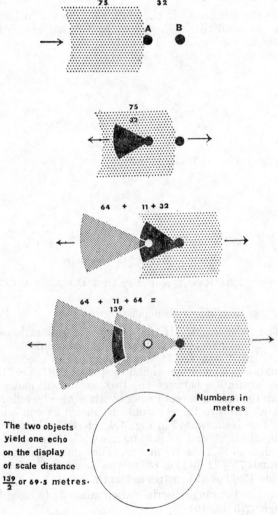

The two objects yield one echo on the display of scale distance $\frac{139}{2}$ or 69·5 metres.

Numbers in metres

FIG. 3.4 RANGE DISCRIMINATION AND PULSE-LENGTH

Bearing Discrimination

Bearing discrimination is the ability of the set to discriminate between two small objects close together at the same range, i.e. the ability to produce separate echoes (Fig. 3.3 (b)).

The bearing discrimination is governed by:

(a) The horizontal beam-width (which depends on the horizontal dimensions of the scanner and the wavelength used). The narrower the beam-width, the better the bearing discrimination.

(b) The spot-size. Again, the influence of the spot-size is less when a lower range scale is employed.

Here again, as with bearing accuracy, the horizontal beam-width plays an important role. If the beam-width (between half-power points) were too wide, there would be a short moment when the two objects are contained in the same beam and send out returning pulses simultaneously. Conversely, the scanner picks up signal pulses with the same divergence in bearing as the beam-width and the echo-paints on the display coming from two objects at different bearings but at the same range, will join up or overlap.

Bearing discrimination can be improved by correct focusing and brightness, by reducing the sensitivity or gain and by off-centring on a lower range scale.

Radar sets in the mm. waveband have excellent bearing discrimination.

Further Considerations relating to Beam-width and Pulse-length

When a ship is sailing very near to a coast, or is manoeuvring near a jetty or a quay, her echo on the observing vessel will merge together with echo of the coast, jetty or quay. This is due, of course, to the limitations in bearing and range discrimination. It is one of the reasons why cm. radar cannot be used for berthing unless a wide aerial is employed. Knowledge of this limitation was used during the Second World War. Taking station near a cliff made radar detection by enemy surface craft very difficult.

Let us now consider the shape of an echo on the display in relation to the shape of the target. The shape of the leading edge (towards the scanner) of a target can only be determined from the screen if the width of the target is large in comparison with the horizontal beam-width. Movements of ships, for the same reason, can only be observed without any appreciable time interval from the screen when they are at a very short range (we are referring to cm. radar). Only in such cases is the beam able to " feel " the individual parts out of which the target is made up, one after the

C

other, and to paint their correct positions relative to each other on the screen.

Let us go back to our example (Fig. 3.1) where we illustrated a beam sweeping over a thin vertical pole. The echo on the display is shaped something like a rectangle whose width is determined by the horizontal beam-width and whose radial length is a function of the pulse-length. What we really see on the display is the image of the horizontal cross-section of half the pulse. Besides this, the dimensions of the echo will be increased in all directions by half the spot-width.

Small isolated targets or targets which are small in comparison with the horizontal beam-width, whatever their real shape, are recorded on the PPI as echoes which possess dimensions which are mainly functions of the angular horizontal beam-width and the pulse-length. Such targets are, as it were, " drowned " in the beam. For wide and deep targets such successive echoes may overlap and a rough outline is displayed on the screen.

In order to make some deeper study of what the echo of a small isolated target looks like at a short and a long range *using the same range scale*, let us assume that the scanner has a radiation pattern as shown in Fig. 1.7. Suppose that the target is at a distance of 7 miles and produces an echo 6 db above the minimum detectable, then the scanner will pick up the signal over an arc of ·6°. If a 30 cm. PPI is used with a spot diameter of ½ mm. and the range scale is 12 M then the *width* of the echo on the display represents $7 \times ·6/57·3$ (arc of signal) $+ (1/20)/15 \times 12$ (spot diameter) $= ·07 + ·04$, i.e. ·11 nautical mile or 204 metres.

The *same* target at a distance of 1 M produces an echo 40 db (i.e. $(10 \log 7^4 + 6)$ db) above the minimum detectable (the echoing strength is inversely proportional to the fourth power of the distance) and consulting Fig. 1.7 it is seen that the scanner picks up the signal over an arc of 1·5°, representing a *width* of the echo on the screen of $1 \times 1·5/57·3$ (arc of signal) $+ (1/20)/15 \times 12$ (spot diameter) $= ·025 + ·04$, i.e. ·065 nautical mile or 120 metres.

If the pulse-length used with this range scale is ·5 micro-second, then this corresponds to a length of 150 metres of the radiated pulse. The echo pulse is also 150 metres long but the spot registers an echo representing 75 metres. This is so because the scale on the PPI indicates half the distance travelled by the radar pulse (out and back). The echo therefore depicts the target, both at the 7 M and the 1 M position as having a *length* of 75 metres + spot diameter = 75 + 74 metres, i.e. 149 metres.

Hence at 7 M the apparent dimensions of the target seems to be 204×149 metres while the *same* target at a distance of 1 M seems to measure 120×149 metres (see Fig. 3.5).

This consideration explains why, on the same range scale, ships at the shorter ranges appear to be end-on and at the longer ranges beam-on. Tube edge distortion may also increase the " beam-on " effect at the circumference of the screen. Thus no inferences should

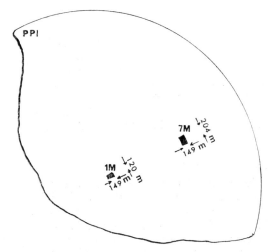

FIG. 3.5 ECHO DISTORTION

be drawn from the shape of an echo about the relative or true movement of the ship's echo over the screen. Neither can the aspect be read off directly (at least not for cm. sets). It is one of the reasons why plotting is necessary in the observance of motion of other vessels.

Vertical Beam-width

The vertical beam-width for different sets can vary between 15 and 30 degrees (between half-power points). It is wide in order to allow for rolling and in order to pick up low-lying objects at a minimum distance of 50 yards (45·7 metres). Close objects, however, can be detected well outside the boundaries of the strong centre radiation. *Hence the minimum range is not limited to the main beam.*

Aerial Gain

An omni-directional aerial radiates energy in all directions, but a scanner has its radiation concentrated into a beam. If one could compare the field strength within the beam at a given distance from the scanner with the field strength the same distance away from an omni-directional aerial which has the *same* power output, then it should be clear that the ratio between these should be a number greater than unity. This ratio is known as " Aerial Gain " and is expressed in decibels. It obviously depends on the solid angle of the beam, i.e. on the horizontal and vertical angular beam-widths.

Scanner Rotation

British marine radar sets have a continuously rotating scanner, maintained at a uniform speed. The speed for different types varies between 20 and 80 r.p.m. This rate must be maintained in relative high wind speeds and sometimes for this purpose balancing fins are fitted to the sides of the reflector.

In some American sets, working in the 10 cm. band and designed for navigational purposes primarily, the rate of rotation is much lower.

Pulse-Repetition Frequency

As already remarked in Chapter 1, most radar sets use short pulses for the short range scales, but the p.r.f. is high. Longer range scales employ long pulses combined with a lower p.r.f.

The brightness of the echo on the screen is also determined to a certain extent by the number of pulses striking the target when the beam sweeps over it, and also upon the number of pulses returning. There is an upper limit to the storing of these pulses on the screen and hence to the brightness. Most radar sets are designed so that at least 10 pulses are returned from the target (having a suitable aspect) during a single sweep. This number of pulses depends on the horizontal beam-width, the p.r.f. and the scanner rotation speed.

Example. The horizontal beam-width is 2°, the p.r.f. 750 and the scanner rotation speed 22 revs. per minute. A rotation of 360 degrees takes 60/22 seconds and the sweep of the beam lasts 2/360 × 60/22 or 1/66 second during which 1/66 × 750 or 11 effective pulses strike the target if the target was at such a range that it could be *just* detected by half-power.

Receiver Noise

The word " noise " is borrowed from the background noise in a radio loudspeaker when we turn the volume control up too much. This " noise " will also become visible on the screen of a radar set, especially when the sensitivity setting is too high. It is caused mainly by thermal agitation of the electrons in conductors and vacuum tubes, resulting in random movements. On the A-Scan it will show up as " grass " covering the trace, on the PPI a speckled background appears. When the noise is too strong, it will be understood, weak echoes and echoes from *maximum detection ranges* will be difficult to detect. Hence the ratio $\frac{signal}{noise}$ for a particular set is extremely important. The factor is expressed in decibels and stated in the manual. When the signals are strong in relation to the noise, the ratio is high and the operation efficient.

It can be shown that an increase in the sensitivity of the receiver results in a greater increase in range than that obtained by a large increase in the transmitter output.

The local oscillator (Klystron) is quite a "noisy" component and during the mixing process will inject noise in the I.F. signal. To overcome this difficulty rader sets have now two mixers in parallel. At one of the crystal mixers the local oscillator signal and the R.F. signal are in phase ; at the other they are 180° out of phase. This means that the I.F. signal at one mixer diode will also be 180° out of phase compared with the I.F. signal at the second mixer diode. By subtracting the two outputs via a push-pull transformer, wound in the proper sense, the signals currents (in opposite phase) are added and the noise currents (random phase) are subtracted.

This arrangement is called "Balanced Mixers" and gets rid of most noise generated by the Klystron.

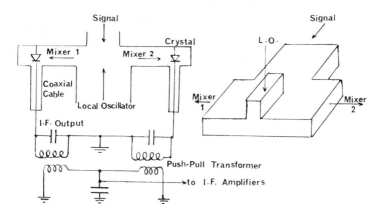

FIG. 3.6   BALANCED MIXERS

It should be remembered that speckled background on a PPI does not necessarily mean that the receiver is working efficiently. The receiver can, however, be tested and the next section will deal with this.

Testing or Checking the Performance

It may happen that after switching on the radar set in fog, no echoes or sea clutter are seen on the display and one starts wondering whether there are no targets in the vicinity or whether there is something wrong with the set itself. To test the action of the set, an artificial echo which has the shape of either a plume or a sun, can be produced on the screen.

The artificial echo is produced by energy collected and re-radiated from a small box (echo box) incorporated in the rotating part of the scanner unit (sun pattern) or mounted on a special stand, funnel or main mast (plume pattern). A paddle rotates at a high speed inside the box and its function is to prevent distortion of frequency (pitch), so that the outcoming signal has the same frequency as the collected pulse (the cavity of the echo box is sensitive to temperature changes). If the paddle does not work, the pattern on the screen is broken up or divided into sectors. The broken effect of the sun pattern can also be produced when the automatic tuning device, which some sets have, fails to "lock-on" correctly.

The " ringing " of the box lasts for about 12 seconds, leaving an echo on the display of radial length $\dfrac{12/2 \times 300}{1852}$ i.e. about 1 N.M. For more details see Appendix III.

If the artificial echo is a permanent feature of the screen, i.e. it cannot be removed, then the echo box must be placed in a blind sector, i.e. a sector, where, due to ship's obstructions in the path of the radar beam, no other echoes are recorded.

By measuring the extent of the artificial echo for given gain and clutter settings and comparing this with the value given in the radar manual or radar log, one can obtain an independent check on the action of the *transmitter, receiver* and *overall reliability*.

Decrease in the length of the plume, or a diminishing in the filling out over the screen of the sunray pattern, may indicate a decrease in transmitting power, weakening of the amplification system, mistuning or the presence of moisture in the waveguide.

Some radar sets are equipped in such a way that transmitter and receiver can be tested separately. The echo box is replaced by a neon tube. When the scanner sweeps over the tube, the gas inside the tube is ionised and causes the plume to appear on the display. The extent of the plume gives a check on the transmission. Alternatively, the current through the neon tube is converted into a wave form whose peak output is indicated on a meter. If the efficiency of the transmitter aerial and waveguide is not below standard, the meter reading should exceed 100. When the receiver is tested, a sun pattern is displayed on the PPI. The energy causing the sun pattern is collected by a tuned cavity inside the transmitter unit. See Fig. 3.7.

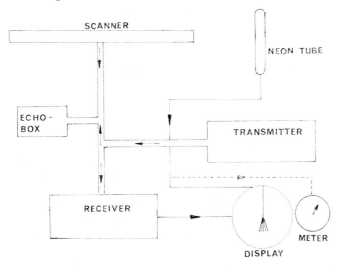

FIG. 3.7   PERFORMANCE MONITORING

It is sometimes difficult with the sunray pattern to see where it actually ends. The plume echo yields more contrast and one can compare it with the speckled background either side of it.

Instead of the neon tube recent radars have a transponder (Chapter 11) which transmits a *coded* reply if the signal is of sufficient magnitude. At a certain calibrated level with degrading signal the code will cease, and for greater degrading the display pattern will gradually disappear.

It is good practice to test the set every time when switching it on and thereafter at frequent intervals so that the observer can be assured or warned of the proper or defective working of the installation.

This testing is also known as *Overall Performance Monitoring*.

CHAPTER 4

FUNCTION AND ADJUSTMENT OF CONTROLS

Main Controls

Main On-Off Switch

Starts the set. The waiting period, before the filaments are warm and high tension is supplied, is two or three minutes. When the set is ready for action, the modulator (see Appendix III) starts firing, which often is accompanied by a high-pitched tone. Sometimes the deflection of a needle on a meter indicates the start of transmission.

When the radar set is fitted with a stabilized tube for True Motion Head-Up facilities, the tube is rotated during the warming-up period to automatically clean the slip rings.

Scanner On-Off Switch

Starts or stops the scanner. There are sets where the scanner is stopped during the warming-up period, but automatically starts when transmission commences.

One of the main reasons for separate switching is that one can run the aerial for de-icing purposes when the remainder of the radar set is not in use.

Stand-by/Transmit Switch

Offers an opportunity to the operator to stop transmission and high tension supply while keeping the filaments warm. During "stand-by" very little power is consumed and h.t. (high tension) components do not wear. The set is ready for immediate use and this can be essential when sudden fog-banks are expected. Another advantage is that if the operator switches to stand-by when not viewing then his set is causing no radar interference with other radar sets of other ships (see: Interference, Chapter 8). *If all radar users had and used this stand-by device in clear weather then there would be a great decrease in radar-to-radar interference.* There are sets which are automatically on "Stand-by" and a "Press to View" switch has to be depressed which makes the equipment go into operation immediately for a few minutes, after which it returns automatically to "stand-by".

On a dark night, however, in an area where fogbanks can be expected the set should be left on " transmit " most of the time as the vessel could easily be steaming along near to a fogbank which is giving no visual indication of its presence.

Brilliance or Video Control

Regulates the brightness of the picture. The screen and the brilliance circuits are so constructed as to keep the brilliance below a certain value, the saturation value. If this was not done, there would be an extremely great difference between the brilliance of different echoes, for example, between the echo of a small buoy and that of a large ship. Too strong a brightness would also produce a halo effect (blooming) resulting in de-focusing. It would at the same time make the *afterglow*—the retaining property of the screen to " hold " the painted echoes during at least one revolution of the scanner—too strong and a long afterglow on a moving ship with a relative motion display will yield a blurred appearance of the picture. Despite the *video limit* control, the operator should not set the brilliance too high. The visible rotation of the trace is tiring for the eye and it will reduce the life-time of the tube. Some sets are equipped with a supplementary control which can adjust *contrast* of intensity against the background.

On *some* sets the brilliance should be turned down when the scanner is stopped or the set is on " stand-by ".

Focus Control

Focusing will be assisted if the brilliance is turned up slightly and it can be done on the range rings, speckled background or heading marker. Later on this extra brilliance can be reduced. In some radars the focus control is pre-set.

Centring (Horizontal and Vertical Shift) Controls

These controls are used to centre and align the radar picture linearly until the electronic centre (the starting spot of the trace) is in the geometric centre (the centre of the bearing scale). This electronic centre represents own ship's position and the geometrical centre forms also the pivot of the mechanical curzor. Bearings taken by this curzor can only be accurate when the two centres are exactly superimposed. The electronic centre should be checked each watch and after an alteration of course. Due to internal magnetic disturbances, changes in the earth's magnetic field (magnetic latitude change), changes in the ship's magnetic field (alteration of course) and vibration (alteration in the setting of the centring potentiometers), the direction of the electron stream in the CRT will be affected and the electronic centre displaced. In some cases mumetal screening is employed to guard against magnetic influences.

If the geometrical and the electronic centres are displaced, radar bearings taken with the mechanical curzor will be wrong unless the bearing coincides with the line which connects the two centres. Perpendicular to this line, the error will be maximum.

The horizontal component of the displacement of the two centres is responsible for the deviation of the heading marker from her zero position. Such an error is sometimes known as " Radar-A ", as it is comparable with the A-coefficient of the compass. Radar-A, for example, will vary several degrees (depending on the size of the tube) during a voyage from London to Australia, if the horizontal shift is not adjusted.

In some sets, designed for River Navigation, the electronic centre can be purposely off-set by means of a switch. This is done to make more use of the available display. In such a case the mechanical single bearing curzor cannot be used and a parallel index or an electronic curzor should be provided.

More than one method can be used to achieve the coincidence of the two centres.

(i) Put the bearing curzor on 0° or 180°, i.e. right up and down. By means of the horizontal shift control, bring the electronic centre under the line of the curzor. Then turn the bearing curzor to 90° or 270°, i.e. in a horizontal position and use the vertical shift control to bring the electronic centre once again under the curzor. The two centres should be superimposed now. A final check can be made by rotating the curzor round a couple of times. It is advisable during the adjustment to have the viewing hood placed over the screen and have one's face pressed well up against the visor, so that no error due to parallax is introduced.

On a short range scale the electronic centre may not be visible due to the dark transmission mark, or one may have an old tube with a centre burn. In these cases the following method can be employed:

(ii) Put the bearing curzor on 0° and align the heading marker parallel to the bearing curzor. Use the horizontal shift control and bring the heading marker on 0° under the curzor. Turn the curzor to 90° and by means of the " Picture Rotate Control " swing the heading marker until it again is parallel to the curzor. Adjust the vertical shift in such a manner that the heading marker indicates 90°, i.e. it is exactly under the curzor. This procedure can be repeated.

Picture-Rotate or Turn-Picture Control

This control swings the picture round so the heading marker can be set to indicate zero or the true course of the vessel on the bearing scale. The latter setting is best when the display can be gyro-stabilized. Switch on the compass repeater inside the display unit when the ship is on course. The Picture-Rotate Control will then be automatically de-clutched from the gearing which rotates the deflection coils.

If the picture is rotated until the heading marker is on zero of the edge scale the bearings of echoes will be given relative to the ship's course whereas if the picture is rotated until the heading marker indicates the ship's course on the edge scale the bearings of echoes will be true.

Auto-Trim Picture Alignment or Compass Repeat Control

Sometimes this control is provided which automatically re-orientates the picture by switching it on or by depressing and turning a knob.

Some sets are equipped with a partly automatic control, known as the *Heading Marker (Line) Coarse Switch* which will make the heading marker (and picture) revolve. Operate this switch until the heading marker is in the vicinity of the intended position, then use a second control, the *Heading Marker Fine Control*, to make the final adjustment of the picture alignment.

Gyro-Stabilized Bearing Scale

Some American, Russian and Japanese sets possess two bearing scales (or azimuth rings). The inner one is marked relatively, the heading marker indicating zero. The outer ring can be gyro-stabilized, so that the true course can be read against the outer ring. All bearings taken by means of this outer ring will also be true. See Fig. 10.1.

Heading Marker-Off Switch

It is good practice to switch the heading marker off frequently just to make sure that it does not conceal small echoes of buoys or small craft. The switch is *spring-loaded* and when it is released the course line of the vessel becomes visible again. In some sets the intensity of the heading marker can be regulated.

Secondary or CRT Controls

Brilliance, Focus, Centring, Picture-Rotate and the Heading Marker-Off Control are sometimes called Secondary or CRT controls.

Gain or Sensitivity Control

This control shapes the overall picture. It brings a varying quantity of detail on to the screen and its setting governs the amount of amplification of the signal, but unfortunately also of receiver noise The speckled background should be just visible. It can be recognized more easily on the longer than on the shorter range scales. On the short range scales the time-base is comparatively fast and the speckles are elongated into thin lines which are difficult to detect. Hence adjust this control when on the medium or long range scale.

Some radar sets have a switch on their display unit marked LIN.LOG., which selects the linear receiver circuit or the logarithmic receiver circuit. See Appendix III and Fig. AIII.10. The logarithmic receiver can, in certain cases, display the difference in echo strength of two strong signals which otherwise would yield the same echo intensity on the screen (see graph, Fig. AIII.10). Examples are echoes of ships in strong sea clutter or two echoes, seen simultaneously, one from a small object near the ship and a second one from a strongly reflecting object a distance away from the ship.

The LOG. setting can be very effective when used in conjunction with STC or FTC (see next sections).

Anti-Clutter, Clutter, Sensitivity-Time Control (STC) or Swept Gain

This variety of names is given to the control which enables the observer to reduce the gain of echoes from nearby targets without affecting the gain of echoes of the more distant objects. The Gain Control should be adjusted before the anti-clutter is applied. When the Anti-Clutter Control is increased, then the *reduction* in gain becomes strong when the spot leaves the centre of the PPI, but gradually the gain is restored to normal for a range determined by the setting of the Anti-Clutter knob. The maximum range at which this occurs does not exceed 4 or 5 miles.

The *purpose* of the Anti-Clutter Control is the elimination of sea clutter or sea return, i.e. echoes caused by reflection against the sloping sides of the waves. These echoes are strongest from waves near the vessel and anti-clutter application keeps in step with them.

It is important to reduce the clutter for the following reasons:

(a) Echoes of targets nearby " drown " in sea clutter and cannot be picked out;

(b) Continuous strong sea clutter will reduce the sensitivity of the coating near the centre of the tube;

(c) The bright patch in the middle of the screen is distracting to the observer.

The amount of anti-clutter should frequently be varied. *Too much suppression might create extremely dangerous situations* as echoes of large targets will be suppressed in intensity and might not be visible. Too much clutter on the screen also is not advisable as small navigational echoes, for example, those coming from buoys, will merge with the clutter echoes and cannot be picked out any more. However, *some clutter should always remain on the display*. As long as the operator can detect echoes of ships and buoys from amongst the clutter, no further suppression is desirable. A short range scale affords the best and easiest arrangement for accurate anti-clutter control. If side echoes are present a compromise should be struck between the suppression of these echoes and the disappearance of echoes coming from small navigational targets.

The Anti-Clutter Control should never be left on a fixed setting. It should be handled *continuously*. Weather conditions and local conditions in connection with the detection of objects in the vicinity of the ship vary continuously. Echoes of ships should be traced inside the clutter and the area near the ship should be investigated by frequent manipulation of the Anti-Clutter knob.

Rain Switch, Anti-Rain Clutter, Differentiator or Fast-Time Control (FTC); CP Switch

These names are given to the control which alters the rectangular waveform of the echo pulse to a peaked one. It does not actually reduce the gain but reduces the amount of paint on the screen. Only the leading edge of echoes can be seen. Masses of echoes due to rain, hail or snow are broken up and this may make it possible for echoes of ships or other targets to be picked out amongst the clutter. In several radar sets the setting is pre-determined and when using the rain switch it is sometimes advisable to increase the gain slightly. In other radar sets, however, the FTC control is variable.

Another type of Rain Switch makes use of so-called *Circular Polarization* (CP). Strong reduction of echoes will then only take place when there exists symmetry in the striking surfaces of the targets. As this is the case with rain drops, only their echoes will be prominently reduced. For more details see Appendix III, page 341.

In this system, the horizontally polarized wave is transformed into a circular polarized wave of a given direction (clockwise or anti-clockwise). The scanner will only accept wave signals which are polarized in the same direction as the transmitted wave and this depends on the number of reflections taking place at the target. If this number is odd, the direction of polarization is reversed and the echo rejected.

The rain switch should be used with caution and not be left on continuously. Echoes of small targets often disappear and when the CP switch is on, symmetrical objects, such as spheres and corner reflectors, may suffer strong attenuation.

Clearscan

DECCA Radar Ltd. some years ago, introduced two video processors—highly sophisticated miniature Video Processing circuits —VP1 and VP2, resulting in a practically clutterless noise-free display with intensified and bigger echoes. Other Marine Radar manufacturers have since added similar facilities.

VP1 suppresses sea and rain (snow and hail) clutter *automatically*, i.e. it assesses the extent and intensity of the clutter and *then* applies a gain level based on this assessment. In other words a gain level proportional to the extent and intensity of the clutter signal is constantly and rapidly applied (this is known as an "adaptive" signal). A further signal is required to suppress clutter effectively within the first mile of radar range. This signal is derived from sea clutter returns experienced during *previous* radar transmission cycles and is added to the adaptive signal to form a composite signal, providing automatic operation over the entire radar range (wind and rain clutter). Normal echoes from ships, navigational marks and coastlines are little affected, but Racon signals are suppressed. There is normally still some slight clutter speckle around the electronic centre and this will increase with strong winds and high scanners.

VP1 is now a standard unit for the ordinary-sized DECCA Radar displays used on ships. Normal sea and rain clutter controls can still be used, but if Automatic Adaptive operation is required, the sea clutter control is lifted—showing a green inset marked VP1—and this action electronically disconnects the normal sea and rain clutter controls.

VP2 is an optional unit, operated by the same control as VP1. With the sea clutter control lifted a red inset, marked VP1/VP2 will be shown.

VP2 suppresses interference from other ships' radars when their pulses or signals pass into own ship's receiver when the T/R cell is open. First of all a practically noise-free video signal is produced by eliminating signals below a pre-set threshold level. This signal produces a dark background for good contrast and yields improved afterglow trails on the display.

The distinct analogue signals are then quantized, i.e. converted into digital signals which are discrete signals of rectangular wave form and uniform intensity (full brightness level on the PPI). This means that weak signals, which would be difficult to see and might otherwise be missed, paint at a brightness level similar to that of strong echoes.

A radar interference free display is produced by the use of pulse correlation technique, which compares echoes during the current transmission with stored echoes from the previous transmission. Only echoes received at the same range from two consecutive pulses are shown. This virtually filters out all radar interference from other ship's transmissions as such interference pulses do not normally occur at the same range on successive transmissions.

The echoes can also be made longer (stretched) so that smaller echoes are more easily detectable. This latter facility is only available on the 12-, 24-, 48- (or 60-) mile range scale where echoes outside 2·5 miles (the clutter range) and exceeding 0·5 micro-seconds are stretched an additional 1, 3 and 7 micro-seconds respectively. In order to improve range discrimination, the stretching of echoes on the 12-mile range scale can be discontinued by switching over to short-pulse transmission.

In some radar sets VP2 is standard equipment and the processor is then denoted by VP3.

The processors discussed are available both for 3- and 10-cm. radars.

Pulse-Length Switch

A long pulse-length can be selected for long range detection, a short pulse-length can be chosen for a sharp picture, good range discrimination and better observance of an echo trail (tadpole tail). The latter can also be used as an alternative for the Anti-Clutter Control or the Anti-Rain Switch. It also gives better minimum range.

Nowadays, multiple pulse-lengths are introduced, for example, 0·05, 0·15, 0·5 and 1·2 micro-seconds. On the shortest range scale the shortest pulse is automatically selected by the range selector switch but on all the other scales a long or a short pulse can be chosen by the operator.

Multiple pulse repetition frequencies are also automatically selected as the pulse-length is changed. The very high p.r.f. is available on the shortest range scales to give the brightest picture possible, yet to minimise any showing of second-trace echoes (Chapter 7).

Range Selector Switch and Scale Indicator

When changing scale this control should be turned quickly and deliberately from one range scale to the next. Slow movement may cause sparking at electrical contacts when the p.r.f. is changed. Use step-by-step movements. Switching, for example, from range scale 1 to range scale 3 in one go, may cause bad contacts. We have seen already in Chapter 3 that for the sake of bearing accuracy, on many sets, the scales are multiples of the shortest scale, for example $\frac{3}{4}$, $1\frac{1}{2}$,

3, 6, 12, 24 and 48 nautical miles. Use the shortest range scale possible when measuring the bearing and range of an object, to obtain the highest accuracy. The *working range scale* when radar is used as a collision warning aid is 10–15 miles. Such range corresponds roughly to the clear weather visual range and provides time for planning avoiding action. Ships at close distances can be followed regularly with intervals on the shorter range scales which yield better discrimination. One of the first things one should do when coming on watch and the radar set is on, is to observe *which range scale is in use.* This is especially important in fog when no land echoes are visible. A clear radar screen, unknowingly on a short range scale, may lull the Master into the false security that everything is safe. However, *do not change the range scale from a short scale to a longer one when the echo of a dangerous target shows up at close range.*

The brilliance may slightly fluctuate when switching from one range scale to another and this should be regulated (see Maladjustments).

Switch (Control) for Fixed Range Rings. Range Ring Interval Indicator

The number of range rings must be sufficient to ensure good range accuracy. Too many rings, however, can cause confusion. The rings should be thin. On the short range scales, due to the comparatively fast speed of the spot along the time-base, the rings are thicker than on the longer range scales where the time-base is comparatively longer and the spot moves a shorter distance during a generated range pip. Therefore use the rings on the longer scales for focusing purposes. Switch the rings off when they are not needed because they may obscure echoes of small objects. Sometimes the intensity of the rings can be varied.

Variable Range Marker Switch, Control and Scale

Its use is handy for interpolation of a range of an object whose echo lies between two fixed range rings. The Range Marker should be calibrated against the fixed range rings or calibration rings (see Chapter 3). The range at the moment of observation can be read digitally or by means of a pointer which moves over a sliding scale. With certain hoods and visors, it is handy to employ the Range Marker. Its use is also excellent for navigational purposes, making a landfall for example, when an accurate range is required. It is also useful for plotting ships at great distances. However, range finding by means of the Variable Range Marker is not such a quick process, and therefore, the range rings are preferred for the determination of ranges of close objects and especially of close *moving* objects. Finally, too much use of the Range Marker can cause loss of confidence in the taking of ranges with the fixed range rings.

Sometimes the Variable Range Marker (or Strobe) is used, together with the Bearing Curzor to pinpoint an echo on the screen. The Range Marker is then brought *through* and the cursor *over* the echo. The motion of the echo (relative motion of the object) can then be deduced by observing into which quadrant the echo moves.

Variable Range Delay (Two Switches)

This unit has been introduced by Decca, and it can increase the displayed range by amounts from 1 to 99 miles. This is achieved by a delay mechanism which paints echoes on a display whose centre represents the maximum range of the previous range scale (i.e. one scale down from the present range scale set by the operator).

Suppose that land is expected at a range of 60 miles and the maximum range scale on the radar is 48 miles. A 24-mile delay is set on the unit and then switched in. Instead of say, six range rings representing 8, 16, 24, 32, 40 and 48 miles, they will now represent (24 + 8), i.e. 32, 40, 48, 56, 64 and 72 miles. The expected land echoes would show up about half-way between the fourth and fifth ring. When selecting the range-scale and for range delay to be used it should be remembered that bearings can be measured more accurately with the echoes nearer the edge of the screen, and that ranges of echoes can be interpolated more easily on an expanded scale.

There is distortion in shape compared with the normal echo appearance and this can lead to difficulty in echo identification.

The unit is separate from the radar display and mounted adjacently to it. It operates between 0 and 99 n.m. in 1-mile increments. The first switch sets the additional range (or delay) which has to be added to the maximum range of the scale in use. The second switch (sometimes spring-loaded) brings the additional range on the display and operates a red (or flashing) warning light to remind the operator that an equivalent distance is missing from the centre of the display.

Range Calibration Switch

Some radar sets are not provided with independent switches for the Range Rings and Range Marker. There may be only one switch, e.g. " up " for the Rings, " down " for the Marker. For calibration both rings and marker must be brought on the screen simultaneously and this is the function of the Range Calibration Switch.

If there is no such switch and there exists a dual control for Range Rings and Range Marker, the following method can be tried:

Switch on the Performance Indicator—we assume that the artificial echo is not permanently placed on the screen—and this will also bring the Range Marker on the display. Put the dual switch on " Range Rings " and reduce the Gain and Clutter to get rid of the artificial echo. Calibration can then be started.

Performance Monitor or Performance Indicator Switch

This switch is used when the artificial echo is not a permanent feature of the display. When depressed the Range Marker will also appear so giving the observer the opportunity of measuring the extent of the pattern. The testing can then be proceeded with as described in Chapter 3.

Tuning Control

Tuning the set to the correct frequency (Appendix III) helps to yield the best picture obtainable. Fine and coarse control is sometimes available. Turn the control(s) until the best response is revealed. Various methods can be employed. Tune on:

(1) Sun or plume pattern caused by the echo box.

(2) Sea clutter.

(3) Outstanding echo or echoes. If no echoes are available, an empty drum could be thrown overboard.

(4) The A-Scan or oscilloscope. Stop the scanner and adjust until maximum height of the echo pip. If an A-Scan is part of the equipment, this is one of the most sensitive methods of tuning.

(5) The echo trail of the wake of own ship while making a turn. Use a short range scale.

(6) Meter reading. The manual will provide the details.

(7) Tell-tale light or " Magic Eye ".

Manual tuning should be done at frequent intervals. There are radar sets which have no Manual Tuning Control near the display; instead it is incorporated in the transceiver (mixer unit). In such a case an Automatic Frequency Control (AFC) unit keeps the tuning adjusted. See Appendix III.

Mechanical Curzor, Cursor Control and Bearing Scale

The curzor consists of a double hair-line and can be adjusted to bisect an echo, by means of the Curzor Control. The Bearing Scale Ring, graduated in degrees, around the screen, provides facilities for reading bearings and for aligning the picture with the aid of the heading marker.

Illumination Scale (Variable)

The Bearing and Range Scale can be illuminated by means of a Brightness Control, which also makes the curzor visible. It is better not to leave the scale brightness on full when not occupied with the taking of bearings. This light can be distracting for the eyes and obscures the speckled background so that proper adjustment of gain is difficult.

Parallel Index

Consists of equidistantly spaced parallel lines engraved on a transparent screen which fits on the PPI and can be rotated. The distance between two successive lines should be calibrated against the range rings of the scale in use. This device can be extremely handy for estimating distances, anchoring technique, following clearing lines, determination of the nearest approach of a ship and the laying-off of a course directly from the screen when the display is gyro-stabilized, etc. A fuller treatment will be given in Chapters 9, 10, 13 and 15.

Electronic Bearing Marker Switch (On/Off), Scale and Control

The Electronic Curzor is a bright line, sometimes dotted, emanating from the electronic centre and can be used with stabilized and unstabilized presentation. Neither parallax is introduced when taking a bearing, nor is the reading affected by centring error.

In most sets correct bearings can only be taken when the trace passes over the curzor position. This makes the process slow, especially during the procedure of setting-up alignment with the echo. This may require a number of successive attempts, each needing one revolution of the aerial to show its effect. A solution is found in one radar set by connecting the electronic curzor mechanically to the Parallel Index lines so that the Electronic Curzor and the Parallel Index are always parallel to each other. By aligning the Parallel Index lines roughly to the bearing line through the echo, there is no need to wait for successive paintings of the curzor.

Sometimes the curzor can be used in conjunction with the Variable Range Marker and the radius at which it begins to paint is the radius of the indicated range. This avoids confusion between the Electronic Bearing Marker and the Heading Marker.

Electronic Curzor (Interscan)

The Interscan Curzor is displayed as a bright line originating from the electronic centre. *Its brightness is uniform throughout the revolution period of the scanner, in other words, it is independent of the position of the scanner.* This property eliminates the disadvantage in connection with the taking of bearings with the electronic bearing marker mentioned in the previous section. The length, from the centre outwards, can be varied and a measurement of range can be obtained. *Coarse and Fine Ranging Controls* are provided.

An *Interscan Bearing Control* rotates the Interscan on the PPI and its brilliance can be adjusted. A scale indicates bearing and range by Interscan. Scales should be calibrated from time to time.

Bearings and ranges can be taken rapidly as there is no need to wait for the trace to sweep over the particular echo under observation.

A very brief description of the Interscan Technique is given in Appendix III.

Reflection Plotter and Fixing Studs. Illumination for Plotter

This plotter is described in Chapter 14. Illumination can be provided only when the plotter is fitted.

Hood, Visor, Magnifier and Fixing Studs

Hood and visor screen the display from daylight intrusion, the magnifier enlarges the picture. Some hoods offer opportunity to view the picture by two people simultaneously and are equipped with a polaroid filter which helps to eliminate light which is not intrinsic to the tube. The hood can be useful for reducing parallax errors.

Specific Controls

Many ships, nowadays, have True Motion facilities incorporated in their radar sets. By feeding direction and speed information into the display, the electronic centre, which represents own ship, moves across the face of the PPI at a rate and in a direction representing own ship's movement. A true motion plot is produced. Moving targets yield an echo with a "tadpole" tail on the screen while stationary targets (no current is assumed) give an echo which has no echo trail. The length and direction of the tail informs the observer about the speed and the course of the target. The range rings move together with the electronic centre and for taking bearings the electronic curzor or Parallel Index must be used. The display must be gyro-stabilized to prevent blurring and discontinuity in movements. After some time the display has to be re-set. This is done sometimes automatically (with an overriding control), sometimes manually and a warning signal may be given when re-setting is essential. One can compare the movement of the electronic centre over the screen with the movement of the ship's position over the chart (Chart Plan). By moving a switch, the movement of the electronic centre can be halted and relative motion presented, though generally the display is then off-centred. Conventional relative motion can also be selected.

Two displays are really advisable, each, if possible, with a separate time-base unit, one showing the true, the other showing the relative motion.

True motion is only provided for the short and medium range scales. The " tadpole " tail for longer range scales is so small that not enough information can be obtained.

Controls are the following:

Speed Switch

Speed information can be fed in from the ship's log, shaft speed (plus slip) or by a so-called artificial log. The last is a manually operated control which should be set according to the estimated speed of the vessel and re-adjusted when speed is altered.

Presentation Switch

Selects True Motion, Relative Motion Stabilized (North-Up) or Relative Motion Unstabilized (Head-Up) Presentation.

In some sets the *cathode ray tube* can be stabilized so that Head-Up True Motion can be selected.

Re-set

Engages North/South and East/West re-set controls so that the electronic centre representing own ship can be brought back to a suitable place on the screen. In other sets the control is set to the reciprocal of the mean line of advance of the ship. Sometimes automatic re-setting is arranged so that the heading marker goes through the centre of the picture after re-setting, thus making use of all available space on the screen. This can be a disadvantage when the tube has a centre-burn.

An automatic warning signal (visual or audible) is given when the display needs re-setting.

Tidal Correction

Estimated speed and direction of tide can be fed in so that Ground Stabilization can be obtained.

Sometimes a " *Course Made Good Correction* " *Control* is provided and this in conjunction with the Speed Switch can remove movement over the display due to tide.

Gyro-compass Alignment

This control has the same function as the " Rotate-Picture " Control. It can be done manually or automatically.

Compass Repeater

A separate Repeater which must be synchronized with the Gyro-Compass. It is responsible for stabilization of the display. Its function therefore resembles the Gyro Compass Alignment, but includes an additional component on which the course of the ship—and the direction of the heading marker which is synchronized with it—can be read off at a glance.

In some cases the electronic bearing cursor is duplicated on the compass repeater and its reading on the compass scale indicates the bearing of the target.

Zero Speed, Check or Hold Switch

This switch stops the movement of the electronic centre on the display. A warning light is shown when the switch is switched on.

When the display is switched back from True Motion to Relative Motion, or when it is switched over from True Motion to a long range scale where no true motion can be introduced, the picture sometimes needs re-centring.

Colour Displays

At present information is displayed in six different colours :—

Red, indicating echoes of strong intensity.

Yellow, indicating echoes of medium intensity.

Green, indicating echoes of weak intensity.

White, showing heading marker, variable range marker (dotted) and electronic bearing curzor (dotted).

Black, showing the plot (history track).

Blue is the background colour of the screen.

Symbols for Controls

For pilots going from one ship to another one, and also for officers joining a new ship, standardization of controls on marine radar equipment is highly advisable. This, of course, requires a great degree of co-operation between manufacturers, and although this state of standardization has not yet been achieved, IMO has issued a list of symbols with which the various controls should be marked so that these controls can be identified on foreign (for example, Russian) vessels. These symbols are illustrated in Fig. 4.1.

1. Radar Off
2. Radar On
3. Radar Stand-by
4. Aerial Rotating
5. North-up Presentation
6. Head-up Presentation
7. Heading Marker Alignment
8. Range Selector
9. Short Pulse
10. Long Pulse
11. Tuning
12. Gain
13. Anti-Clutter Rain Minimum
14. Anti-Clutter Rain Maximum
15. Anti-Clutter Sea Minimum
16. Anti-Clutter Sea Maximum
17. Scale Illumination
18. Display Brilliance
19. Range Rings Brilliance
20. Variable Range Marker
21. Bearing Marker
22. Transmitted Power Monitor
23. Transmit/Receive Monitor

ARPA Controls

Automatic Radar Plotting Aids are sometimes separate units connected to the main radar display, but in other cases they are incorporated in the main display. If one studies the ARPA Performance Standards drawn up by IMO (see ARPA Appendix I (a)) one realizes that a great number of extra controls have to be included to fulfil the requirements of the Specification. Moreover some of the conventional controls such as gain, anti-clutter (wind and rain) do not work in the same way as they do in analogue or "raw radar" displays. The reason is that the input signal to these displays (sometimes called Videos) is digitized with the signal either 'on' (level 1) or 'off' (level 0). Echoes which reach a certain threshold voltage are accepted as viable echoes and are amplified to a '1' level driving the CRT to limit level. Echoes and noise signals which do not reach the threshold voltage are treated as '0' and do not paint.

Owing to the varying strength of echoes more than one threshold level is employed. Before the signals are applied to the CRT some sorting out in amplitudes of the binary coded signals to two or more discrete levels is carried out. In many cases two levels are adequate and with this arrangement the video signal is fed to two threshold circuits in parallel. The upper threshold is set so that only strong signals will exceed it and those which do produce a logic level output. The lower threshold is set so that all signals above noise or clutter will exceed it and produce a similar logic level output. Note that with such an arrangement saturation or "blooming" of the display is avoided since the upper threshold is set for a maximum level which cannot be exceeded regardless of signal strength.

In ARPA displays the tracking of echoes of movable targets is one of the most important functions and as noise signals are never uniform and their general shape or envelope is dependent on external

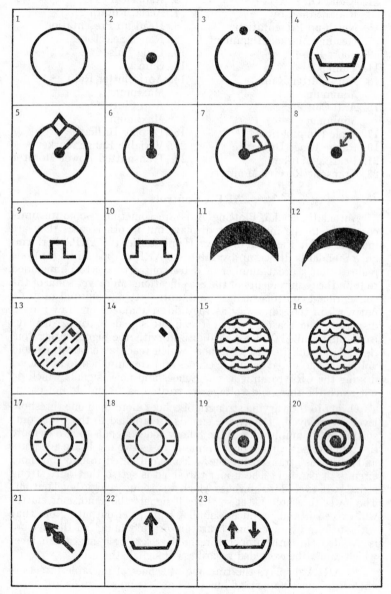

FIG. 4.1 SYMBOLS FOR CONTROLS

and internal conditions, a fixed lower threshold level may invite the risk that echoes cannot be acquired or tracked by the computer, or that echoes are lost during the process. It is for this reason that Racal Decca in their ARPA have applied automatic *control of gain* (sensitivity) *for each of their* 20 *echoes of targets being tracked.* Owing to the logarithmic amplifiers (Appendix III) being used, there exists a wide dynamic range in the video signal (the range of measurable signal levels between an echo which is just visible in the noise and one large enough to saturate the receiver). *Whenever an echo of a target is acquired,* two threshold levels, separated by a fixed amount, sweep through the dynamic range to examine the signal strength at the lower level just above the noise and also at the upper level. For example, suppose that in a *particular window* an echo is strong enough to be detected at the higher (control) level, the computer will automatically raise the level of both thresholds during an aerial revolution. This process is continued until the echo can only be seen at the lower (elevated) level. Thereupon the computer starts lowering the two thresholds progressibly until the target appears at the higher (control) level. In this way the gain is automatically controlled so that no tracked target, no matter how strong its echo-reflecting intensity, presents a signal to the system greater than the difference between the two thresholds. What, in fact, this means is that *signal/noise ratio is continually measured* and echoes *to be tracked* are projected on the screen with equal brightness irrespective of their original signal strength.

The ARPA picture differs from the conventional radar picture in the following ways:—

1. The gain control, as we have already seen, is mainly determined by the circuitry although some small adjustments can be made by the operator. Independent adaptive video-thresholding for individual echoes of targets are sometimes applied in the tracker unit.

2. To set the threshold levels, *logarithmic amplifiers* have to be used. The largest signal received by a marine radar from nearby targets is 10,000 times as great as the smallest signal. In order to drive the CRT the receiver must amplify the smallest signal one million times to an amplitude of the order of a volt. The largest signal would then become 10,000 volts! The amplifier must operate over the very wide dynamic range of 10,000 to 1 without distortion of the essential shape of the pulses or their amplitude relationship.

A logarithmic amplifier ensures that, while the dynamic range of the input signals it can handle without distortion is 10,000 to 1, the amplitude range at the output is compressed to a range of four to one, although the smallest signal is still amplified to one volt. Thus the smallest signal appears at the output as one volt, and the largest signal as four volts giving a dynamic range of four to one which is the logarithm of the input range of 10,000 to 1.

3. Automatic sea and rain clutter suppression can be applied automatically by means of adaptive circuits. When entering harbour or sailing close to land, it is advisable to switch over to manual control as strong returns of land-based targets may give rise to over-suppression and result in loss of echoes of small targets. See also the Section on Clearscan.

ARPAs without automatic clutter suppression need some protection against automatic acquisitioning of wave and rain echoes and consequent overloading of the tracking computer (minimum acquisition and tracking range controls are provided).

4. Echoes appear on a constant time-base for all range scales, i.e. its velocity is the same, corresponding to the velocity at the normal 6 or 12 mile (depending on the make) range scale. The video-return signals are received and stored (or "written" as one often hears) in real time into the memories, and then redistributed (or "read-out") at a speed according to the range scale in use. If, for example, the standard time-base is the one used for the normal 12-mile range scale (duration 148 micro-seconds), then on the 3-mile ARPA range scale echoes from targets one mile apart which were written with a time interval of 148/12 micro-seconds are read-out on the time-base with an interval of 148/3 micro-seconds. It will be seen, therefore, that on range scales shorter than 12 miles, the spot will move at a much slower rate over the display compared with the conventional radar display. This will result in making fuller use of the energizing properties of the CRT phosphor and hence a brighter display than normal. This, exactly, is one of the purposes of using a standard time-base, i.e. to obtain a *bright radar display* so that no hood or curtain has to be used in daylight conditions, and more than one person can watch it together. If the brightness is not sufficient on the 12-mile and longer range scales so that echoes have to be "touched up", this can easily be achieved (because memories are available with the ARPA unit) by re-cycling the echoes during the interscan period. Alternatively one can have echo-stretching (see Clearscan section) applied, if the facility is available with the longer pulse-lengths.

5. In ARPA displays the video signals are processed and written into the memory; it is possible therefore to compare and correlate one set of video returns with the video signals received during a previous transmission. Anything that appears during one transmission and not during the next one is liable to be suspect and could be a random noise spike or could be due to trans-

mission of other radar(s) (external interference). To get rid of such unwanted signals, MOON (M Out Of N) correlation techniques are used, meaning that if M identical echoes are received out of N successive transmissions, there is a reasonable probability that the M responses come from a realistic target deserving its echo-place on the screen. For example, M/N could be 1 (say 2 out of 2), or 0·6 (say 3 out of 5).

6. The rotating CRT coil has been replaced by fixed coils. These represent no problems for slow time-bases. With fixed coils it is a simple matter to put up interscan information-electronic bearing markers, collision markers and all the symbols that go with a modern automatic radar plotting aid display.

After this introduction it should be easier to understand the ARPA controls and their operation. Although there are several stand-alone indicators, let us assume that the ARPA is incorporated in the main display. Most of the basic radar controls have already been discussed, but some extra ones will be mentioned.

I. Basic Radar Controls

Presentation Switch. Besides the *Course-up* (stabilized) and *North-up* (stabilized, relative or true motion), there is a *Head-up* (stabilized, relative or true motion) *presentation*. Both the picture and bearing scale are stabilized (double-stabilization) and during course alterations no blurring of echoes will take place while the heading market remains in the "Up" position showing the ship's true course on the bearing scale. If the ship, for example, heading originally 000°T, turns to starboard, the heading line moves clockwise around the display while the stabilized display turns anti-clockwise simultaneously in order to keep the heading marker in the upright (i.e. pointing in the forward direction) position. This is in contrast with the *Course-up* display where the operator has to bring the heading marker back manually to the "Up" position after each alteration of course.

Echo Ref (Ref. Point or **Auto–Drift).** This switch is part of the True Motion Speed Input (Manual/Log) switch. In this mode a known stationary target is chosen as the ground relative velocity reference, and its echo is tracked by the computer so that own ship's ground velocity (i.e. speed *and* course) can be derived. Its direction and magnitude can be shown as a vector and can be read out from a digital display.   The direction and rate of the current/tidal stream can also be obtained, though, if there is wind, this direction and rate includes the effect of the wind.

To do the operation, position the computer circle by means of the Joystick over the selected stationary echo and press the button 'select echo'. At the same time the word 'REF' is written on the PPI next to the selected echo. The display is now 'ground-stabilized' as opposed to 'sea-stabilized' with log or manual input. The facility is sometimes also available for Relative Motion displays so that accurate *True Ground Track vectors* can be produced.

The same facility can be used when own ship is at anchor and is then known as 'anchor watch'. Own ship and other nearby ships can be monitored so that if own ship or any other selected ship (done by Joystick and 'select echo' button) moves over the ground more than a pre-set distance, an alarm is generated. The distance selected can be either 0·2, 0·4 or 0·6 n.m. and is set on a control switch indicator and scale. When the limit is exceeded, an audible alarm sounds, a warning light starts flashing and a symbol starts flashing over own ship's position on the PPI or the echo of the offending target. If the 'acknowledge' button is pressed, the audible alarm stops, the symbol and the visual alarm stop flashing (but stay on).

Picture Shift or **Offset.** When this button is pressed the control of shift is transferred to the Joystick control. The picture (on all presentations) can then be off-centred up to two-third radius in any direction. When the button is released, control is removed from the Joystick and the picture will stay in the position set. The button is sometimes situated on the display panel, but in other cases is incorporated in the Joystick.

Electronic Range and Bearing Line (E.R.B.L.). The distance of the line is controlled by the range knob, its direction by the bearing knob and digital readouts are available near the PPI.

Several modes are usually available :—

Centre Mode. The origin of the line is fixed at the centre of the PPI.

Fixed or Own Ship Mode. The origin of the line remains at the Own Ship's position on the PPI.

Free or Drop Mode. The origin can be shifted (either by horizontal and vertical shift controls or by the Joystick) to any position on the PPI. This mode can be useful for measuring range and bearing between two navigational marks (equivalent to parallel rulers and dividers) and for parallel index techniques on relative motion stabilized displays.

Carry or Off-set Mode. The origin is subject to the course and speed input set on a True Motion display and is *carried* across the display as if attached to the position marked by own ship, i.e. its bearing and range from own ship remain constant. If the origin is superimposed on the echo of another vessel and the line is rotated such that it passes through the position marked by own ship on a True Motion display then it will become immediately obvious to an observer if there is danger of collision (echo remains on the line). If the echo moves away from the line and the line is rotated to pass through the new position of the echo, then its direction reveals the apparent motion line of the target and the nearest approach or CPA can be estimated.

Pulse-boost Control. Its function is the same as the *"stretched echo"* facility discussed in the Clearscan section.

II.   Basic ARPA Controls

Before discussing the controls a few terms have to be explained first.

Acquisition. This procedure informs the computer about the position of an echo on the PPI. It initiates the tracking of the vessel showing the echo, and on some ARPAs starts the determination of the Point of Possible Collision or PPC. Acquisition can be done *manually* by bringing the computer ring on or round an echo and then pressing the *'Acquire'* button. Once the target has been acquired, its echo will show a dot, or have a small circle round it, or a small line on each side of it (known as "eyebrows"). On some ARPAs the automatic acquisition is done by the computer according to a priority programme, based on relative bearing and range, CPA and TCPA. In other ARPAs the echoes of the targets have to pass through a range ring (known as 'guard ring') or enter a zone or a sector before automatic acquisition takes place. Much more will be said about this later.

Tracking. This is the continuous calculation and updating of the various parameters which are important to be known by the operator.

(The IMO Specification states that the ARPA should be able to automatically track, process, simultaneously display and continuously update the information on at least :—

(*a*) 20 targets, if automatic acquisition is provided, whether automatically or manually acquired ;

(*b*) 10 targets, if only manual acquisition is provided.)
These parameters are :—
   1. Present range of target ;
   2. present bearing of target ;
   3. predicted target range of the closest point of approach (CPA) ;

4. predicted time of CPA (TCPA);
5. calculated true course of target (often denoted by 'True track');
6. calculated true speed of target.

The parameters can be required in alpha-numerics via a data channel operated in parallel with the PPI display.

In some sets the position of the PPC (*P*oint of *P*ossible *C*ollision) and the PAD (*P*redicted *A*rea of *D*anger, an area to avoid heading towards in order that own ship will always stay a minimum, safe distance—as chosen by the operator—away from another vessel if such vessel keeps her course and speed and own-ship maintains speed) are also calculated and shown on the display.

To read the six items in alphanumeric form, the operator has to reposition the circle by means of the Joystick over the echo and then presses the Joystick button or a "Read" button. On DIGIPLOT during the manipulation of the Joystick a dual switch is set to "Tag" and the information is displayed in pairs (range/bearing, CPA/TCPA, course/speed) by pushing in a series of data selector buttons. On the SPERRY ARPA all the information is displayed on a combined operator control/data display panel. The RACAS has an extra push button to read the Bow Crossing Range and the Bow Crossing Time.

In some ARPAs the placing of the circle over acquired echoes of targets in order to obtain tracking information, is done simply by pressing a *"select target"* button. Each time the button is pressed the circle jumps to a different echo—according to a programmed selection following the original acquisition sequence—and the data of its target are displayed.

Vectors and Tracks. Targets being tracked should display a vector on the display (time-adjustable or having a fixed time-scale) in a true or relative motion form which clearly indicates the target's predicted motion. The ARPA should also be able to display on request at least four equally time-spaced past positions of any targets being tracked over a period of at least eight minutes. This latter facility is denoted by several names : *Tracks, Trails, Plots, Past History* or *Track History*. It will indicate recent manoeuvres of targets if true tracks are depicted and provides a method of checking the validity of the tracking system.

Collision Assessment Display or Intercept Display. Such display shows the Possible Points of Collision (PPCs or PCPs) of own ship with all targets being tracked. In other words, each such point is that point where interception with a target would take place if the target keeps its present course and speed and if own ship heads towards it at its present speed. In the SPERRY ARPA the point is surrounded by a perimeter showing an area based on a safe distance of approach from the target chosen by the operator. Such an area is known as a Predicted Area of Danger or PAD.

NAV-Lines. These are lines which can be drawn electronically on the PPI for navigational purposes. They can be used for the practice of Parallel Index Techniques or the construction of a simple "electronic" map superimposed on the radar picture. Such a map could show courses and leading lines, deep channel limits and way-points.

After this introduction we can return to the basic ARPA controls.

Joystick (Target Selector). This can be operated at different speeds (from very slow to fast) and can determine the system response by means of positioning a symbol over the echo and operating an instruction button. Some of these responses are :—

Acquisitioning of targets, using the Acquire button.

Cancelling the acquisition, using the Delete, Release, Cancel or Drop button.

Production of PPCs and PADs.

Reading of data, using Read or Data Selection button.

Ground stabilization using Echo Ref plus Select Echo button.

Anchor Watch, using Anchor Watch Panel switch plus Select Echo button.

Picture Offset (all presentations).

"Time to Go" to any position on the PPI.

Electronic Bearing Line Offset in conjunction with Free Mode switch (only on some ARPAs ; on others there are separate shift controls).

Drawing of sector zones and barrier lines.

Drawing of NAV-Lines, using Map panel.

Vector Length Control (Future Position Control). This thumb-wheel control is set by the operator in minutes of time, resulting in instantaneous vector length on the display. A selector switch in conjunction with this control gives the choice of true motion vectors, relative motion vectors or true motion vectors plus history plot. One can bring, for example, relative motion vectors on a True Motion display, or, true motion vectors on a Relative Motion display (Sperry ARPAs only use the latter display with $\frac{1}{2}$-radius offset). Vector lengths can be related to range scale : 12 minutes on the 12-mile scale, 6 minutes on the 6-mile scale, etc.

Set CPA; Set TCPA. Both are useful controls for clear and restricted weather conditions. The operator adjusts the CPA control

(increments 0·1 n.m.) to what he considers is a minimum safe
distance to pass off another vessel. But he does not need *unnecessary*
early warnings (for instance 0·2 miles in 2 hours). To avoid undue
alert, the operator selects e.g. 2 miles, 30 minutes for CPA and TCPA
respectively. The setting of the 'Set TCPA' control prevents alarms
going off for vessels which might not become dangerous targets. A
visual and/or audible signal is given when a target is predicted to
close to within the minimum range *and* time chosen by the observer.
At the same time the warning is clearly indicated on the display
itself by echo-flashing, vector-brightening or a distinct symbol. An
audible warning can be silenced by an "acknowledge" button. Note,
by the way, that targets must have been acquisitioned before a
response can be achieved.

It is, of course, quite possible that small targets when first
detected have a TCPA less than the selected value. This might
result in giving the operator very little or no warning at all.

III. Less Used ARPA Controls

Guard Rings. Generally there are two, one for a long range to
detect the bigger targets, one for a short range for the smaller and
less reflective targets. They are either fixed, or one is fixed while
the other one can be adjusted in range, or both can be adjusted.
In fact, it would be better to speak of guard *zones*, for looking closely
at the 'rings', it can be seen that they consist of two circles close
together. This allows the echo to accumulate data related to the
target before the alarm is sounded. Or, as with some ARPAs, the
first level (outer ring) of operation will set off the audible alarm.
At the second level (inner ring) the intruder is acquired automatically; thereafter Level Two operation will alarm only.

The function of these zones therefore are two-fold :—

(*a*) They act to initiate a long-*range* warning sign to the operator
(compare with the TCPA-alarm for long-time warning) and can be
a very useful device in open sea in clear weather. Alarms are audible
and visual and often a symbol appears on the display near the echo
of the offending target. And/or,

(*b*) they start automatic acquisitioning procedure (auto-watch).
Near land, echoes or land-based objects may be acquired and some
desired targets may be cancelled owing to the accumulation of more
than 20 targets. To overcome this problem, RACAL-DECCA
ARPAs have a facility to draw sectors on the screen by means of
the Joystick and circle marker. Echoes entering this sector zone
will be automatically acquired in the tracking system.

If the zones are used for both functions, then as soon as the vector is produced, the zone alarm will go out and the symbol near the echo disappears.

In the KELVIN HUGHES ARPA, Anticol, search *areas* with adjustable forward and beam range limits are introduced. These extend to 24 n.m. maximum and any echo entering such areas is automatically acquired. These areas can be restricted and bounded by broken lines on the PPI, electronically generated screening lines against unwanted echoes, as for example, from land masses and rain-storms. Priorities in tracking and display of vectors depend on the following classification : *Early interest*, depending on CPA and TCPA ; *Range interest* ; *Shorter range* and *CPA interest*.

Auto-acquire. ON/OFF switch.

Barrier Lines. Exclusion Zone Boundaries (EZBs).
Area Rejection Boundaries (ARBs).   Area Elimination Boundaries (AEBs).

To prevent the computer becoming saturated near land when targets are *automatically* acquired, these electronic lines can be placed on the PPI either by means of translation and rotation controls or by using the Joystick. Such barriers can be stabilized (i.e. they remain in the same position with respect to the coastline echoes) by using a stationary reference target (see Echo Ref.).

Similar controls to prevent overloading by sea clutter and false echoes are :—

Minimum acquisition range. Could be, for example, 2 miles.

Minimum tracking range. The tracking of targets within this range is discontinued to prevent target swop and computer overload. In fact, within this range all targets are erased from the memory.

Minimum detection range. Inside this area targets are neither acquired nor tracked except in the case when targets enter this area from outside.

Blind sector area. This area is introduced to prevent acquisition and tracking of false echoes.

Other less used controls are :—

Mark targets. Echoes are marked on the PPI. Some ARPAs show the letter 'N' when tracking is done but no vector is displayed.

Mark Selector. For taking a bearing and range of a fixed point, which is useful for position finding. Instead of Course and Speed, Direction and Rate of Tide are calculated and displayed on the read-out screen.

D

Man Overboard button. When this button is pressed, a small circle detaches itself on the display from the point marking own ship, indicating the point where the mishap took place thus giving a reference point to return to, e.g. when contemplating the Williamson turn.

Centre Circle button. This button is used as a quick way of centring the circle marker over own ship's position, instead of using the Joystick. If the 'read' button is then pressed, the read-out will indicate own ship's course and speed.

IV. Alarms

Alarm signals are displayed so that one can hear and see them. The audible alarm is generally acknowledged and silenced soon after hearing it. A volume control is also provided to regulate the sound intensity.

System or Equipment Alarm. The equipment, including the computer, is self-tested at regular intervals and a fault condition will activate the alarm. This alarm is probably the most important one, because an ARPA is not of much use if the calculations are wrong.

Test Programme Warning. It is possible to bring echoes of artificial targets on the screen, so that the unit acts as a radar simulator. As the movement of these echoes is determined by the test operator, it provides an overall check on the tracking system. When this presentation is selected, normal operation of the ARPA is prohibited.

CPA and TCPA Warning. Operator is warned when own ship is predicted to pass within a pre-selected CPA distance *and* when own ship is predicted to pass CPA within a pre-selected time interval. On some sets a small 'T' symbol (for "threat") is shown on the display.

Possible Collision. A warning that a target with a CPA less than the selected CPA, has a TCPA of 18 minutes. Such a target will have its PAD symbol in contact with the 18 minute position on the heading marker.

Collision. Warning that a target with a CPA less than the selected CPA, has a TCPA of less than 12 minutes.

New Target. New target is detected in the search area.

Zone Alarm. A new target is detected in the guard zone of sector.

Bad Echo. Owing to wind, rain conditions, shadow sectors or poor reflecting properties, it is difficult to track the target.

Lost Echo (*Target*). Tracking system has lost the echo, most probably due to conditions mentioned under "Bad Echo".

Anchor Watch Alarm. Ship has drifted through a pre-selected distance.

Automatic Stranding Alarm. This alarm is provided in the CAS II (Sperry) ARPA when "Channel Navigation" Lines are incorporated and a ground-stabilized digital map is held in correct registration on a sea-stabilized relative motion display.

Tracks Full. The tracking system is saturated. DIGIPLOT shows a crossed line on the Heading Marker. Sperry CAS II display flashes the acquisition symbol when the 18th target is acquired (maximum capacity is 20 targets).

Manoeuvring Time. A lamp will start flashing when the time has arrived for a simulated (trial) manoeuvre to be carried out. See later, this section.

Reset. This is the usual alarm to tell the operator that it is time to reset a True Motion display. Or, in some ARPAs, it places the acquisition symbol back at the geometrical centre of the tube, unseen, but available for subsequent use.

In the following cases the echoes on the display themselves will convey a warning to the operator by echo-flashing, vector brightening, showing of distinct symbols (squares, triangles, circles etc.) or a combination of such signals.

Target on collision course.
CPA and TCPA limit setting have been exceeded.
Echo has entered a guard zone or sector.
Target has altered course and/or speed.
Bad echo.

V.   Trial Manoeuvre or Simulation

This is a facility which is mandatory according to the IMO Performance Standards for ARPA and is useful for the operator to predict and pre-judge the effects of avoiding action and to be able to make a decision about the best choice available. It, of course, must be assumed that targets maintain course and speed.

Three variables have to be fed in : Trial course, trial speed and trial delay. The latter is the estimated time interval between the start of the simulation and the time when the *actual* course or speed alteration is expected to be implemented. The simulation on the display starts when a switch or button is depressed, stating **Simulation, Trial Manoeuvre** or **Selection and Interrogation.** At the same time a symbol on the display shows clearly (large letters

S or T) that the simulator mode is in operation so that the observer cannot interpret the picture as the real situation. On some sets the system reverts back to the normal mode after one minute and the operator has to press the Trial Manoeuvre button again.

Simulation can be shown on True Motion and Relative Motion Presentations. Switches, Controls and Alarms operate the same as when in Normal Mode so that, for example, the operator can change the vector lengths, note when targets are expected to come within the CPA limit and is warned of collision dangers. The data read-out panel(s) can also be observed but data are now given for the trial manoeuvre(s).

Static trials show the vector change of *own ship*, allowing for time delay and also for ship dynamics during the turn when the display is on the *True Motion* presentation. With the PPI in *Relative Motion* mode the newly calculated vectors of *all* acquired *targets* are shown as a result of the proposed course and/or speed changes, and taking into account the delay time. Various courses and/or speed changes can be entered while the trial switch is depressed and the results observed both on the display and the digital data read-out.

The Dynamic trial shows generally the *True Motion* display of a particular course and/or speed change which includes own ship's dynamics. The manoeuvre is observed on the display and the digital data read-out in *quick time* (1 to 30 speed-up). The trial delay time allows a delayed turn to be displayed, starting 1 to 9 minutes in future.

There are certain variations in the controls, depending on the make of ARPA ; in some of them the *change* of course (starboard, port) and/or speed (slower, faster) is incorporated. Also, with displays showing PPCs and PADs, effects of proposed course alterations are obvious and only speed simulation is incorporated (watch the PPC and PAD movements).

VI. NAV. Lines (Fairways) ; Radar Map

These electronic lines can be used for the application of Parallel Index Techniques—which will be fully discussed in a later Chapter, and also for electronic map drawing, e.g. marking traffic separation schemes, anchorages, etc.

The drawing itself is a straightforward method. The map panel section is switched on. The Joystick is positioned with the circle marker over the starting point of the line and the **'start/end'** button is pressed. When the Joystick is now moved, the circle marker pulls a line out behind it as if it were a piece of elastic. After the second point has been entered by means of the same 'start/end' button, the circle marker can be removed leaving the desired line in position.

If a dot is required (for example to mark the position of a buoy), the circle marker of the Joystick is placed over the required PPI position and the button **'point'** is operated. A dot will appear on the PPI surrounded by a symbol. The latter shows the observer that the dot is a synthetic mark and not a radar target. Generally up to 15 elements (lines + points) are available. When the stock is exhausted, a notice **(MAP-FULL)** will appear.

During this process X and Y coordinates in n.m. (northings and eastings) are shown on the PPI and also recorded in the store. The starting point of the map is the reference point (0,0). During the drawing of the map one can always use the **E.R.B.L.** facility in the same way as one would use parallel rulers and dividers when drawing lines on an actual chart.

There are *three* levels of producing an electronic map.

1. The map, consisting of leading lines, deep channel limits, shoal boundaries etc., is drawn right over the top of the radar picture. Temporary storage is available, but only so long as the radar set *remains switched on.*

2. The map is prepared well beforehand on the radar display, say when the ship is in open sea. A navigational mark, being a good radar target, is chosen as a reference mark, (0,0). The chart is then consulted for the X and Y coordinates required for the various parts of the electronic map. The latter is then drawn on the radar display by watching the X and Y coordinates of the marker circle on the radar display. The map is next given a reference number and stored in a permanent memory by pressing the **'store'** button. Later, when required, the map is recalled **('select map'** button), aligned (press **'alignment'** button) with the Joystick bringing the reference mark (0,0) of the electronic map over the chosen reference mark on the actual radar display.

3. The map is prepared by the manufacturer and stored by means of tape or disc. It is then fed into the radar display by means of a reader, aligned, and latitude and longitude of the reference point are set in by means of a keyboard. This is the method applied when the ARPA unit forms part of an Integrated Navigation System.

For map presentation a True Motion Ground-Stabilized display is of course ideal and one can leave the system alone.

With a Relative Motion display the situation is more complex and a DR/EP programme, driving the electronic map, should be included. In this case there is a need for continuous comparison by the computer between the coordinates of the electron map reference point and the coordinates of the reference point on the radar display itself.

Several ARPAs have the facility to feed in—by means of a keyboard and the Joystick—the latitude and longitude of a salient navigational point, or of the position of own ship when berthed alongside. During the voyage, when pressing the button marked 'Own Ship', the vessel's position will be displayed, either as a "Fix" or a D.R. Position, depending whether the display is in the Ground-stabilized or Sea-stabilized mode respectively.

This then concludes the discussion of the controls of an ARPA quantized display. At a certain range scale setting—on a time-base where echo signals are recorded in *real* time—one, sometimes, can find a dual switch, giving the opportunity to switch over to "Raw Radar" if something has gone wrong with one of the digital circuits.

Switching On. Testing
Setting-Up Procedure
(Analogue Displays)

The first and foremost aim is to obtain a good navigational picture so that the radar set can be used to the best of its ability as an aid to navigation and anti-collision. The sequence in switching on and employing the controls can be varied. One way of doing it is set out below.

1. Make sure that the scanner is clear, the Scanner Switch is On, the Standby Switch on " Transmit " and the gyro-repeater is disconnected.

2. Set Gain (Sensitivity) Control to minimum, Brilliance and Anti-Clutter Control to zero position. Switch to a medium or long Range Scale.

3. Switch on. Wait a few minutes till transmission starts.

4. Turn up the Brilliance Control till the trace is just visible. Switch on the Range Rings. Focus on one of the rings situated half-way between the centre and the edge of the display.

5. Switch off the Range Rings. Turn the Brilliance Control down until the trace is *just barely* visible.

6. Check that the electronic centre is precisely under the centre of the face of the tube. If necessary adjust with Centring Controls.

7. By means of the Picture Rotate Control turn the picture until the heading marker indicates zero or the ship's true course on the bearing scale.

If the display can be gyro-stabilized, connect the compass repeater to the display at that moment when the reading of the

Heading Marker on the bearing scale is the same as the reading of the gyro-compass.

8. Adjust the Gain Control until the speckled background is just visible. Note the setting of the Gain Control if it is calibrated.

9. Switch to the shortest Range Scale. Turn the Gain Control to Maximum (or to a setting stipulated in the manual) and test the set by the Performance Indicator Switch. Try tuning by means of the Tuning Control if the test is unsatisfactory. After testing (and tuning) put the Gain Control back according to the setting found in (8).

10. Select range scale required and adjust clutter as required.

With a *True Motion* radar presentation, then, in *addition*:

(*a*) Switch the Presentation Switch to " True Motion " (North-Up or Course-Up).

(*b*) Use the " Re-set " Control to off-centre the electronic centre to a suitable place on the screen.

(*c*) Feed in the ship's speed either from the ship's Log, Shaft Speed or Artificial Log (manual setting of estimated speed).

(*d*) If desired, correct for current. This may be useful for pilotage. All small echo trails should then be removed from echoes caused by stationary objects.

The operator can check if the correct speed has been fed in by turning the lines on the Parallel Index at right angles to the Heading Marker and then note the time it takes for the electronic centre to traverse between two or more lines. It is important in this connection to check whether the range rings have the correct diameter (the Manual provides information about the outer range ring). If the scale of the particular Range Scale is not exactly correct, the speed of the electronic centre across the screen will then be too small or too great and this will affect the motion of all the echoes on the display. If the diameters of the range rings need slight adjustment then the velocity of the time-base should be varied internally by means of a potentiometer and the Manual can be consulted for the necessary information.

Note. If the Range Scale is altered from medium to long, where True Motion is not provided, put the Presentation Switch back to " Relative Motion Stabilized " before changing Range Scale, otherwise the centring will not be right.

If the *Relative Motion Stabilized Off-Centred* display is wanted, then instead:

(*a*) Switch the Presentation Switch to " True Motion ".

(b) Put the Zero Speed Switch on zero.

(c) Off-centre the electronic centre by means of the " Re-set " Controls.

Both for True Motion display and Relative Motion Off-Centred display the mechanical curzor and the edge bearing scale should be disregarded. Instead the electronic bearing marker should be used. Alternatively the parallel index can be employed in conjunction with the edge scale to obtain quick but not very accurate bearing observations.

Switching Off

Switch the Range Scale to a short scale (three or six miles) in order to be ready for an emergency. Then switch the mainswitch to " Off ".

In the case of an emergency, just switch on again. If there is plenty of time for switching on, turn Gain Control, Anti-Clutter Control and Tube Brilliance down before setting the mainswitch to " On ".

Switching on and off repeatedly during a watch is bad practice and not good for the set. It is then better to keep the set on " stand-by " or running fully operational if a stand-by switch is not available.

Maladjustments

Maladjustments should be recognised instantly. Look for the speckled background on the longer range scales. Reduce the gain if the speckled background is too bright. If the centre of the screen is too bright or too dark on the medium or long range scale, the Anti-Clutter Control should be adjusted. *Always check the Gain with Anti-Clutter down before adjusting the Anti-Clutter Control.*

If the picture is dull and the Heading Marker seems faint, the tube Brilliance is probably too low. To check on this, turn the Gain Control fully down, then adjust the trace by the Brilliance Control in such a way that it just is *barely visible*. Finally restore the Gain to normal.

When the complete trace can be seen rotating, then the Brilliance setting is too high. Turn the Gain Control down again and do as described in the previous paragraph. This suppression of gain or sensitivity, when adjusting the brilliance, is especially essential when there are many echoes on the screen. These echoes light up when the trace sweeps over them and an apparent trace is created which makes proper adjustment of the brightness of the trace difficult. The apparent trace disappears when the Gain Control is turned down.

Check the focus occasionally. If the picture is not sharp, bad focusing may be the reason.

Switch Range Rings and Range Marker off when no ranges are to be taken. Dim the illumination of the Bearing Scale when not taking bearings.

Check tuning frequently, especially during the first ten minutes after switching on.

Check the course input and the slight hunting of the Heading Marker when the display is stabilized.

Correct course and speed input must be fed in when the True Motion display is used. *Errors in these will affect the tracks of echoes of other vessels* (see Chapter 14: Plotting Errors). When the speed is fed in manually, close attention should be paid to any alterations in own speed. One must also remember that the speed of a vessel is seldom uniform and adjustments have to be made from time to time.

With a *new* reflection plotter it is advisable to check the position of the plotting mirror (see Fig. 15.2) in order to avoid parallax effects. To do this remove the cover of the plotter while it is in position on the display so that access can be obtained to the three adjusting screws.

With a chinagraph pencil mark the plotting surface at three equally spaced positions, adjacent to each of the three adjusting screws, on a circle at half the radius of the display. With the display operating adjust the plotting mirror by means of the three adjusting screws so that the reflections of the three pencil marks appear to be in the same plane as the display picture. Care must be taken that *all* three pencil marks appear at this position.

Whilst maintaining the setting of the adjusting screws by means of a screwdriver, lock the adjusting screw nuts.

After a final check, replace the plotting cover.

Setting-Up Procedure
(ARPA Displays)

The controls found on conventional, analogue displays are also provided on ARPA displays and many do not need further discussion, as, for example the display presentation switch. The three main differences are :—

1. The gain does not need any setting up although slight adjustments occasionally may be required to get better contrast. The display is quantized (conversion from a continuous signal to a set of discrete ones) and an echo is seen when its intensity is above a certain level and is not seen when its intensity falls below that level.

In fact, ARPA displays have two video levels, one for strong, and one for weak, echoes ; in one particular ARPA these two levels—at a fixed interval—sweep through the dynamic range of the echo-signal, i.e. from noise to saturation level.

2. Clutter adjustment is done for a reason different from the analogue display. Besides automatic clutter control (Clearscan), separate clutter controls are sometimes provided to prevent the acquisition and false tracks of clutter echoes when the display is in the automatic mode.

3. Speed input must *always* be fed in irrespective of the display presentation, because vector/graphic calculations are based on it.

One way of setting up an ARPA could be as follows :—

1. Select the range scale and adjust brilliance and contrast.

2. Select presentation mode (Relative or True Motion).

3. Select orientation—Head-up, North-up or Course-up.

4. Input speed. This can be done manually, or via the log sensor (sea speed) or via an ARPA facility called **Auto–Drift** or **Echo Ref,** whereby the computer keeps the radar picture (true or relative motion) in the ground-stabilized mode. In the latter case, the input is not "speed", but "velocity" and can be represented by a vector, having magnitude *and* direction.

5. Select the acquisition mode—Manual or Automatic. For manual acquisitioning the Joystick and circle-marker is used.

6. Set minimum CPA and TCPA limits.

7. Set Vector Mode—True or Relative Motion.

8. Set 'vector length'—in minutes. 'History track' can also be selected, but it will take some time after acquisition before the track is shown on the display.

9. Watch for alarm signals and compare the graphical presentation shown by acquired targets with the digital read-outs.

Let us go back to point no. 3, **speed input.** For navigational purposes, a ground-stabilized picture (in the true or relative motion mode) referenced by the ARPA computer is ideal and nobody will dispute this. But even for anti-collision purposes this type of velocity input is the best choice, and for two reasons :—

(*a*) Even with an accurate log sensor one can never be 100 per cent sure that the true motion track of a target generated on an ARPA display yields accurate information about the aspect of the other vessel. The reason is that it is never possible to calculate the leeway angle for another ship (depends on type, size, windage, etc.) and secondly, it should be remembered that there are many places in the world where the current strength and direction in one section of the area covered by the display for a particular range scale, is not the same as in another section of that area.

With ARPA ground-stabilized displays (Echo Ref) on the other hand we can be quite certain that the vector direction of own ship and targets, represent the true ground tracks.

(b) The only cases where the type of avoiding action depends on the aspect are related to Rules 14 to 17 (incl.) but these Rules apply to the conduct of vessels in *sight* of one another when the aspect can be deduced by watching the ship or her navigation lights.

In restricted visibility Rule 19 has to be obeyed. Here the type of avoiding action depends only on the knowledge if ships on opposite courses, crossing or overtaking, are forward, abeam or abaft the beam ; in other words the relative bearing must be known. In addition we must know if we are overtaking another vessel, but here Rule 13, although written for clear visibility, should give guidance for restricted visibility : "When a vessel is in any doubt as to whether she is overtaking another, she shall assume that this is the case and act accordingly."

What is really said here is that under conditions of restricted visibility a knowledge of accurate aspect (if this were available) is not required for the application of Rule 19.

There are ARPA displays working in the *Intercept Mode*, showing PPCs and PADs, which operate only in the Relative Motion presentation (ground-stabilization is not provided). If there are errors in the log speed input, then the PPC of ships on collision courses will move up or down the heading marker. The sub-divisions on the heading marker are influenced in a compensatory manner by the same speed error, so that the time interval to the PPC (TCPA) remains the same. If the other vessel is not exactly on a collision course there will be also an azimuthal displacement of the PPC, further away from the heading marker when the speed input is too low (Fig. 15.23).

Speed error also involves all forms of trial manoeuvre predictions. In the case of the PPC/PAD type of display, this will manifest itself in minor errors in the values of Time Interval and Heading to Steer (to clear a PAD), the principal information of the PAD symbol.

Now regarding point 5 : The acquisition mode. If the mode is automatic, it will relieve the observer of some of the load and there will be a lot of information available in busy traffic, perhaps too much. A programme is sometimes written for the computer to attach certain weightings to the targets and does the sorting out for the tracking priority. The I.B.M. Bridge System had the degree of collision categorized according to five priority steps :—

   (i) Based on range set by operator ;
  (ii) Based on CPA/TCPA limits set by operator ;
 (iii) TCPA greater than limit ;
 (iv) CPA greater than limit ;
  (v) Range increasing, i.e. target has passed CPA (TCPA is Negative).

This priority Diagram or Threat Profile is shown in Fig. 4.2.

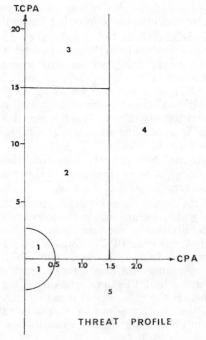

FIG. 4.2 THREAT PROFILE

In the IMO Publication "Minimum requirements for training in the use of Automatic Radar Plotting Aids (ARPA)" (ARPA Appendix I (a)), it is stated that the training for masters, chief mates and officers in charge of a navigational watch should be "a knowledge of the criteria for the selection of targets by automatic acquisition".

A word of warning about the Threat Profile shown above. No

allowance has been made for the relative bearing of the target and targets coming from astern have the same priority as targets coming in on the bow at long- and medium-range with the same CPA and TCPA ('Priority 3' section).

In the **auto-acquisition** mode one has to watch for warning signals that the computer system does not become saturated when new acquisitioning will stop and even the tracking of some targets may be dropped. *Warning signs* are :—

"*Tracks Full*" notice.

Flashing signals indicating "*bad*" or "*lost*" echoes.

A dotted ring indicates the area in which automatic acquisition still takes place. This ring expands or contracts depending on the degree of saturation.

Remedies are :

Switch to the manual mode of acquisition.

Use ARB (Area Rejection Boundaries) lines to screen off areas of land echoes and rain clutter.

Use the controls to set a "Minimum Acquisition Range" and/or "Minimum Tracking Range", in that order. Within the latter range, targets (echoes) are completely erased from memory.

In all cases consultation of a raw radar display should not be neglected.

In **Manual Acquisition** Mode CPA and TCPA values are not immediately available and range is the only guide to base priority on. One therefore starts the acquisitioning procedure from the nearest ship-echo outwards. Even in this mode after acquisitioning "bad" or "poor" echoes can be recorded. In such a case the lower video level might need some adjustment and the radar engineer or radio operator have to be called in.

Before we finish this section on the adjustment and interpretation of ARPA displays, some *final* remarks are given :—

1. When targets are close together, vectors may exchange places or one vector is lost, wrong information is displayed on the digital read-out panel and the history track is broken up into two parts, containing initial and final information of each target. This disturbance is known as "Target Swop" and the operator has to re-acquire the targets.

What, in fact, has happened is that echoes came within the same gate-width and computer discrimination was lost. To reduce this risk, several ARPAs have a facility incorporated to reduce the size of the gate (Harbour Mode), and there are also computers which employ "rate-aiding". In that case the computer calculates the position ahead of an echo for a short period of time. The proper vector is only drawn if this position is confirmed later.

2. In many ARPAs the tracking is based on *relative motion* information and tracking errors may occur in relative motion vectors of all targets when own ship alters course or in the true motion vector of a target during its course alteration. In some ARPAs therefore tracking is discontinued during these manoeuvres.

If, however, the position and velocity of tracked targets is stored in *true motion* format, as is sometimes the case, it means that target vectors do not need to be re-established following the change in relative motion which takes place each time own ship alters course and/or speed. Vector data continue to be updated throughout own ship's manoeuvres even when these follow each other continuously. A change in the target's true motion will be quickly detected and updated.

3. Targets at *short* range sometimes produce vectors which are directionally unstable or the tracker may lose the target (echo). This is due to the very rapid bearing changes which may occur. Also, at short range, when target manoeuvring becomes extremely important for own ship, smoothing constants are sometimes reduced deliberately.

4. When using intercept systems with PPCs and PADs, note the following points :—

(*a*) The line joining the echo of the target to the PPC is a True Motion *direction only line*. By itself it does not give any information about the speed of the target. However when comparing its length to the distance from Own Ship to the PPC, it will indicate if the target's speed is slower, equal or greater than Own Ship's speed.

(*b*) The PPC is not usually at the centre of the PAD. In Sperry CAS II display the trackline—mentioned in (*a*)—terminates at the *geometric centre* of the longitudinal axis of the PAD. If a selected CPA of zero range is called up, a minimum PAD is displayed of 300 yards radius to compensate for any residual error influences. In this case the actual PPC is not shown on the PPI. This presentation differs from the Raycas display where the precise PPCs locations can be brought on the screen.

(*c*) To determine the expected CPA and TCPA, use the alpha-numerical read-out display panel and not the PPI presentation.

(*d*) Sometimes two collision points and two PADs are displayed for the same target. Providing the Collision Regulations are obeyed there is no objection to setting a course with the heading marker intersecting the line between the PADs.

(5) The "Trial Manoeuvre" facility or simulation, is useful but it is not simply a "Trial and Error" device. The Collision Avoidance Rules should be complied with.

Owing to the specific design of the C.A.S. Sperry System, the trial manoeuvre for course alteration is not required here as the display presentation offers solutions directly. Owing to the distribution of the PADs there is sometimes the temptation to alter course during reduced visibility in a direction which contravenes Rule 19, and/or to make small alterations of course contrary to the requirements stated in Rule 8.

Abbreviations

AC	Anti-Collision.
AEB	Area Elimination Boundary.
AFC	Automatic Frequency Control.
ARB	Area Rejection Boundary.
ARPA	Automatic Relative Plotting Aid.
C-Band	4,000—8,000 MHz.
CP	Circular Polarization.
CPA	Closest Point of Approach.
CRT	Cathode Ray Tube.
DR	Dead Reckoning.
EBC	Electronic Bearing Curzor.
EBL	Electronic Bearing Line.
EP	Electronic Plotting, or Estimated Position.
ERBL	Electronic Range and Bearing Line.
EZB	Exclusion Zone Boundary.
FTC	Fast Time Constant (for rain clutter suppression).
IC	Interscan.
IVST	Indirect Viewing Storage Tube.
Log-Amp	Logarithmic Amplifier.
MOON	M Out Of N selection procedure.
PAD	Predicted Area of Danger.
PCP	Possible Collision Point.
PPC	Possible Point of Collision.
PPI	Plan Position Indicator.
RM	Relative Motion.
S-Band	2,000—4,000 MHz.
STC	Sensitivity Time Control (for wind clutter suppression).
SWG	Slotted Waveguide.
TCPA	Time of Closest Point of Approach.
TM	True Motion.
X-Band	8,000—12,500 MHz.

Range and Bearing Measurement

The range rings should be checked as described in Chapter 3 (Range Accuracy) and the Range Marker should be calibrated against the Rings.

When determining a fix by ranging, select objects well apart in azimuth, moderately high and steeply descending in the direction of the observer (Chapter 9: Coastal Navigation).

The *range accuracy* of a radar set is generally high.

In contrast to range accuracy, the *bearing accuracy* is generally not so high. This is because the bearing accuracy is governed by so many factors related to inherent characteristics of the set and these have been fully discussed in Chapter 3; secondly, if the observer is not very careful, he may easily overlook simple adjustments and precautions connected with the bearing accuracy which can prevent large errors.

Two errors falling under the first category will be repeated here:

(1) Beam-width distortion, which can be partly eliminated by reduction in gain;

(2) Heading marker error, which can be determined by various methods.

Other errors falling under the second category are:

(3) Centring error;

(4) Error due to yawing of own ship;

(5) Error due to parallax when viewing the display.

Errors (3) and (5) can be important when we use the mechanical curzor, error (4) can come into existence when the display is not gyro-stabilized.

The error caused by a displacement of the electronic centre and the geometrical centre of the face of the tube will vary in sign and magnitude. The distribution is semi-circular. The magnitude depends on (a) the distance of the echo from the centre of the tube, and (b) the angle between the bearing of the echo and the line connecting the two centres.

This is illustrated in Fig. 4.3 where a set of curves is drawn showing the errors in different zones of a 9-inch (23 cm.) and 15-inch (38 cm.) PPI for 2·5 mm. displacement of centres. Comparing the 9-inch and 15-inch displays respectively, it will be noticed that the errors amount to 1·3° and ·8° at the edge, 2·5° and 1·5° at ½ radius and 5° and 3° at ¼ radius. For displacements greater than 2·5 mm. the error will be increased in direct proportion. The error is smaller at the edge of the screen than near the centre and it is zero in the direction GE and maximum in the direction perpendicular to GE. *It is important to note that bearing errors, especially when the target yields an echo near the centre of the display, can be quite large for a displacement of the centres, which is unlikely to be noticed unless the display is checked regularly.*

The same figure shows that a very dangerous situation can develop when another vessel is on a collision course and the Master is under the impression that the bearing opens out. His plot would show a wrong nearest approach and collisions have occurred from

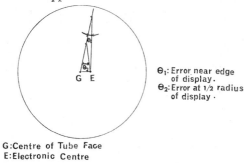

Θ₁: Error near edge of display.
Θ₂: Error at 1/2 radius of display.

G: Centre of Tube Face
E: Electronic Centre

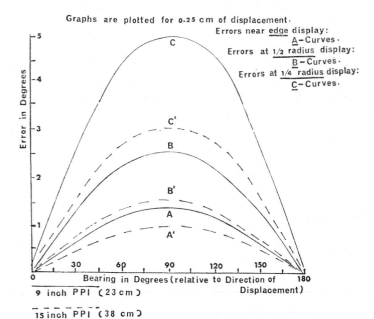

Graphs are plotted for 0.25 cm of displacement.

Errors near edge display:
A - Curves.
Errors at 1/2 radius display:
B - Curves.
Errors at 1/4 radius display:
C - Curves.

Error in Degrees

Bearing in Degrees (relative to Direction of Displacement)

9 inch PPI (23 cm)

15 inch PPI (38 cm)

Fig. 4.3 Bearing Errors due to Displacement of
Electronic Centre

this cause. As a slight discrepancy between the two centres may remain even after adjustment due to centre-spot wander, it is always advisable to take bearings of echoes near to the edge of the display. As we have seen before, the range scales may be altered to get the echo near to the edge of the display, or the display can be off-centred.

To decrease the possibility of errors due to parallax, the visor can be put on the display when taking a bearing or the inner and outer lines of the mechanical curzor should be kept superimposed.

Centring error and error due to parallax can be eliminated by using the electronic curzor and for this purpose the specially provided electronic bearing marker scale should be employed.

Then there is the error due to yaw when the display is *not* gyro-stabilized. If the heading marker is at zero, bearings of objects are read off as relative bearings. If the heading marker is placed at the true course of the ship and the display is non-stabilized, then true bearings are indicated by the cursor *only when the ship is on her correct course*. It is essential therefore to note or to ask for the heading *at the instant* when a radar bearing is being observed. It is the same procedure which is followed when taking a D.F. bearing and the goniometer is not connected to the gyro-repeater. Only when the display is gyro-stabilized or when the bearing scale is stabilized (Fig. 10, 1) can the true bearing or true course be read off directly without noting the ship's head. Otherwise yaw must be allowed for.

As yawing will cause blurring on an Unstabilized Display, no bearings should be taken during the yaw. The bearing should be taken when, for a short moment, the ship's heading is steady while another officer should sing out the heading of the ship. In general, the habit of taking bearings of the afterglow instead of the echo, should be avoided.

To summarise, when taking a bearing:

(i) Adjust the centres correctly. Select the shortest range scale possible to put echo in the outer half of the display. Alternatively, use the electronic curzor and scale.

(ii) Apply the heading marker error if there is one.

(iii) If the display is not stabilized, check on the compass (and true) course.

(iv) Reduce the gain, so as to reduce the beam-width distortion.

(v) View cursor and bearing scale from the correct position to avoid parallax. Alternatively, use the electronic curzor and scale (calibrate the latter from time to time).

It is assumed that with stabilized displays gyro errors are known and that the true course is fed into the display.

CHAPTER 5

INTERPRETATION OF DISPLAY (1)

The following four chapters deal with the interpretation of the display.

The shape and the strength of echoes on the screen and the detection range of targets depend on:

(i) Certain characteristics of the set itself;

(ii) Certain characteristics of the target;

(iii) Various effects of atmospheric and weather conditions;

(iv) The existence of blind and shadow areas and sectors.

In Chapter 3 we discussed the characteristics of the set which affect echo strength; Chapters 5 and 6 consider the characteristics of the target; Chapter 7 is devoted to the influence of atmospheric and weather conditions and blind and shadow sectors. The final chapter in the series looks into the cause of unwanted echoes and mentions the remedies which can be adopted.

Effect of Aspect, Shape, Composition and Size of Target on Echo Strength and Detection Ranges

Aspect

This is the most important factor of a target in relation to echoing strength and detection range though it is related to surface texture. By aspect is meant the *angle of view*, both in the horizontal and the vertical planes which the target presents to the observer. For example, a horizontal aspect of 90° means " beam-on " and a vertical aspect of 90° means " perpendicular ".

It will be easily understood that a target which is steep and beam-on will return an excellent echo, especially when its surface is smooth. Such an extremely strong reflection is called *specular* reflection, as it can be compared with reflection when facing a mirror. If the aspect is not 90°, for example, a sloping surface or an

85

object which does *not* present its face towards the observer, the echo response can still be good, provided the surface texture is rough and causes a certain amount of scattering. This is illustrated in Fig. 5.1.

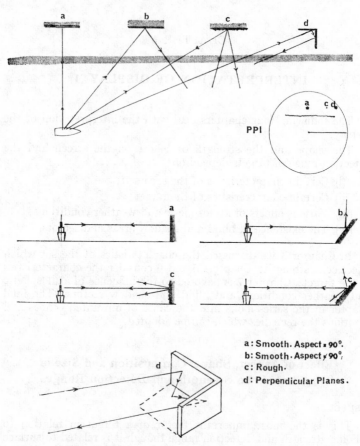

a : Smooth. Aspect = 90°.
b : Smooth . Aspect ≠ 90°.
c : Rough.
d : Perpendicular Planes.

FIG. 5.1   ASPECT AND SURFACE TEXTURE

Shape and Surface Texture

The shape is a permanent feature of the target in contrast with its aspect which can change as the ship and/or target change their relative positions.

(1) *Plane* (Fig. 5.1)

    (a) Perpendicular and smooth.

There is excellent response only when facing it, i.e. specular reflection.

Examples: Some cliffs and isolated buildings.

(b) Perpendicular and rough.

Fair reflection from all directions. (a) and (b) can be compared with the illumination of a road by the headlight of a motor-car. The road scatters light back in all directions, while the " Cats' eyes " only provide specular reflection. Examples: Craggy cliffs and vertical faces of icebergs which are broken up into many facets.

(c) Sloping and smooth.

There will be very little response.

Examples: Sloping faces of icebergs, growlers, sand and mudbanks, gently sloping hills not covered with vegetation, dunes and foreshore.

(d) Sloping and rough.

There is fair response.

Examples: Sloping hills covered with trees, undergrowth or boulders.

(e) Three mutually perpendicular planes (Fig. 5.1 (d)).

Extremely good reflection from all directions. It can be shown by simple geometry, that from whatever direction the beam comes, the reflected ray will return in exactly the opposite direction (compare with " cats' eyes " on the road or the prismatic reflector buoys in the Suez Canal). Radar reflectors (see later) are constructed on this principle. In general, steel or iron structures which consist of perpendicular connections, planes or lattice work yield very strong echoes.

Examples are radio and television transmitter masts, groups of buildings, steel structures on buoys and beacons.

(2) *Spherical* (Fig. 5.2 (a), (b))

A sphere is a poor target. Scattering takes place in all directions. There is only one point where, upon reflection, the radar energy returns to the scanner. This point is hit when the beam is directed straight towards the centre of the sphere. This is true for a perfectly smooth sphere; in practice, due to the roughness of the surface, slightly better response will be received.

Example: Spherical buoy.

(3) *Cylindrical* (Fig. 5.2 (c))

A cylinder possesses the properties of a plane and a sphere combined. Upright and vertical it presents a fair target and its aspect remains the same when viewed from any point in the horizontal plane passing through it.

Funnels, gasometers, oiltanks, can buoys and chimneys, provided they are not tapered, give reasonable response.

If a cylinder with its axis horizontal is observed, however, then its aspect changes when it is observed from different directions in the horizontal plane. When its axis is at 90° to the beam its response will be the same as for an upright cylinder but with any other aspect its response is poor.

Examples are cylindrical mooring buoys which are sometimes detected but in other cases are not recorded at all, depending upon whether they are facing the scanner beam-on or end-on.

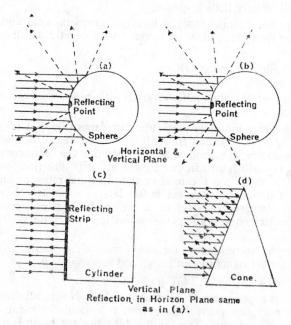

FIG. 5.2 SHAPE AND REFLECTION PROPERTIES

(4) *Conical* (Fig. 5.2 (*d*)).

A cone standing on its base is a very poor target. Theoretically, all the radiation will be deflected away and none will return. The surface, in practice, is not completely smooth and some, though little, radiation will come back.

Examples: Conical buoys, lighthouses, tapered chimneys and church steeples.

Composition

The reflection coefficient (ratio of the amounts of reflected radiation and incident radiation) depends on the intrinsic electrical properties of a particular substance.

Metallic conductors are good reflectors for radar waves. The reflection coefficient of steel is nearly 1. Water also reflects well; in fact it reflects the pulse away and we cannot see an echo of it unless the sea surface is rough and sea clutter is recorded. The reflection coefficient of seawater is approximately ·8. Ice—with no salts to conduct electricity—is a much poorer reflector and its reflection coefficient can be as low as ·32. Stone and earth do not give such a good response; here again, it depends on the composition and vegetation. Ground covered with short grass (reflection coefficient ·7–·8) yields much better echoes than ground covered with long grass (reflection coefficients ·1–·4). If the ground is covered with scrub or trees, the reflection coefficient will be lower than ·1. Metal ore, for example, in mountains improves the echo strength. Wood is a poor reflector. Wooden boats are difficult to detect by radar and so are boats made of fibreglass.

Aspect, of course, plays a very important part. The edge of sheet ice can be clearly visible as an echo on the screen. Sails, if the aspect is right, may yield a fair echo and dhows have been reported at a distance of 9 miles.

Size

(a) *Width.*

If the target is wide in comparison to the horizontal beam-width then the radar beam prods along the target, cutting it in small sections, one after the other. These sections have the width of the beam, about one to two degrees in azimuth, depending on the target. In such a case the echo strength depends on the strongest echo of each individual part and *not* on the total width. A wide object may cause a long, but not always a strong echo on the screen.

If the target is small in comparison with the horizontal beam-width, then the echo strength not only depends on its aspect, shape, surface structure and material but is also proportional to its actual width. The *shape* of the echo is a function of the horizontal beam-width, the pulse-length and spot-size (Chapter 3).

(b) *Height.*

Here, the vertical beam-width does not offer a limitation as is the case with the horizontal beam-width. The vertical beam-width is large and will span any surface target. But there are other reasons that the highest mountain does not necessarily yield the strongest echo.

The first reason is obvious. A high mountain will re-radiate a lot of energy, but whether this energy will return to the observing vessel is another matter which will be decided by the aspect, shape and surface texture. The second reason is that the reflected signal from a high and sloping mountain does not return at the same time and hence does not contribute to momentary echo strength. The higher parts of the pulse return later and their electromagnetic fields may be in or out of phase with the neighbouring lower parts of the pulse. The echo on the screen, radially, is then longer, but not stronger (see Fig. 5.3).

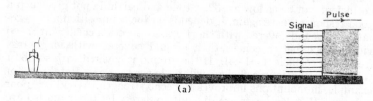

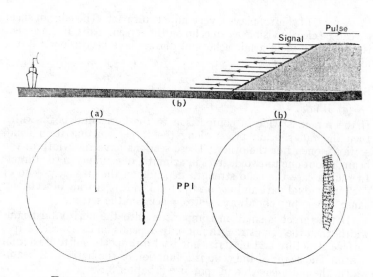

FIG. 5.3   ASPECT IN THE VERTICAL PLANE AND ECHO STRENGTH

The vertical lobe pattern (see Chapter 1) will also have its effect on high targets at a *distance*. Near the transmitter the maxima are close together and only targets with a very small reflecting surface will suffer fluctuations in echo strength. At a good range away from the transmitter the maxima are separating out and quite large reflecting surfaces will pass slowly through successive maxima and minima when the range is decreasing. This explains why a good reflecting part of a high mountain may first yield a strong, but later a very much weaker echo.

A high sloping mountain may not give such a strong echo as a steeper cliff which is not so high. When making a landfall, the echoes which come through first may not be the echoes of the highest landmarks as is the case visually. Careful study of the chart for aspect and local knowledge is essential.

(c) Depth

The radar pulse cannot " feel " the off-side of a solid object. Only the leading edge of such a target is recorded on the screen in the correct position (Fig. 5.4), the side away from the scanner lies in a shadow. The islands, illustrated, will yield exactly the same echoes on the PPI if their " presented " width is the same. This is the reason why, when taking ranges, the *inner* edge of the echo should be measured against range rings or range marker.

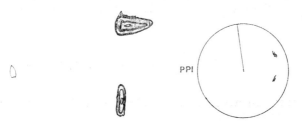

FIG. 5.4   ECHO OF LEADING EDGE

Appendix

Radar Cross-Section (Equivalent Echoing Area)

If an object, intercepting the power of a radar beam and scattering it uniformly in all directions (isotropic re-radiator) has a cross-section of unit area (usually 1 m²) measured perpendicular to the beam, then this object can be used as a standard against which we can measure the effectiveness of re-radiation of other objects, which is returning to the scanner. Hence the radar cross-section may be defined in terms of the equivalent cross-section of an isotropic scatterer which would give the same signal at the scanner position as the actual object with which we are concerned.

The shape of most practical targets is too complex for the radar cross-section to be calculated, but for objects of simple geometircal design, the calculation can be carried out and some results are shown below.   In the calculations it is assumed that the wavelength is very small compared with the dimensions of the object.

Object and its Aspect	Radar Cross-section
Metal sphere, diameter d	$\frac{1}{4}\pi d^2$
Metal plate, area A. Aspect 90°	$\dfrac{4\pi A^2}{\lambda^2}$
Cylinder, radius r, length l. Axis parallel to electric field	$\dfrac{\pi r l^2}{\lambda}$
Triangular corner reflector, length of short side l	$\dfrac{4\pi l^4}{3\lambda^2}$

The radar cross-section of an upright smooth cone would be zero.

Suppose that we want to compare the echo power of a metal sphere and a metal plate placed at right angles to the radar beam. Their ranges are the same, their actual cross-sections one square metre and the wavelength used is 3·2 cm.   Then the relative echo power of the plane is

$$\frac{4\pi \times 100^2 \times 100^2}{3·2 \times 3·2}$$
$$\frac{}{100 \times 100}\ \text{i.e. } 12,000 \text{ (approx.)}$$

Note that in many cases, the radar cross-section is dependent on the wavelength used.

CHAPTER 6

INTERPRETATION OF DISPLAY (2)

The foregoing chapter gave an account of the aspect and features of the target related to echo strength and we follow this up now by making a closer study of the echoes from objects in which the navigator is interested. The objects are divided into three categories, namely, Shore Targets, Navigational Targets, and Ice.

Shore Targets and Topography

Groups of buildings return excellent echoes. They form perpendicular surfaces with the ground between them as a third plane. A prominent part here is played by hillside towns and seaside resorts. Sometimes they are picked up at a distance of 25 miles or more. The echoes of such targets, even well inside land, will often appear earlier on the screen than the echo of the coastline itself when approaching the coast.

Very good echoes are returned for the same reason, from steel structures when they are rectangular in cross-section or where steel girders are employed. Aerial towers and masts, dockside cranes, coaltips, metal lamp standards, pylons, scaffolding, funfairs, etc., yield excellent echoes.

To pick out strong echoes, reduce the gain momentarily. These echoes will come out clearly while the weaker echoes will disappear. Strong reduction in gain will bring out those targets having specular reflection. Most of the echoes remaining on the screen come from vertical faces which are facing the scanner or from three mutually perpendicular planes. Sometimes specular reflection can cause a distorted echo. A well-known example is the overhead power cable in the Straits of Messina which is depicted as a single spot on the screen and can be mistaken for a ship. An alteration of course simply moves the echo in the direction of the cable. The true bearing of the echo from own ship remains the same.

Cliffs yield good response. Their surface texture is not too smooth and they are broken up, so that reflection of energy in one direction only is prevented. Vertical cliffs show up on the screen at a good range (up to 20 miles). They give a thin strong echo. Sloping cliffs on the other hand, yield a thick weak echo and they are not detectable at a long range.

Gently sloping hills and mountains which are bare, are very poor targets. Their size, as we have seen, does not influence the echo strength. Low foreshores, sand dunes (coast of Holland, West Coast of Australia), sandy hills (Persian Gulf) fall under the same category. In approaching these coasts, strong targets far inland, may be picked up much earlier than the coastline. Such sand or shingle foreshores with low-lying land or dunes in the background will show up on the screen as a thinly drawn line. The painted echo is smoother and has more evenness than those echoes of vertical cliffs. They are never detected at long ranges (from 1 to 5 miles).

Sloping hills covered with trees yield a fair response and the radar picture of a coast shows a clear distinction between the parts which are wooded and those which are not. Steep hills and mountains covered with vegetation can be detected at distances between 15 and 40 miles.

Marshes produce weak echoes, but mangroves and other thick growth may yield a strong echo. Swamps may return either strong or weak echoes, depending on the density or the size of the growth of vegetation.

Where land and water border each other, clear echoes can be expected at several miles distance. The echoes on the screen of the boundaries of a river are often clearly depicted. The mudbanks themselves do not often give an echo—unless one is very near—but what is seen are the water clutter echoes caused by the rippling of the water against the sloping bank.

Lock entrances and docks are detected at a good distance (up to 5 miles). Sheds along the quays, ships in drydock and bridges over rivers produce shadow sectors. The arches of the London bridges can be picked out clearly when sailing on the Thames.

Piers and breakwaters show up at medium range (5–10 miles). When a breakwater is scanned end-on, on the screen, the end sometimes seems disconnected from the remainder of the breakwater. This is due to the shadow sector, caused by the bulky end, the more so if a substantial lighthouse and a signal station are situated at the same spot.

Chimneys and towers, provided they are not tapered, provide good targets at medium range.

Streets show up well on the screen. This is due to the echoes of the buildings, walls or fences on both sides. The same holds true of country roads, which can be traced on the screen by the reflection against the hedges or trees bordering the road. A railroad comes out very clearly on the PPI (1–3 miles) and is painted as a series of regularly spaced dots. These dots are caused by the echoes from stations, overhead bridges, signal posts and signal boxes.

In Chapter 3, it is mentioned that the shape of a target can only be determined from the screen if the size of the target is large in comparison with the horizontal beam-width.

Even when the target is large in relation to the beam-width, the shape, as it appears on the screen, is often distorted. Smaller details of shape are not shown as there is not enough range and bearing discrimination and what is really seen on the display is a chain of the best reflecting parts. Echoes, for example, of large cranes and coaltips, show a well-defined, often rounded boundary of the side towards the scanner, but the other side of the echo is ill-defined and frayed.

To summarise, here are some points which may well be remembered when studying topography from the radar screen.

(i) The highest mountains do not always yield the strongest echoes. If the tops are smoothly rounded and bare, they may only be recorded due to good reflection properties lower down. Nor are the highest mountains detected first. Their detection may be much later than that of vertical cliffs which have much less height.

(ii) Vegetation increases the echoing property of smooth sloping surfaces, but it decreases the echo strength of a vertical rock when it is facing the scanner.

(iii) High land shields targets behind it and casts radar shadows.

(iv) The background of a coast moves past the foreground in the direction of the ship's movement. The same effect is noticed visually in clear weather. We see the landscape in the background move past any objects in the foreground. Echoes will behave in a similar way on the screen.

(v) The screen of a hilly and mountainous coast changes as the observer's position changes. Some reflecting parts become more favourably situated, others less favourably.

(vi) The shape of the echo of small or isolated distant objects is determined by the characteristics of the set (Chapter 3). The shape of the echo of large objects only resembles the true shape of the object on the side towards the scanner.

(vii) The state of the tide can affect the radar picture. In some areas more echoes will be shown on the radar display at low water than at high water because as the tide falls more objects, such as rocks, mudbanks and breakwaters, become exposed.

Since February, 1956, an appendix has also been inserted in the Sailing Directions, entitled *Reported Radar Ranges* giving ranges in miles at which radar echoes can be expected from ordinary navigational land features shown on the chart.

Navigational Targets

General Remarks

Water acts like a plane mirror when the water surface is smooth and it therefore subjects the radar pulse to specular reflection. No response is recorded. The fronts and the backs of waves also reflect energy specularly and the probability of returning energy when the sea is choppy is much higher at the shorter ranges than at the longer ranges. Hence sea clutter is restricted to the shorter ranges ($\frac{1}{2}$–4 miles depending on wind and sea).

Due to clutter tide-rips can be seen on the screen at short ranges. Strong tide-rips could be mistaken on the screen for the coast. The same is true when relatively strong echoes are returned from surf produced by waves breaking on a sandbar, beach or surface rock. A false coastline echo may appear on the screen. This is disconcerting.

Crests of swell often show up on the screen distinctly as straight line echoes. Their detection range varies up to 3 miles, depending upon the height of the crests. Their echoes pass through the centre of the display and by timing—using a short range scale—the distance travelled from a point forward of the beam to a point abaft the beam, and allowing for own speed, one can obtain an estimation of their speed. By measuring the distance between a number of these line echoes, one can deduce the wavelength of the swell. Such observations, for example, could be carried out on a Weather Ship.

Sandbanks, sand spits and mud flats produce weak echoes, but often a mud flat is detected earlier on the radar screen than by eye in clear weather. This is often the case with water–land boundaries.

Sandbanks and mudbanks a short distance below water can be discovered due to clutter on the screen, though these echoes are ill defined.

Reefs with breakers over them show up quite well at ranges of 1 to 2 miles.

Icebergs give a good response mainly, but their echo strength is weak when they have a sloping surface. Growlers provide very poor targets and should be carefully watched for, the more so when their echoes move with the sea clutter. Large icebergs have been viewed on the radar screen at distances of 12–15 miles. More details are provided later on in this chapter.

Occasional bright spots on the screen may be caused by waves, flocks of birds, whales, porpoises, etc. breaking the surface. It is needless to say that these echoes are only observed at very short ranges.

Lightvessels

The maximum detection range of lightvessels varies between 6 and 10 miles. Their stated light range gives a fair estimation.

Lighthouses

Lighthouses are poor targets generally. They are often made of stone, tapered in shape and surrounded by a landmass.

Buoys

The echo strength of buoys is not powerful unless they are equipped with a radar reflector. The following list of different types of buoys arranges them, in descending order, according to their echo strength: Radar reflector, pillar, can, spherical, conical, spar buoy. Detection ranges vary approximately from 6 to $\frac{1}{2}$ mile respectively. A light structure on the buoy will improve the echoing strength.

At medium ranges, the echo strength of a buoy fluctuates. This is due to:

(*a*) Change in aspect as a consequence of the wind, sea and swell. Conical buoys without radar reflectors yield a better response when their heads are pointing towards the observer.

(*b*) Breaking-up of the radiation in the vertical plane into many lobes due to sea reflection (Chapter 1). This can be especially the case when approaching a radar reflector buoy in a smooth sea. The reflector, when own ship is moving, will pass through a series of maximum and minimum radiation and the echo is sometimes very bright and sometimes may disappear altogether.

Another way of explaining this is by starting at the buoy end, as illustrated in Fig. 6.1. The *signal*, besides travelling via the direct route, returns via a reflecting surface, as for example, a smooth sea.

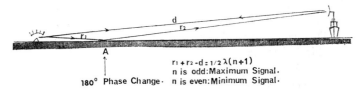

$$r_1 + r_2 - d = 1/2\,\lambda(n+1)$$
n is odd: Maximum Signal.
n is even: Minimum Signal.

180° Phase Change.

FIG. 6.1   VARIATION IN ECHO STRENGTH OF RADAR REFLECTOR

For a ship under way, the electromagnetic fields of the two waves can be alternately in phase, so reinforcing each other, resulting in a maximum signal, or in anti-phase, so cancelling each other out and yielding a minimum signal. The impression one gets by looking at the echo on the screen, is a kind of " twinkling " effect. This property is only confined to small directional targets.

(c) Scintillations caused by local variations in the refractive index of the atmosphere. This can be the case near the water surface when temperature differences are in existence. Compare, for example, with the scintillations caused on a hot summer day above an asphalt road.

At short range the buoy will paint a " firm " or " solid " echo on the screen. The brightness of the paint on the screen is then saturated.

Ships

The echoing strength of ships depends on their aspect, shape, size and type of material exposed to electromagnetic radiation. Steel ships are good targets, especially when they are in ballast and beam-on. Hull and deckhouses play a prominent part in the reflection. Funnel and masts are much poorer reflectors. A small coaster, beam-on, in light condition may give a stronger echo than a larger loaded vessel end-on. Wooden vessels are very poor targets and great caution has to be exercised when small wooden craft are expected in the vicinity. Fibreglass boats are very poor reflectors, but many new lifeboats of this type are made nowadays with fine meshed-metal netting moulded into the sides of the hull so that a fairly strong echo is displayed on the radar screen. Tankers beam-on yield an excellent echo.

Below follow some maximum detection ranges (height scanner 50 feet):

Small wooden boats . . .	$\frac{1}{2}$–4 n.m.
Lifeboats	Up to 2 n.m.
Drift-net vessels . . .	3–5 n.m.
Trawlers	6–9 n.m.
Tugboat and towed barges . .	Up to 7 n.m.
Ship 1000 tons . . .	6–10 n.m.
Ship 10,000 tons . . .	10–16 n.m.
Ship 50,000 tons . . .	16–20 n.m.

In Chapter 3, it was already remarked that ships on the shorter ranges appear to be end-on and on the longer ranges beam-on if the *same* range scale is employed. It should be remembered that the relative motions over the screens can only be deduced by observing ranges and bearings and noting how they change over a period of time.

Only at short ranges, when the size of the vessel is large in relation to the horizontal beam-width, can the leading edge of a ship be seen as an echo on the PPI. A large ship, beam-on and very nearby, will be seen on the screen as a long thick line, not necessarily straight. The echo of a deeply loaded coaster will be broken up, showing the reflecting parts of the elevated fo'c'sle head and poop. The contour of the echo of a vessel, end-on, is sharply defined towards, but is irregular away, from the centre of the screen.

Buoys and Ships

There are several ways to differentiate between the echo of a buoy and that of a ship.

(*a*) Local knowledge or evidence from the chart. A paint in a certain position, for example, can only mean a ship's echo if it is known that no buoy is anchored there.

(*b*) Detection range. For example, an echo detected at a range of 10 miles in open water is unlikely to be a buoy's echo. It is probably a ship's echo.

(*c*) Reduction of gain. When the two echoes are shown near to each other, the echo of the buoy will disappear first, while the echo of a ship will remain longer on the display. Even at lowest gain the echo of a ship may still be visible on the screen. Alternatively, the logarithmic receiver may be switched into action if incorporated in the equipment (see Fig. AIII. 10, Appendix).

(*d*) Plotting decides the identification of a *moving* vessel and a stationary buoy. Effect due to current must be taken into account.

The reduction of gain (*c*) will also assist in deciding what echo represents an ordinary buoy and what a radar reflector buoy. The " twinkling " character (maxima and minima) of the latter's echo can also be of help for identification of a radar reflector buoy. This holds, of course, only when own ship is moving.

It is often not possible to decide by studying an echo whether it depicts a radar reflector buoy or a ship belonging to the smaller or medium tonnage class, which is at anchor.

Detection of Movement

The appreciation of movement can be deduced from plotting (see later). However, when the echo of a ship is strong and moves moderately fast over the screen, the successive echoes of the ship, held by the afterglow, paint a trail on the screen, commonly known as " tadpole tail ". This tail, of course, is best seen on the shorter range scales. Observers should be aware that this trail indicates the *relative* movement of the vessel on the relative motion display. Only further plotting will reveal the true movement. The direction of the " tadpole tail " is the *relative* direction and the length will give a

E

rough indication of the relative speed of the object, but only after one has had long experience with a particular set.

Stationary objects like buoys will also cause a trail on the relative motion display when own ship is moving and the length of the " tail " depends on the speed of own ship. Yawing will cause kinks in the tail if the display is not stabilized.

The wake of a vessel, however, will indicate her true movement. It can be seen on the PPI when fast moving ships pass at close ranges. It shows up as a trail of clutter like the echoes of tide-rips, or as a black ill-defined line caused by the smoothing out of sea clutter. At long ranges it is not visible. At short ranges in fog, when ships possess a reduced speed or are stopped, no wake is revealed.

The depiction of the wake of own ship on the screen can be employed for estimating the diameter of the turning circle.

True Motion displays are ideal for the detection of movement, especially when a reflection plotter is available. The echo track of a vessel indicates the true motion of the vessel which *took place* during the observation. Stationary objects, for all practical purposes, yield stationary echoes (some drift, due to current or internal causes might be introduced, but it is, in general, small). The tadpole tails of echoes from ships on opposite courses are not so long as they would have been if observed on a Relative Motion display. The best range scales on which true motion can be detected without the aid of a reflection plotter are the shorter range scales with the six mile range scale as the upper limit on a 12 inch display.

Turning Circle. A stabilized Relative Motion radar display, with a reflection plotter may be usefully employed to obtain a reasonably accurate trace of the turning circle of own ship. To do this, if there is no set or drift, an echo of a buoy, or similar isolated target, may be observed and plotted on the reflection plotter at the same instant as the helm is put over, and at short intervals (about 30 seconds) while the vessel is making her turn. The circle traced out by the echo will be a replica of the ship's turning circle but the initial course will appear to be 180° in error. The shortest range scale possible should be used. The plot can be recorded on tracing paper; the range rings should be switched on to obtain a scale.

If there is an appreciable current an echo of a freely drifting object should be used instead of the buoy, e.g. a raft or a large oil drum cast overboard just before the trial.

Ice

General Remarks

Smooth, flat ice, field ice and floe ice give no echoes or hardly any. Only the edges or edge can be seen as echoes on the screen. This is partly due to the smooth surface texture and partly due to their electrical properties; ice possesses a rather low reflection coefficient as compared with seawater. Even wet ice—*fresh* water on top—can have a reflection coefficient considerably less than seawater.

Ice clutter is caused by ice floes frozen together. In contrast with sea clutter, it is stationary, with well defined boundaries.

Pack-ice and rugged ice surfaces yield clear echoes on the screen. The use of gain control is useful for detection of other targets.

Tracks made by ships can be detected easily even after being frozen over and covered with snow. The piled-up ice on both sides of the track give excellent echoes. Do not mistake shadow sectors cast upon the ice for cracks or tracks in the ice.

The duller the ice echo on the screen, the thinner the ice is. The brighter the echo, the thicker the ice.

The coastal contour changes when ice drifts against a coast. A false coastline appears as a line-echo on the display and this should be kept in mind. Employment of other navigational aids should not be neglected.

Icebergs

Vertical faces of icebergs give excellent returns. Sloping faces yield a much poorer response. Antarctic icebergs having sheer sides are therefore favourable, as compared with Arctic icebergs. In general, an iceberg as a target compares favourably with a landmass of similar size.

The echo strength of an iceberg will vary due to change in aspect and sometimes a poor response is recorded. Seldom, however, do large icebergs come within two miles of the ship before being detected by radar and this is because the sloping sides of icebergs still possess many facets.

Growlers are very dangerous to shipping. Their surface is smooth and rounded owing to the action of the sea. Due to their shape their response is very poor. Besides this they move up and down with the sea and their echoes get lost in sea clutter. The detection range of growlers in calm weather varies between 2000 and 3000 yards, but when the sea surface is rough, this range is considerably reduced.

Ships trading in ice-infested waters should be equipped with a wide scanner so that a very narrow horizontal beam-width can be obtained. The scanner must be sited in such a way that no shadow sectors dead ahead are created; a position forward of the fore-mast would, in this case, be advantageous. It should not be placed too high, otherwise too much clutter may be recorded.

The most important radar control used for growler detection is the anti-clutter control which must be manipulated continuously and applied in such a manner that the intensity of the growler echo is stronger than that of the sea clutter. The receiver is then held below saturation level.

Other suggestions have been forwarded:

(*a*) Turn the Gain Control up for several revolutions of the scanner, then turn the gain right down. A faint smear will be painted, showing the echo of the growler. This is so because the afterglow of the echo of the growler is stronger than the afterglow of the changing sea clutter.

(*b*) Scan-to-scan integration. If it is possible to reverse the direction of rotation of the scanner, as can be done in some sets, scan clockwise and anti-clockwise over a certain sector by manipulating the reversing switch.

Still, clutter suppression is the most essential.

It should also be stressed that sub-refraction (Chapter 7) can be expected in polar regions where ice and fog are not uncommon. For example, in the winter, Arctic winds of −15° C pass over water of 0° C in the Hudson Bay, or cold air from the Labrador current passes over the Gulf Stream on the St. Lawrence–Bell Isle run and these conditions may favour sub-refraction. As a result of sub-refraction the lobes are lifted and this may make it very difficult or sometimes impossible to pick up echoes of *low* targets such as growlers.

Poor detection of growlers is mainly due to the limitations of radar and is sometimes made worse by atmospheric conditions. Growlers often have a weight amounting to 100 tons; most of it is below the waterline and collision with them can cause *extensive* damage to the ship's stem and hull. Hence, even with the radar set working at full performance, *speed must be reduced during fog and during the night when growlers are reported*.

A *short range scale* is essential for growler detection. Growlers calve from icebergs and where growlers are observed larger icebergs can be expected in the vicinity. Hence the observer should *frequently switch over to a medium or long range scale* for short intervals so that he can beware of surprisal by a large iceberg at a short range.

The best solution would be to have two displays each with its own time-base unit so that the two can operate on different range scales. To obtain a strong picture on the short range scale, the master display should be switched to the longer range scale and the slave to the shorter scale (see Chapter 2: Interswitching System).

CHAPTER 7

INTERPRETATION OF DISPLAY (3)

Atmospheric Factors affecting Echo Strength and Detection Ranges

Refraction in General

Refraction affects detection ranges. The phenomenon of refraction should be well-known to every navigating officer. On many occasions he employs corrections in order to allow for the refraction of light rays.

Refraction takes place when the velocity of a wave is changed. This can happen when the wave front passes the boundary of two substances of different densities. One substance offers more resistance to the wave than the other and hence the velocity of the wave will change. The substance where the resistance is greatest is called the denser of the two. When the wave meets the boundary at right angles, no change in direction will take place, but the wave will be accelerated or slowed down bodily. If, however, the boundary is hit at an oblique angle, a change in the direction of the wave will follow. In the case that the oblique angle is very large, and the wave travels from a dense to a less dense substance, the change in direction of the wave front may be so large that the wave will return to the denser substance instead of going forward. We say then that *total reflection* occurs. If the change in density is increasing or decreasing gradually, so will the direction of the wave front gradually change and it will follow a curved path. Note that the hollow part of the curve is always towards the denser substance. A practical example is the apparent " uplift " of a stick held below water.

A light ray or radar ray passing through a substance which is gradually increasing or decreasing in density will describe a curved path as shown in Fig. 7.1 (*b*). The hollow side of the curve will be towards the increasing density.

The extent of refraction is given by Snell's Law which states that for two given media, $\dfrac{\sin \theta_1}{\sin \theta_2}$ is a constant, where θ_1 is the angle of incidence and θ_2 is the angle of refraction (Fig. 7.1 (*a*)).

The constant equals the ratio n_2/n_1, where n_2 and n_1 are the refractive indexes of the two media and n is defined by the ratio

$$\frac{velocity\ of\ radar\ wave\ in\ vacuo}{velocity\ of\ radar\ wave\ in\ medium}$$

When the angles of incidence and refraction are measured with reference to a normal to a sphere, Snell's Law is modified and becomes

$$\frac{\sin \theta_0}{\sin \theta} = \frac{nr}{n_0 r_0}$$

where r and r_0 are the surface radii of the first and final media (Fig. 7.1 (b)).

In cases where the refractive index decreases strongly, as can take place in the earth's atmosphere, then the angle of refraction, increasing all the time, will finally reach the value of 90°. At this height, it is said that the angle of incidence has reached its *critical value* and from then on the wave begins to be refracted *downwards*. It is in such a case where " ducting " of a radar wave occurs as will be seen later in this chapter.

It can be proved that the refractive index of the atmosphere is a function of the pressure, temperature and humidity.

Warm air has a smaller density than cold air at the same pressure and radar and light rays will turn their hollow side towards the cold air.

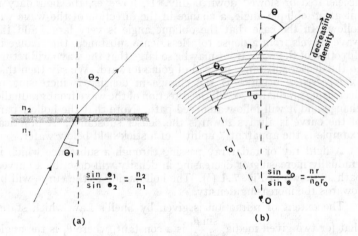

FIG. 7.1   REFRACTION

There are two cases of abnormal refraction which cause abnormal radar " visibility ".   They are *super-refraction* and *sub-refraction* but before dealing with these, we will say something about the radar horizon.

Radar Horizon

Generally, radar rays are restricted in the recording of the range of low-lying objects on the screen by the radar horizon. The range of the radar horizon depends on the height of the scanner and on the amount of bending of the radar ray. The bending is caused by diffraction—a property of the E.M. wave itself—plus refraction, which is governed by atmospheric conditions. There is, therefore, a definite radar horizon. This is in contrast to certain waves which are used in radio. Such waves can by-pass the horizon by reflection against ionized layers in the upper atmosphere. These reflections cause signals to be recorded far beyond the range of their horizon. Radar waves, however, have too much penetrating power and they pass through the ionized layers and are then lost in space.

In the table produced, the distance of the radar horizon for micro-waves is tabulated against the height of the scanner, which acts as a radar eye.

The distance to the ordinary horizon, ignoring refraction, is expressed by $1 \cdot 92 \sqrt{h}$, the distance to the visible or optical horizon by $2 \cdot 08 \sqrt{h}$ and the distance to the radar horizon is given by $2 \cdot 21 \sqrt{h}$. The result is in nautical miles and h is the height of the eye or aerial in metres. See Fig. 7.2. It does not include diffractional effects which are (relatively) small for 3 cm. radar.

Height in		Distance of radar hor- izon n.m.	Height in		Distance of radar hor- izon n.m.
feet	metres		feet	metres	
18	5·5	5	215	66	18
24	7·3	6	240	73	19
32	9·8	7	265	81	20
42	12·8	8	320	98	22
54	16·5	9	380	116	24
66	20·1	10	445	136	26
80	24·4	11	520	159	28
95	29·0	12	595	182	30
111	34	13	680	208	32
130	40	14	770	235	34
150	46	15	860	262	36
170	52	16	960	293	38
190	58	17	1060	323	40

We thus see that the range of the radar horizon is greater than that of the optical horizon, which again is greater than that of the geometrical horizon. This result could have been expected from our considerations in Chapter 1 where the spreading out of waves (diffraction) was discussed. The wavelength of light is much shorter than the wavelength of a radar wave. The ratio of the size of the Earth to the wavelength of light is much larger than the corresponding ratio with respect to a radar wave. The radar wave therefore has a greater inclination to follow the Earth's curvature than the light wave. The longer the wavelength, the greater is the tendency to bend round objects. The formula given above is valid for the 3 cm. wave. The horizon related to 10 cm. waves would have a greater range, because diffractional effects would have to be included.

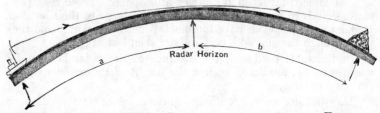

FIG. 7.2   RADAR WAVE PROPAGATION AROUND THE EARTH

The very first condition for an echo to be seen on the screen is that the object must be above the radar horizon. If the object is 60 metres high and the aerial 12 metres high, then the maximum distance at which the echo of the object can be spotted on the screen is $17 \cdot 25 + 7 \cdot 75 = 25 \cdot 0$ miles, provided, of course, the reflection conditions are favourable and the echo falls in a suitable time-base. In Fig. 7.2, the figures quoted above correspond to b and a respectively.

The second condition is that sufficient portion of the object must be above the radar horizon to provide enough reflecting surface for the pulse. It can happen sometimes that the navigation lights of a vessel are observed visually but the echo on the screen is still absent as not enough of the hull is yet above the radar horizon.

Super-refraction
The distance to the radar horizon is extended (Fig. 7.3).

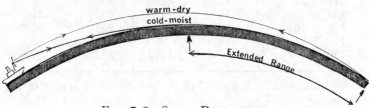

FIG. 7.3   SUPER-REFRACTION

Conditions which favour super-refraction are:

(i) Decrease of relative humidity with increased height (moisture lapse) or (and)

(ii) Warm air situated over cold air.

In many instances the two conditions go together. A typical example is a rise of 7° C in temperature and a fall of relative humidity of 30 per cent over a height of 100 metres.

The weather must be calm and there should be no turbulence, otherwise the layers of different densities are mixed and boundary conditions disappear.

Super-refraction frequently happens in the tropics where cold sea currents are prevalent (for example Gulf of Aden) or where warm landwinds come out over the sea as in the Red Sea, the Mediterranean, and, on summer days, in the English Channel. In some regions, super-refraction is a more normal occurrence than normal standard refraction. It is especially noticeable on the longer range scales.

Sub-refraction

The distance to the radar horizon is reduced (Fig. 7.4).

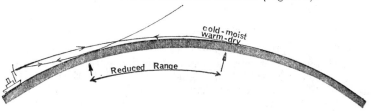

FIG. 7.4   SUB-REFRACTION

Conditions which favour sub-refraction are:

(i) Increase of relative humidity with increased height or (and)

(ii) *Abnormal* decrease of temperature with increased height.

Often the two conditions go together. As the surface layer cannot be dry, sub-refraction at sea must be primarily due to an excessive lapse rate of temperature. Weather conditions must be calm and no turbulence should occur. It is not so common as super-refraction.

Sub-refraction can happen in polar regions where Arctic winds blow over water where a warm current is prevalent. In such a case a cold layer of air is situated over a warm layer. The warm layer will be extremely shallow and a high proportion of the temperature fall is confined to the first 50 cm. or less above the sea surface. The radar scanner of a ship is not situated in this layer and hence it is very unlikely that the distance to the radar horizon is reduced to a very short range. If very poor radar detection is observed, *the reading of the performance monitor should be checked*, as it is improbable that it is due to sub-refraction only.

Sub-refraction may involve an element of danger to shipping. It has been explained that it is unlikely that the reduction in the distance to the radar horizon contributes much to this element. The danger lies in another direction. As a result of sub-refraction, the lobes are lifted and this will affect the minimum range. Also, due to this upward trend of the radar beam, low targets, for example, small ships and ice may be " missed " and not recorded.

Sub-refraction near land, especially when it first appears, can be very noticeable to the observer who is studying the screen. It will give him the impression that something is slightly wrong with the tuning of the set. Echoes of buoys and low land disappear and only echoes of higher land come through.

The conclusion is that one should be very watchful in an area where sub-refraction and the presence of small craft can be expected and during poor visibility speed should be reduced.

For recording purposes (radar log) one should note the temperature of air and seawater.

Anomalous or Freak Propagation

Anomalous propagation can take place due to:

(i) Intense super-refraction so that air ducts are formed and the radar wave is trapped in a waveguide, shaped by the Earth's surface and the slightly higher warm layer. The pulse will be guided along this layer and may be reflected by an object well below the radar horizon. In the same way it can travel back and record an echo. It has been reported frequently in tropical regions such as the Red Sea and Gulf of Aden. This phenomenon, known as *surface ducting* is illustrated in Fig. 7.5.

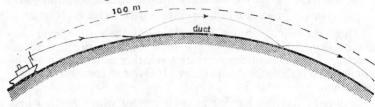

FIG. 7.5 SURFACE DUCT

(ii) A *reflecting layer* of warm air, well elevated above the Earth's level, for example at 250 metres. In other words, a temperature inversion exists. The atmosphere in that region must be calm. Trade wind regions can offer such conditions and it can also happen near land, when on a late summer afternoon warm air drifts slowly from land over a cooler sea surface.

Total reflection of the radar wave takes place against the warm layer and this can only happen when the angle of incidence is large and a glancing one. In cases like this, more than one discontinuity in temperature might occur and the radar energy is conducted along a *duct* well *elevated above the Earth's surface* as shown in Fig. 7.6.

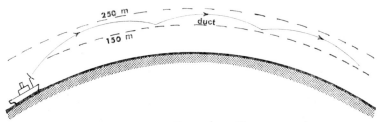

FIG. 7.6   ELEVATED DUCT

Anomalous propagation does not usually occur in *all* directions from the scanner. The guiding and reflection of pulses also depends on the wavelength used. Sometimes echoes coming from objects a great distance away, were observed on a 3 cm. set and not a 10 cm. radar set and vice versa.

Many observers are not familiar with freak propagation. This is because when ships are in regions where it can occur they are often enjoying fine weather and are many miles from land so the radar is not in use. Many reports, however, have been received from shore stations which possess radar sets with a powerful pulse.

During the summer, echoes from the mountains of Sardinia and Greece were recorded by a radar set, located at Malta, working in the 50 cm. band. The Dutch coast, though low, was recorded from East Anglia on a summer day. During the East Monsoon, echoes of ships were seen on the display of Bombay Radar Station at distances ranging from 200 to 700 miles and frequently, the coast of Arabia (1000 to 1500 miles) was observed.

Non-Standard Propagation in General

The mean lapses of refractive index with height depends on the season and the geography of land and sea. However, the mean lapse in the *lower* level of the Earth's atmosphere in temperate latitudes has been designated as " standard " and this amounts to a decrease of refractive index of 4 parts per million per 100 m. Variations in meteorological conditions give rise to variations in the refractive index. Where the refractive index is increased above standard, propagation of radar waves will become above average (super-standard refraction); where the refractive index falls below standard, propagation of radar waves will drop below average (sub-standard refraction). An illustration is shown in Fig. 7.7 where graphs are drawn showing the change in refractive index against the height in metres.

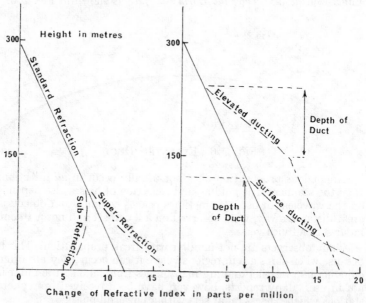

Fig. 7.7   Change of Refractive Index with Height

Second-Trace Returns

The question naturally arises how is it possible to observe an echo from a target, for example, at a range of 100 miles while the maximum range scale is only, perhaps, 48 miles. Such an echo, however, can be recorded, providing a trace is sweeping the face of the tube when the echo pulse returns. This trace is not the original trace which commenced at the same time as the transmission of the pulse that caused the echo. The echo is recorded during one of the subsequent traces and will appear at the wrong distance on the screen. Hence the name *Second-Trace Return* or *Multiple-Trace Return* is used as the echo may arrive back during the second, third or even fourth trace. Sometimes the name " Spilled-Over

Echoes " is used. Exactly the same could happen with a stopwatch when recording the time interval between transmission and reception of a certain signal. Suppose that the watch has only one hand, the pointer indicating seconds, then the reading 15, for example, could mean that the time interval is 15 seconds, or 75 seconds, or even 135 seconds. If the signal returns after 75 seconds and we take the reading as 15 seconds, then the calculated distance is $7\frac{1}{2}v$ and the true distance $37\frac{1}{2}v$, v being the velocity of the signal. In other words, due to the missing of one sweep of the hand of our watch, our distance, as supposedly recorded, is less than the true range by a distance of $30v$.

Suppose that the p.r.f. is 2000, so that the time between the starting of two successive sweeps is $\dfrac{1,000,000,}{2,000}$ i.e. 500 micro-seconds corresponding to a travelling distance of a pulse, out and back of $\dfrac{500 \times 300}{1852}$ or 81 miles (pulse travels 300 metres per micro-second and there are 1852 metres in a nautical mile). If the signal returns when the next transmission takes place, then the range of the target causing the echo is $\frac{1}{2} \times 81$ or $40\cdot5$ miles.

Let the range scale in use be 12 miles. In other words, no echoes are recorded from objects between 12 and $40\cdot5$ miles, because the spot on the PPI is having its rest period. In Fig. 7.8, second-trace returns are illustrated, depicting a coastline at a range of about 45 miles. Second-trace returns appear on the display on the same bearing, but their distances are shortened by $40\cdot5$ miles as observed on the screen, i.e. $ac = bc = 50 - 40\cdot5 = 9\cdot5$ miles and $dc = 43 - 40\cdot5$, i.e. $2\cdot5$ miles. The picture of the coast is distorted towards the centre. If the distance to an object becomes less than $40\cdot5$ miles, its echo will disappear. It is absorbed by the centre at $40\cdot5$ miles and may re-appear again at the edge of the PPI when the distance to the object is 12 miles.

If instead of there being a coastline, a ship was sailing from A to B, then for a stationary observer at C, her echo would follow a curved track on the screen from a to b. Her speed as observed on the screen will be much slower than the real speed. If, however, a ship is approaching on a steady bearing, then second-trace returns will paint a true picture of her course and speed.

In the case stated above, echoes can be picked up when anomalous propagation takes place between the ranges of $40\cdot5$ to $52\cdot5$ n.m. (second-trace), 81 to 93 n.m. (third-trace), etc.

If the p.r.f. is 500, so that the time between the starting of two successive sweeps is 2000 micro-seconds (which corresponds to a travelling distance of a pulse, out and back, of 324 miles) and the range scale in use is 24 miles (say), then second-trace returns will be recorded on the screen if the distance to the object lies between 162 and 186 n.m. By changing the p.r.f. from 2000 to 500 the second-trace echoes illustrated in Fig. 7.8 would disappear altogether.

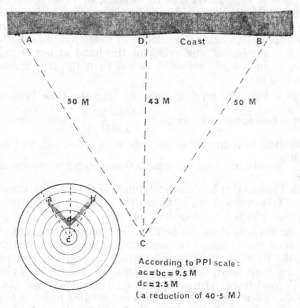

FIG. 7.8   SECOND-TRACE ECHOES

To summarise, multiple-trace returns can be recognised by:

(i) Inability to identify the echoes at first sight. Land echoes appear to come from a position where no land actually exists or from a place, where due to aspect, shape or shadow sectors, no echoes are usually recorded.

(ii) General distortion of echoes towards the centre of the display.

(iii) Abnormal movements of targets, as observed from their echoes on the radar screen.

(iv) Disappearance and re-appearance of echoes due to leaving the second-trace and later re-entering the first-trace, or atmospheric disturbances.

(v) Changing of the range scale (and p.r.f.) will make the echoes either disappear or appear at a different range.

A second or multiple-trace return is generally of no use to navigation. It can be distracting and may mislead the observer.

Elimination of Second-Trace Returns

Some manufacturers produce radars whereby an automatic periodic variation is applied to the p.r.f. so that it varies over a pre-determined range of values (a wobbulated signal).

Suppose the p.r.f. variation is over the range 2025 to 1975 in 1/100 sec. If we go back to our example on page 111 and substitute in the calculation 2025 and 1975 instead of 2000, it is found that on the 12-mile range scale during a period lasting 1/100 sec. a second-trace echo appears respectively at (50-40), i.e. at 10 miles and at (50-41), i.e. at 9 miles (instead of one echo at 9·5 miles). In 1/100 sec. the scanner has turned through about 1° which is approximately the horizontal beam-width between half-power points for X-band radar. During that time a maximum of 16 echo-pulses could return back from the target (assuming the scanner turns at 20 revs. per minute).

As the p.r.f. is varied continuously these 16 pulses (probably far less) will be spread out over a display range of 1 mile, and become elongated and lose their intensity. An observer would not be able to see them and for practical purposes these echoes are eliminated.

Ghost Echoes

Strong bands of echoes, sometimes 15 miles long and 2 miles wide, often with a " sausage-shaped " appearance, have been recorded on occasions in the Red Sea, near Cape Guardafui during the S.W. monsoon and also preceding precipitation echoes of an extended non-frontal line of convective showers or thunderstorms. The air was clear over the ship when they were observed to be over-head. The nature of these echoes is not quite clear, but it is believed that these can be caused by steep changes in lapse rate, 30 to 150 metres above sea level.

Weather Factors affecting Echo Strength and Detection Ranges

Wind

Sea clutter echoes are caused by reflection of the pulse against the sea waves. The reflection is specular and conditions for the pulse to return to the scanner are favourable near the ship. At longer ranges the beam will be deflected away from the ship. The steep front of the wave faces the scanner in a better position than the sloping back. The result is that the clutter on the screen is not distributed symmetrically. From the pattern on the screen it is not difficult to deduce the direction of the wind (see Fig. 7.9).

The echo strength and the extent of the sea clutter depends on (a) the size and aspect of the wave, and (b) the height of the aerial.

Higher aerials give more sea clutter. The latter effect forms a great draw-back on the high bridges of V.L.C.C.s, and it is advisable in these cases to have an additional low-sited scanner on the fo'c'sle head.

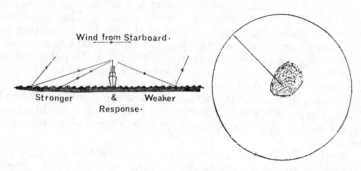

FIG. 7.9   SEA CLUTTER

Blind sectors due to superstructures on own ship can sometimes be observed as dark sectors against the sea clutter while a large target, as for example, a ship nearby, also may cause a shadow.

The boundary of the wave clutter, naturally, changes continuously except when there is an abrupt ending due to, for example, a sand- or a mud-bank. This sudden ending can be an indication of the existence of such a bank which otherwise would return very little echo. Conversely, waves breaking over a shallow patch can cause sufficient clutter to enable the navigator to identify the shallow area.

Echoes of steel ships and most buoys possess a greater amount of reserve echo strength than the sea clutter and the application of the Anti-Clutter Control will show them up. They emerge from a confused mass of echoes.

How often it is that one reads about collision cases where a ship was detected at a range of 10 miles and at the Enquiry the Master stated: " Later I lost the echo on the screen." The whole purpose of the Anti-Clutter Control is for tracing echoes near the centre of the PPI and to minimise the danger of such echoes being lost. Chapter 4 dealt fully with this control, which must be used frequently and carefully for detection of close objects.

Generally the Anti-Clutter Control does not allow for the un-symmetrical distribution of the clutter, but some experimental sets have been designed in which account of it is taken automatically. Research into the subject is being continued, the object in view to find a final and perfect solution. (See 'Clearscan', page 48).

The signal-to-wave-clutter ratio for different wavelengths in the microwave band is a complex subject because so many factors are involved and the scatter calculations do not always follow the same law (it depends on the size of the particle to the wavelength used). The radar cross-section of waves and spray is not only dependent on the frequency and grazing angle but varies also with the plane of polarization (and so does the coefficient of reflection). It has generally been found that clutter echoes have a stronger intensity when the wave is vertically polarized than when it is horizontally polarized. However, even in this field, exceptions have been recorded in the X-band under moderate wave conditions.

Most radar installations working in the X-band have scanner design based on horizontal polarization, whereas many S-band radars have scanners designed to give polarization in the vertical plane. The signal-to-clutter ratio will (theoretically) increase when the wavelength is increased, say from 3 to 10 cm.; on the other hand the change from horizontal to vertical polarization can introduce an opposite effect. It is, however, a fact which can be easily verified from photographs—that in general, sea return on the display of a 3 cm. radar set is of much stronger intensity than sea return on the display of a 10 cm. radar set, except that at very short ranges sea waves can still cause very strong clutter echoes on the display of a 10 cm. radar installation.

In practice, small targets do not show up too well on 10 cm. radar displays, and this should be remembered by a navigator. He should therefore consult both 3 and 10 cm. radars—if provided—when strong wind conditions are prevalent in areas where such small targets might be present.

Precipitation in General

Centimetre and millimetre waves are absorbed by atmospheric gases such as water vapour and oxygen. Figure 7.10 shows a graph where the attenuation by water particles is plotted against the wavelength and it is clearly seen that short waves in the cm. and mm. band suffer a great deal of attenuation. This is the reason why a mm. radar is not suitable for detection of targets at medium and long ranges, although it is of excellent use in river navigation.

Rain, hail and snow affects detection ranges in two ways:

(i) As mentioned above, energy is taken away from the radar pulse and the region beyond the precipitation area will yield reduced response. We could say that there is a certain amount of shadow behind it.

(ii) They cause echoes on the screen which may obscure echoes from targets inside the area of precipitation (something like sea clutter). This effect is known as 'masking'.

Both (i) and (ii) are governed by the size of the particles and the distances between them, in other words, their extent depends on the density of precipitation.

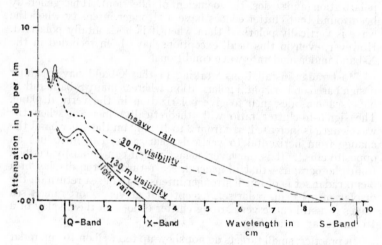

FIG. 7.10 ATTENUATION OF RADIO WAVES BY
WATER PARTICLES

Rain

Rain squalls and showers appear on the screen as a woolly mass. As they may be present in a region up to 10,000 m. or so in height, they are easily detected at a distance. An intense rainstorm can be detected at a distance up to 25 or more miles. Thunderstorms give excellent echoes, but even drizzle affects the radar visibility and may produce light diffuse echoes. If the marine radar operates in the X band (3 cm.), then the further edges of squall line echoes may be indistinct due to radar wave attenuation and these edges present a fuzzy appearance. The echoes of an extended non-frontal line of convective showers or thunderstorms are sometimes preceded by a " sausage-shaped " echo which travels in advance of and faster than the echoes of the main storm belt. The nature of the sausage-shaped echo is not yet quite clear (see " Ghost Echoes ", last section), but it is not due to precipitation as it has been observed to pass over a ship in clear air.

Echoes of ships or land inside the rainstorm will be obscured by rain clutter and targets beyond the rain area will lose some of their echo strength. In some cases, detection range of the latter is reduced to 80% of the normal range.

Continuous rain over large areas will make the centre part of the screen brighter than the rest and the rain clutter, moving along with the ship, looks very much like sea clutter. It can be clearly seen on long range scales. This is due to a gradual decrease in returning power as the pulse penetrates further into the rain area. In fact, it can be shown in such a case that the echo intensity decreases proportional to the square of the distance.

Ships can make use of radar in port to plot oncoming rain squalls in order to give dockers ample time to close hatches or hoist rain tents. A warning in time can prevent extra work for the crew (especially during the night) and damage to the cargo.

At sea where *heavy* rain storms of limited extent can be expected, it is often worthwhile to search for rain storms on the *long* range scales and start plotting their track(s). Meanwhile the observer can begin to investigate if there are echoes of targets inside the rain clutter area by gently reducing the gain. The echoes of such targets should, in most cases, possess more reserve echo strength than the rain echoes. The Gain Control setting should then be restored and reduced again alternately in order to keep an eye on eventual targets outside and inside the rain area. At the same time the master or o.o.w. could contemplate taking avoiding action for the rainstorm (for a variety of reasons, but mainly for safety) and/or consider the application of Rule 6 (*Safe Speed*). He may put the engines on 'Stand-by' or, even start reducing speed. When the storm comes closer, one can start making use of the Rain Switch or Differentiator (FTC) control (see Chapter 4), but the Gain Control must not be forgotten and, quite probably, may have to be turned *up* slightly. When rain actually starts to fall, the combined operation of Gain and FTC control will become quite difficult because, besides the masking effect of important echoes by the rain clutter, attenuation will now take place. The echo strength of the target is now reduced, its detection range is reduced to one or two miles and it becomes more and more difficult to separate the clutter from the echo of a vital target. A very heavy rainstorm, like those sometimes encountered in the tropics, can obliterate most of the X-band radar picture. Signal/clutter ratio then may drop below unity and the echo of the target is "lost".

Many collisions have taken place between ships in heavy rainstorms, mainly because the search for targets started *too late* and owing to the combined effect of masking and attenuation, and not knowing where the target was—although the observer had probably detected it first at a range of 8-10 miles. A long time was spent on the manipulation of Gain, Anti-Rain Clutter and Anti-Wave Clutter Controls. Then the collision came suddenly, with both vessels at full speed!

In all cases, therefore, when fairly heavy rain storms can be expected, start plotting the storms *and* any ships inside them at an *early* stage. Take either avoiding action for the storm or put engines on 'Stand-by' and *reduce speed*, when in heavy rain storms it might no longer be possible to separate the signal from the clutter.

3 cm. radar sets were in use during these collisions. The signal-to-rain clutter ratio is directly proportional to the square of the wavelength (Raleigh's approximation for scattering on the assumption that the diameter of the particle is small compared with the wavelength). Using a 10 cm. radar set, this ratio is increased by $(10/3)^2$ or 10·5 db up. Also, in any given rate of rainfall, absorption (attenuation) decreases as wavelength increases, firstly as a direct function of the wavelength and because a decrease in the dielectric constant (less absorption) takes place as the wavelength increases. This combination, of less scatter (greater diffraction) and less absorption, makes the rain clutter on 10 cm. sets far less dense in relation to the echo-strength of the target. *A 10 cm. set, or just a 10 cm. r.f. section with interswitching facilities with the 3 cm. radar installation, is therefore very advisable for many ships.*

Hail

The echo strength of hailstones is only comparable with the echo strength of raindrops if the rate of precipitation is the same. This will only occur when large hailstones are falling. Generally, there-fore, we may say that the clutter effect of hail is less than that of rain. The Gain Control and/or the Rain Switch (FTC) should be employed as described in the previous section.

Snow

For snow, also, the overall effect of clutter on the picture is less than that due to rain. The clutter will have approximately the same density only if the rate of precipitation is the same. This is unlikely as one inch of rain is equivalent to ten inches of snow per hour. Ice pellets inside the snow increases the echo strength.

In warmer climates, snow may produce stronger echoes as the flake is made up of a number of crystals, whereas in colder regions, it only consists of a single crystal.

Falling snow is only observed on the displays of 3 cm. radar sets, not on the display of 10 cm. radar sets, except when the snowfall is very heavy.

The danger of snowfall lies in the fact that, after its fall, it tends to conceal the shape of the object which it has covered. Electromagnetic energy penetrates snow to a certain extent but also loses some of its power due to absorption. The result is that snow-covered coasts may yield poor echoes on the screen, and caution must be exercised when approaching these coasts. Radar as a navigational aid can then be limited, and where possible, as always, other aids for position fixing, in addition to radar, should be employed.

The pulse will penetrate through thin layers of snow. This can be very useful for ice-breakers when coming to the rescue of a vessel. The track made by the ice-breaker through the ice often becomes covered by frozen snow and so becomes invisible to the eye. The PPI in such cases will show up the track, and the ship can be guided out through this route, which has already been made, relatively easy. Fuel, also, is saved.

Fog and Smog

In most cases fog does not actually produce echoes on the screen, but a very dense fogbank may yield a faint haze on the display. Heavy fog, especially in the polar regions, can reduce radar "visibility" to 60% of the normal radar visibility. Detection ranges are similarly affected. See Fig. 7.10.

Not much is yet known about the effect of smog on radar. Investigations are being carried out.

Clouds

Clouds do not return detectable echoes unless there is precipitation inside them. Cold fronts can be located as a line of cloud echoes. Echoes from warm fronts are extremely variable. A bright band in the echo of a cloud may indicate the freezing level.

Tropical Revolving Storms

Hurricanes etc. are easily identified by echo patterns which are concentrically arranged round the centre of the storm. The location of the centre can be of help to the mariner, but the range of a marine radar set, of course, is not large enough to assist the mariner in taking early avoiding action. Shore radar stations with a range of 150 to 200 miles can be of great value in tracking these storms.

Dust and Sandstorms

There is a general reduction in radar detection in the presence of dust and sandstorms. The impression is that their effect is the same as that of heavy fog. Here, also, few details are available.

Localised sandstorms show up on the screen as nebulous patches.

The Effect of Blind and Shadow Areas and Sectors on Echo Strength and Detection Range

Blind and shadow areas are caused by obstructions on land, by another vessel or by obstructions on own ship in the path of the radar beam. The radar beam is completely cut off in a *blind* area. The beam can bend or diffract round ridges or smaller objects causing an area of reduced intensity, generally called a *shadow* area. See the *Appendix* of this section.

Blind areas caused by cliffs, high coasts, high buildings and sheds will create considerable discrepancy between the radar picture on the display and the view presented to the observer when studying the chart.

Permanent blind and shadow areas in the horizontal and the vertical planes are caused on board by cross-trees and foremast, funnel, samson posts and fo'c'sle head. Such areas are known as *sectors* as they cause dark sectors on the screen extending from the centre outwards to the edge.

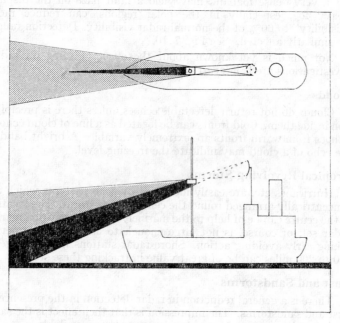

FIG. 7.11 HORIZONTAL AND VERTICAL CROSS-SECTIONS (APPROXIMATE) OF BLIND AND SHADOW SECTORS

The diagrams shown in Fig. 7.11 only give a very rough indication of the boundary of the sectors as the subject of radar shadows is complex. First of all, the scanner sweeps round and has a certain width; secondly radar waves, due to their longer wavelength, possess a much greater tendency to bend round obstacles than do, for example, light waves. One can see, however, from the Figure that there exists a small blind core in the horizontal plane of the complete shadow sector caused by the foremast, while in the vertical plane, the cross-trees are the worst offenders, but due to the comparatively large angular vertical beam-width, only part of the radiation and re-radiation is stopped, while the other part diffracts round. In this connection it should be remembered that the directional properties of the scanner are exactly the same for transmission as well as for reception (for example, $1\frac{1}{2}°$ and $15°$ between half-power points in the horizontal and vertical planes respectively).

Although the detection range of a large ship coming towards own ship inside a shadow sector might be reduced from, say, 12 to 6 miles, the greatest danger of the existence of shadow sectors applies to the small type of vessel, where the reduction of detection range might be in the region from 4 miles (in the absence of a shadow sector) to $\frac{1}{2}$–1 mile (when located inside a shadow sector).

Shadow sectors of foremast, samson post and funnel may vary between $1°$ and $3°$, $5°$ and $10°$, $10°$ and $45°$ in angular width respectively, depending mainly on the distance between the aerial and the offending structure, and the wavelength used. If the obstructing part is near and much larger than the scanner, as could be the case with the funnel, then the blind sector will take up nearly the complete shadow sector. Careful siting of the aerial should prevent or at least reduce this (read Chapter 2: Siting of Scanner Unit). A high fo'c'sle head in comparison with the scanner height may also give rise to a blind sector and increase the minimum range forward.

In order to have a margin of safety, the angular width of the shadow sectors are determined, *not* the blind sectors. This can be carried out by the following methods:

(a) Observation of the shadows against a background of weak sea clutter. If the clutter is too strong, the gain should be reduced. The shadow sectors are revealed as dark sectors on the screen and their extent determined by means of the curzor. The dark sector on the screen should be clear of multiple, indirect and side echoes (see Chapter 8).

(b) Steaming slowly round near a small buoy and noting the bearings between which the echo disappears and re-appears on the radar screen. The buoy should not carry a radar reflector. Give the scanner an opportunity to scan through a narrow shadow sector, hence steam slowly.

(c) Measurements from the ship's plan or by sextant from a place just in front of the scanner. This is a very approximate and the least satisfactory method, because the scanner is not a point object and the rays do not propagate entirely rectilinearly. Besides, the exact boundaries of a shadow sector are difficult to predict. Blocks, cargo gear, ladders, etc. all may contribute.

The emphasis in methods (a) and (b) on weak sea clutter and a small buoy without a radar reflector has a reason. Rough sea surface and a radar reflector buoy can be detected in the partial shadow; their echoes may penetrate the dark sector on the screen and the blind sector instead of the shadow sector is determined. Strong echoes have a lot of reserve power upon which they can draw when their objects are situated in an area of reduced intensity, and hence they are not suitable for the determination of the shadow sector.

After determination of the angular width of the shadow sectors, make a diagram showing their extent and display it prominently next to the PPI so that their existence is not forgotten by the Master or navigator. Record the extent of the shadows in the Radar Log.

In some cases different diagrams have to be prepared for different conditions of trim.

The blind and shadow sectors caused by the cross-trees can be particularly dangerous as they are cast ahead if the scanner is centre-sited. These sectors have been the cause of several collisions, especially with small boats, such as wooden vessels which did not possess much reflecting capability. Such targets may remain un-detected even in the area of reduced intensity unless they are very near. The Master must keep this in mind and it is essential during reduced visibility, *to weave round the course to an extent of at least half the angular width of the dangerous shadow sector* (snaking). An excellent suggestion is to compile a zig-zag diagram and to order the man-on-the-wheel to alter course alternately to starboard and port the required amount of degrees each time the whistle is blown. A zig-zag alarm clock can also be very useful for this purpose and, when plotting, for timing the plotting interval.

It should be noted that the discussion above applies to 3 cm. radar. One of the advantages of using 10 cm. radar—with its longer wavelength and greater diffraction properties—is the quite large reduction of blind and shadow sectors when compared with those observed of a 3 cm. radar.

Appendix

The diffraction effect of radar rays is illustrated in the graph shown in Fig. 7.12, where the effect of diffraction around a sharp ridge is indicated. " A " represents the ship's scanner position and " B " an object behind the ridge. The direct distance between A and B is $d_1 + d_2$ or d. The distance measured via the top of the ridge is $r_1 + r_2$. The differences between these two distances, expressed in wavelengths, is plotted against the loss in echo strength in decibels (scale on right-hand side).

When AB just clears the ridge then there is already a loss of 12 db. There is a slight gain in echo strength when AB is above the top of the ridge and $(r_1 + r_2) - (d_1 + d_2) = \frac{1}{2}$ wavelength. The height of AB above the ridge for this gain to occur, can be calculated from the formula below the diagram.

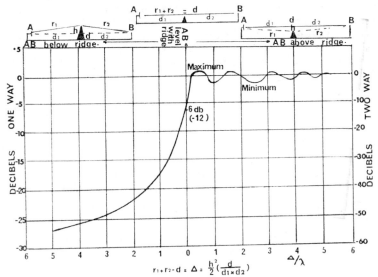

FIG. 7.12   DIFFRACTION OF RADIO WAVES

For example, if d_1 and d_2 are 1 and 4, 2 and 3, $2\frac{1}{2}$ and $2\frac{1}{2}$ nautical miles, then the clearances above the ridge for maximum echo strength of object B are 6·7 m., 8·2 m. and 8·3 m. respectively.

The reasons for these gains in echo strength are the formation of secondary sources of radiation at the top of the ridge. In fact the ridge is responsible for the formation of a new radiation structure in the vertical plane (Chapter 1).

On the other hand, if *AB* is *below* the top of the ridge and the distance *AB* is 5 nautical miles (as in the case above), then with the vessel 4 miles away from the ridge (d_1=4M, d_2=1M), objects with large radar cross-sections, up to about 20 metres below the top of the ridge for 3 cm. radar, and about 36 metres below the top for 10 cm. radar could be detected on the display of ship *A*. It is said in this case that the radar beam has *diffracted* round the ridge.

Conclusion

In Chapters 3, 5, 6 and 7 we have now discussed all the factors on which the strength and the quality of the picture on the display depends.

What it amounts to is in short:

Get to know your own set, study its performance, accuracy and characteristics. Consult the chart carefully for aspect and shape of coastline. Try to predict land shadow sectors, take into account tidal, atmospheric and weather conditions and remember the local shadow sectors caused by obstructions on own ship. Co-ordinate and confirm any obtained information with other navigational aids, if possible.

CHAPTER 8

INTERPRETATION OF DISPLAY (4)

Factors which might cause Faulty Interpretation

Sometimes unwanted echoes or signals appear on the screen. The most common are:

(i) Multiple echoes;

(ii) Indirect echoes;

(iii) Side echoes;

(iv) Interference pips or spokes.

It often happens that a target is recorded by several echoes shown in different positions on the screen. In such cases the echoes which are shown in positions which do not truly represent the position of the target are known as either *False echoes* or *Spurious echoes*, and the other echo which is shown in the correct representative position of the target is known as the *True Echo*.

Multiple Echoes

Multiple echoes are caused by the reflection of the signal *between* own ship and the target before its energy is finally collected by the scanner. The effect on the screen is that, besides the original echo, there are seen, on the same line of bearing, one or more echoes, equidistantly spaced and having ranges of multiples of the true range. They generally happen at short ranges up to about one mile, often when another vessel is passing closely, beam-on or nearly beam-on to own vessel (Fig. 8.1). The shapes of multiple echoes are less defined than that of the original echo and they are weakening in intensity outwards.

Reduction of gain or clutter will remove them before the true echo.

For those who have been sailing with an electronic sounding apparatus, the phenomenon should ring a familiar note. These multiple reflections are also recorded when a ship is sailing in shallow waters.

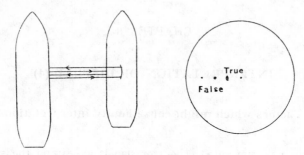

FIG. 8.1   MULTIPLE ECHOES

Multiple echoes are a nuisance when a ship is approaching an anchor road where several ships are riding at anchor. The picture on the screen can become confused due to several multiple echoes of different ships, and it is difficult to distinguish between true and false echoes.

Multiple echoes can sometimes be observed in the dark sectors on the screen caused by the shadow sectors of large sheds, etc., nearby.

Indirect Echoes

Indirect echoes are caused by the reflection of the outgoing pulse against part of the superstructure (funnel or foremast) or against a building, cliff, river bank, bridge or ship etc. nearby. The echo pulse returns the same way, and due to reflection, it paints its echo in the wrong direction on the screen. What really happens is that we see the target via a mirror. See Fig. 8.2.

The most offending parts are the cross-trees of the foremast. Their mirror effect is worse than that of the funnel because the pulse reflected by the cross-trees contains more concentrated energy than that reflected by the funnel.

An excellent remedy is to cover the cross-trees with angle bars or a piece of corrugated metal, so that the reflected energy returns directly back to the scanner when the receiver is paralysed or is scattered in all directions, thereby losing all its energy.

Owing to its curvature, the funnel may distort the echo. An indirect echo, from a ship approaching from ahead, reflected via the

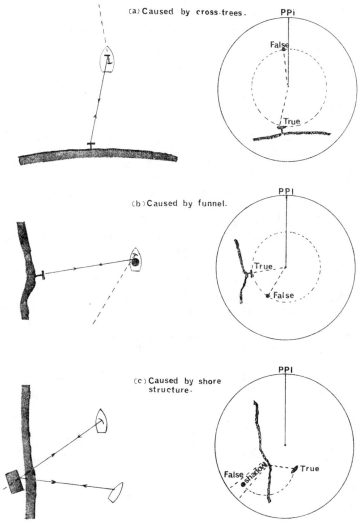

(a) Caused by cross-trees.

(b) Caused by funnel.

(c) Caused by shore structure.

FIG. 8.2   INDIRECT ECHOES

funnel, will show up on the screen as an overtaking vessel. An indirect echo reflected via the cross-trees from a receding cliff or building appears on the screen as a target ahead of own ship, moving away and possessing twice her speed.

Large ships passing close by in confined waters may give rise to part of the picture being re-produced in reverse on the screen. These strong indirect echoes move very fast in a circular path over the display.

A ship dead astern sometimes gives rise to four echoes on the screen. Three of them nearly in one line and perpendicular to the course, show up forward, and one shows up abaft the beam. The three indirect echoes forward of the beam are caused by reflection via the samson posts on the foreship and the foremast, the fourth one is the true echo.

When there are many echoes on the screen, indirect echoes are likely to escape notice. It is when the screen is reasonably clear that one starts wondering about some echo, which one knows, from experience or visual observations, should not be there.

Indirect echoes can appear from targets at quite large ranges provided the reflecting surface has an excellent aspect. They can be wiped off the display by reduction of gain or clutter.

Characteristics of indirect echoes can be summed up as follows:

(*a*) They usually appear in blind or shadow sectors and areas because the obstructions producing these regions of reduced response often themselves act as " mirrors ". See Fig. 8.2.

(*b*) When caused by the ship's obstructions, they will appear on the same relative bearing (foremast or funnel) although the bearing of the target may change. This is also the case when the ship is stationary and the indirect echo is caused by a shore object (Fig. 8·2 (*c*)). If the ship is moving, in the latter case indirect echoes will appear only for a very short time.

(*c*) When caused by an obstruction on own ship the true echo and its image, the false echo, appear at about the same range. This is because the range discrimination is not sufficient to measure the difference in distance between the two routes of the pulse (Fig. 8.2 (*a*) and (*b*)).

(*d*) When caused by an obstruction not on own ship the range of the false echo will be the range of the obstruction from own ship plus the range of the true echo from the obstruction. See Fig. 8.2 (*c*).

(*e*) Movements of false echoes are abnormal if they are compared with the movements on the screen of the true echoes.

(*f*) There is a distortion in shape (via the funnel) and in presentation on the screen as only the *best* reflecting surfaces of a large target can produce a false echo (for example, a false echo of a river bank could be presented on the screen as a single spot).

(g) When caused by obstructions ("mirrors") on own ship an alteration of course will make the false echo disappear although *another* false echo may appear on the same relative bearing.

Side Echoes

Side echoes are caused by the side lobes, which are beams of electromagnetic energy re-radiated by the scanner plate in a different direction to that of the much stronger main lobe. It is impossible to design a scanner without side lobes (see Appendix, Chapter 1) although the construction of an aerial affects the magnitude of the side lobes. Nearby targets are picked up by the side lobes as well as with the main lobes. See Fig. 8.3. As the side lobes are far less powerful than the main lobe, the effect of side echoes will only be observed at short ranges. Large buildings near the river bank, piers or harbour entrances, etc., give rise to side echoes.

When another vessel alters course, side echoes may suddenly flash over the screen. It happens when the other vessel presents her hull beam-on and can be observed up to a distance of maximum five miles.

Side echoes appear as concentric arcs round the centre of the screen. Their radii represent the true ranges of the targets. Near the centre of the screen, the possibility of side echoes naturally is great and this is the reason why, in narrow waters, the centre of the display is often masked by a white circular patch.

By reducing the gain judiciously, or increasing the anti-clutter, the less powerful side echoes will disappear. However, as remarked before in this book, it is not always good practice to reduce the side echoes too much. We do not want to have a tidy picture, but a good navigational picture—and that may simply have to include side echoes so that other important small echoes can be observed.

Many scanners at present—especially slotted waveguide aerials—are for all practical purposes unidirectional and the visible effect of side echoes is almost eliminated.

Interference

This word is used for many unwanted effects which reveal themselves on the screen. These can be caused by internal disturbances in the radar set itself, by disturbances from electrical apparatus on board or by radar sets of other vessels. The first two disturbances may cause an effect known as *spoking*.

The interference from other radar sets is due to the recording of the other ship's pulses and their echo pulses on the screen of own ship. This can only happen when the other vessel transmits in the same frequency band as that of own ship. If the p.r.f. of own ship's radar pulses and the other ship's radar pulses are the same, the interference pips will be displayed in circular patterns, if the p.r.f.'s differ, interference will reveal itself in spiralling pips or traces towards the

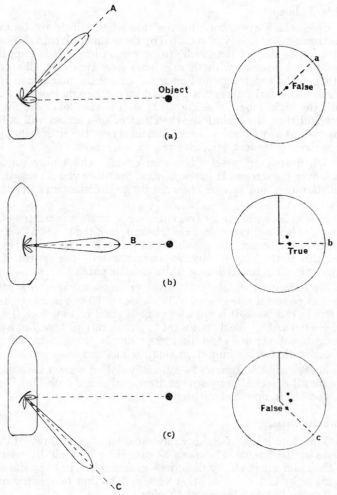

FIG. 8.3 SIDE ECHOES

centre of the display. When there is a large difference between the p.r.f.'s the pips will be distributed haphazardly. Interference can be most markedly seen on the longer range scales. On the short range scales the trace or sweep is so quick that the pips are pulled out into thin lines which are liable to escape notice, unless the interference source is very close.

If very strong interference makes it difficult for the observer to keep an eye on important navigational echoes, he might try, just temporarily, to off-tune the set slightly.

Indirectly, *if all radar operators would make proper use of the Stand-by Switch in clear weather, then there would be a great decrease in external radar interference.*

N.B. Observation of the presence or absence of external interference on the display is of *no use* in determining whether a vessel, whose echo is displayed on the screen, is or is not using radar. Interference could be shown which is caused by the radar of a vessel not yet detected. On the other hand, the vessel which has been detected may be using radar which is transmitting in a different frequency band to that of our own radar, in which case no interference from her radar would be shown on the display.

Some radar sets have quite effective suppressor units for this type of external interference. (See 'Clearscan', page 48).

F

CHAPTER 9

USE OF RADAR FOR NAVIGATION (1)

The following three chapters deal with the use of radar for navigation. This chapter views the subject in general terms and also contains a section for the navigator to appreciate the merits of 3- and 10 cm. radar sets. Chapter 10 deals with some techniques and choice of display presentation, while Chapter 11 studies various types of radar aids.

Types of Navigation

Employment of radar to assist navigation falls into three categories:

(a) When making a landfall;

(b) For coastal navigation;

(c) For pilotage.

Landfall Navigation

Radar here can be very useful, especially when approaching land during bad visibility. Fog or hazy monsoon weather could be among the reasons for the reduced visibility.

However, and this should be remembered, *initial* radar fixes are often not reliable when making a landfall. *Identification at long ranges is difficult.* The reasons should be clear from the contents of Chapters 5 and 6.

Strong echoes mainly depend on the positions of the reflecting surfaces with respect to the scanner position, in other words on the aspect of the targets. Such surfaces, generally, are not charted. When approaching land, the picture may change completely. The positions of the reflecting surfaces of the individual targets are changed with respect to the scanner position whenever the range is altered, and pulses are more favourably or less favourably reflected.

Another reason is that some targets are above and others are below the radar horizon.

These two reasons have the following consequences:

(i) The foreshore, unless a steep cliff coast, is not detected. Low ying coasts are not picked up. Hillside and seaside resorts show up well. So may other targets inland.

(ii) There is an apparent movement of land towards the observer as the lower parts appear above the horizon. This effect, of course, is also true in the visual case, but it is more conspicuous on the radar screen.

(iii) Lighthouses are not usually conspicuous.

(iv) The first detectable echoes are not necessarily the highest targets.

As a consequence of (iv), the table giving the distance to the radar horizon (Chapter 7) can be employed to a certain extent, but its result must be accepted with reservation. An example will make this clear. Suppose an echo is first seen on the screen at a distance of 24 miles. The height of the scanner is $16\frac{1}{2}$m., corresponding to a horizon distance of 9 miles. Assuming the object is just above the horizon, it must span a distance to the horizon of 15 miles corresponding to a height of 46 metres This means that a good reflecting surface, at least 46 metres in height, is above the horizon. It does *not* mean that it is marked in the chart as an object about 46 metres in height. It could be the lower part of a mountain which is very much higher, the top of which, due to its shape or surface texture, is not a good reflector. Still an estimation can be obtained, sometimes good, sometimes not so good.

Apart from these considerations developed in Chapters 5 and 6, it may happen sometimes that the highest peak of a mountain range is obscured by one of the lower peaks. The highest peak, shown on the chart is in the shadow of the lower one (Fig. 9.1).

FIG. 9.1   SHADOW EFFECT

Meteorological conditions as for example, sub- and super-refraction should be studied. The latter may cause second-trace returns. Momentary reduction in gain will show up strong responses. Such strong responses, though not always coming from the highest targets, can offer excellent guidance. Sketches can be made of the picture at long ranges and the making of photographs of the display, if equipment is available, is even better. The identification of strong echoes should be recorded in the radar log, together with their ranges and bearings. Clear weather practice is important so that visual and radar detection can be correlated. All of this is extremely valuable for future occasions, especially when sailing on a regular route.

The Sailing Directions contain an Appendix, entitled *Reported Radar Ranges*, giving ranges in miles at which radar echoes can be expected from ordinary navigational land features shown on the chart.

The 'Echo-stretching' facility and the 'Variable Range Delay' (Chapter 4) can be very useful for this landfall navigation.

Identification sometimes can be combined with the plotting of a fix in the following manner. Take three ranges and bearings of three targets which yield strong and isolated echoes, and plot them on the same scale as the chart, from a point representing the ship, on transparent or tracing paper. Place the point near the D.R. position and move the tracing paper until the plotted echoes correspond with the targets on the chart which are expected to give a strong return echo. The point then will represent the fix and can be marked on the chart.

No position should be assumed before at least three targets have been identified beyond all doubt. Astronomical observations, if possible, the employment of other radio aids and the taking of soundings should be used in conjunction with radar observations.

Coastal Navigation

Better use of radar as a navigational aid can be made the nearer the vessel comes to the coast. Reflecting conditions become more favourable which allow the navigator to extract more accurate information from the display.

When establishing a position by means of radar, one should remember that range accuracy is generally higher than bearing accuracy and this is more true when the mechanical curzor is employed.

A position determined by using three radar ranges as position circles is preferable and can be very accurate (Fig. 9.2).

If possible, select objects well apart in azimuth, each moderately high and steeply descending in the direction of the observer from the object. The coast itself should then be at right angles to the observer's line of sight. The stipulation of steepness and its direction brings two advantages with it:

(i) The radar range can be easily laid-off on the chart and no doubt arises as to the point on the chart from which it has to be laid-off;

(ii) Beam-width distortion error is eliminated and so is an error in the range, which otherwise may be too large or too small.

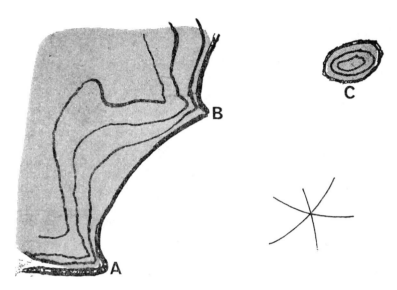

FIG. 9.2 FIX BY RANGING

In cases when the coast on the screen can be seen on different sides of the ship, take radar ranges from objects on each side. If errors are introduced, the plotted position will be displaced along the direction of the channel, which is usually a safer error than a displacement across.

Quite accurate distances between conspicuous radar targets can be measured from the screen by means of rubber-tipped dividers, which then can be laid-off radially against the fixed range rings.

Of course, if the coastline, although steep, is fairly straight without too many salient points, or if one has to rely on the chart contours to lay-off a radar range, then it might be difficult to determine the exact centre of a position circle on the chart. There is also a practical point to consider, namely, that a ship should acquire *beam compasses* which are really necessary for laying-off position circles with large radii.

In some cases the observer might think it better to take radar bearings instead of ranges, but be aware of the fact that errors can be introduced in bearings, especially when a mechanical curzor is employed (Chapter 4). Visual bearings are usually better than radar bearings and a visual bearing together with a radar range can yield a good fix.

If bearings are taken, choose isolated targets of relatively small size. Reduce the gain as much as possible so that a fine echo is obtained and beam-width distortion is reduced. Then bisect the echo by means of the cursor. The echo should be well away from the centre of the screen when a mechanical curzor is used, as the possibility of a slight displacement between the electronic centre and the centre of the bearing scale is always present. Put the visor on the display when taking a bearing. Do not forget to restore the gain later.

When obtaining a fix with radar bearings, at the same time as observing the bearings note the ranges of various conspicuous landmarks on the radar and use them to check the fix on the chart.

Ranges and bearings can be combined to find a fair estimation of the total compass error. Plot the fix first by means of ranges, then take off the true bearings of the objects from the chart. From the relative radar bearings the compass bearings can be determined if the compass heading is known. The mean of the differences between true and compass bearings indicate the total error.

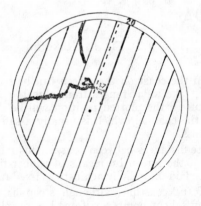

FIG. 9.3   USE OF PARALLEL INDEX

The *Parallel Index*, already discussed in Chapter 4, is an extremely handy device for navigational purposes, especially so when the display is gyro-stabilized. It consists of equidistantly spaced parallel lines engraved on a transparent mask which fits on the PPI and can be rotated (see Fig. 9.3).

By means of the Parallel Index the course can be laid-off directly from the screen though, of course, the chart must be consulted for shoals and hidden rocks.

The instrument can be of excellent use for anchoring technique and for following clearing lines. The distance between two successive lines should be calibrated against the range rings of the scale in use. Once this is done, then distances between fixed marks can be quickly estimated. Intermediate lines can be drawn in chinagraph pencil in order to increase the accuracy of the Parallel Index. More will be said about this device in the next chapter when special techniques will be discussed.

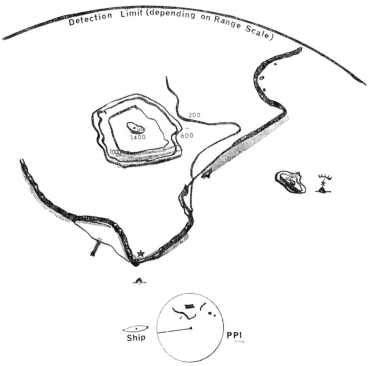

FIG. 9.4   P.P.I. AND CHART

An electronic chart can now be produced on many ARPA displays (Chapter 4, ARPA control: NAV-lines), showing lines, marking channels and bordering shallow areas. This enables navigation to be carried out with less frequent time-consuming consultation of the conventional chart during the actual passage, but careful pre-passage planning is needed for preparation of the electronic chart.

Owing to beam-width distortion, a false picture of a coastline is painted on the screen. Coastlines not at right angles to the line of sight yield echoes which are widened-out, the more so when their aspect is decreasing. The same is the case with the echo of a corner which is extended by an amount proportional to half the beam-width in a perpendicular direction to the trace. The ship's course as seen on the radar screen will seem to run closer to the coast than it actually does. This usually yields a margin of safety.

Figure 9.4 shows diagrammatically some of the points discussed, in particular beam-width distortion, insufficient bearing and range discrimination, and radar shadows.

In addition, it must be stressed that one should *be acquainted with the state of tide*. The radar picture of a coastal area at low water often differs considerably with the radar picture of the same area at high water.

Study the chart for aspect and shadow sectors. Blank areas on the screen can be due to water, poor aspect or shadow sectors.

Coastal navigation by radar is of great value for the mariner when sailing along a coast which lacks salient features for the taking of visual bearings; or steaming at night along a coast or in straits where no light beacons are present or where the lights are weak and difficult to identify (occulting character). Unlighted rocks above water can also be detected.

Pilotage

For navigation in narrow waters radar can be of great service, especially during reduced visibility. Usually the ship's position is not plotted on the chart. The minimum range of the set is important. A short range scale must be used, but periodically the scale should be switched over to a longer one. The existence of radar shadows caused by structures on own ship must be kept in mind. During fog, the derricks ought to be down. False echoes may also confuse the picture.

In narrow channels, rivers, etc., it is difficult to distinguish between the echoes of buoys and ships, the more so when ships are riding at anchor. Local knowledge and the reduction of gain can offer great assistance (see Chapter 6: Buoys and Ships). If a logarithmic receiver is incorporated, then the related control on the display unit should be switched over from LIN to LOG.

The chart should be thoroughly studied for radar echoes, especially the echoes of buoys. This makes it easier to pick out the echoes of ships. Too many buoys can make the picture confusing and it is good practice to check buoys and record them when passing.

Great care must be taken when sailing in narrow waters round Norway and Sweden. The channels there are bordered with many wooden spar buoys (there are about 50,000 of them) which are ineffective radar targets. In fog, corners constitute an especial danger when navigating by the radar screen.

In the past, trials were carried out in laying buoy patterns near sharp bends or junctions. Such patterns consisted of buoys equipped with radar reflectors, grouped in the shape of a T, Y, V, diamond or triangle.

Owing to ground reflection, however (see Chapter 6: " Twinkling " of radar reflector buoys), it was found that echoes of individual reflectors had a tendency to disappear and re-appear one after the other, resulting in loss of the pattern, so defeating its purpose. Also, obscuring of the pattern by large ships took place.

Still, in this connection, it may be worthwhile to remember that any buoy patterns or buoyed channels can be very useful during reduced visibility in picking out the approaches to a harbour from the radar screen.

Spotting of the echo of the pilot boat in fog can also save time.

Radar can be used for anchoring. Having a radar-conspicuous object ahead when approaching the anchorage can be of great help. Put the variable range marker on the distance one requires to anchor from the object. Steam very slowly and just before the range marker makes contact with the echo, drop the anchor. Any dragging of the anchor can similarly be checked.

One can easily check head- or stern-way, assuming that there is no swinging, by placing the cursor or the lines on the Parallel Index perpendicular to the heading marker on the relative motion display. If there are land echoes beam-on, any fore and aft movement can be discerned. The electronic curzor should be used if the presentation switch of the display is on "True Motion".

For coastal vessels, blind pilotage in clear weather is good practice. For example, the wheelhouse can be screened off for the time being, and the Master should try to navigate the vessel solely from his observation of the radar screen. The pilot or the mate would have to stand outside the wheelhouse where he could maintain a watchful lookout lest something should go wrong, in which event he can advise the Master or act accordingly.

Off-Centre Displays, if provided, can be extremely useful in narrow channels. *A good presentation for navigation in narrow waters, especially with many large and small vessels in the vicinity, is the True Motion display set to a short range scale.*

More will be said about these presentations in the next chapter.

Owing to the limited range and bearing discrimination and minimum range of a cm. *shipborne* radar, it is not suitable for berthing. This is a great drawback, especially in fog. Requirements for a set to make it satisfactory for docking and berthing would be:

(i) Large scale or short range scale (up to ½ mile) ;

(ii) Short pulse and hence good range discrimination ;

(iii) Narrow horizontal beam-width. This requires a wide scanner or a radar set in the mm. band ;

(iv) Short afterglow to detect small movements, and consequently

(v) higher number of scanner revolutions.

Modern radar sets possess most of these qualities with the exception of the very narrow horizontal beam-width, mentioned under (iii) and it is just this characteristic which is very essential for a radar set to make it suitable for docking purposes.

Already in several ports, shore radar sets with much wider aerials than that of shipborne radar sets, have contributed to a solution.

3 cm. and 10 cm. Radars (X-band and S-band)

Although various comparisons between 3 and 10 cm. radar installations have already been made in several places in this book, a fuller treatment and summary of this topic is given below :—

I. Detection Range

The maximum range of detection for free space—assuming *equal gain* aerials and ignoring surface reflection and constraint by the horizontal plane—varies as the square root of the wavelength and would therefore be better on a 10 cm. radar (1·8 times the maximum detection range of 3 cm. radar provided the object has the same radar cross-section). There is also less atmospheric attenuation with a 10 cm. wave.

However, on board ships, the gain of a 10 cm. aerial, owing to practical limitations, is less than the gain of a 3 cm. aerial. This, in practice, makes the free space detection range of 10 cm. radar only slightly larger than that of 3 cm. radar.

The influence of surface reflection will widen the minima between lobes for a 10 cm. set and *smaller objects* will be more difficult to detect.

On the other hand diffraction increases the detection range for

10 cm. radar and we may generally say that *larger objects* are detected earlier on a 10 cm. display than on a 3 cm. one.

II. Propagation Effects

The two regions of interest for radar propagation are the *interference region* and the *diffraction region*. The first one is located within the line of sight of the radar, and is the region where lobed radiation is produced.

The second region beyond the interference region and below the radar line of sight is the shadow area where diffraction takes place.

(*a*) *Interference effect*. It was seen in Chapter 1 that owing to interference between direct waves and waves received by ground reflection, the number of lobes between the scanner and the sea are far more when using the 3 cm. wavelength than when using the 10 cm. wavelength (1000 against 300 respectively for an aerial 15 m. high). In other words the lobe structure is far coarser for the 10 cm. radar set, while for the 3 cm. set there is much better sea coverage,

Lobing, by the way, is not as pronounced for a curved earth as for a plane earth (as shown in diagram 1.3); because the waves, when reflected from a curved surface, are subject to divergence so the minima are not as deep, and the maxima have a less extended range.

When the wavelength is increased, the angle between the earth's surface and the centre-line of the first lobe will increase (directly proportional to the wavelength and inversely proportional to the height of the aerial) and it will become more difficult to detect low-lying objects. However, the increase is smaller with vertical polarization than with horizontal polarization and vertical polarization is favoured for 10 cm. installations. Of course, as the sea becomes rougher, the interference pattern tends to be destroyed, especially at the minima, and polarization dependence disappears.

The *Raleigh distance* (Appendix 1.1)-where the radiated field and the azimuthal width of the returning signal is a function of the scanner width and not of the horizontal beam-width (varies directly as the square of the scanner width, and inversely as the wavelength) —will often be smaller with the 10 cm. wavelength as with the 3 cm. wavelength despite the use of wider scanners with the former.

(*b*) *Diffraction Region*. This region, where a certain amount of bending round the Earth takes place, away from the line of sight, is generally neglected for 3 cm. waves, but is taken into account for 10 cm. waves and results in slightly longer ranges than given in the Table in Chapter 7 (distances to the radar horizon). Note that this table is based on calculation of normal *refraction* in the atmosphere and is the same for all micro-waves, irrespective of wavelength.

Diffraction or the bending of e.m. waves round objects, plays an important role also in other aspects. In contrast with refraction, it is a property of the wave itself and does not depend on the medium in which the wave is propagated. The amount of diffraction is a function of the ratio $\dfrac{\text{wavelength}}{\text{size of object}}$, and the greater this ratio, the more bending will take place. Some more explanation of diffraction is given in Appendix 1.1 and at the end of Chapter 7.

Some of the consequences, apart from the bending round the Earth, are as follows :—

(i) The beam-width in the horizontal and vertical plane is wider for the 10 cm. than for the 3 cm. wavelength if the same scanner is used. The beam-width (horizontal or vertical), in fact, is proportional to the ratio

$$\frac{\text{wavelength}}{\text{effective scanner width (in the horizontal or vertical plane respectively)}}$$

It follows that the same beam-width would be produced if the scanner dimensions of the 10 cm. scanner are 10/3 times the size of the 3 cm. scanner. In other words, a 3 cm. radar scanner of effective width of 1·80 m. would produce the same beam-width as a 10 cm. radar scanner effectively 6 m. wide. Such type of scanner for 10 cm. wavelength would be too large for a ship as there is the danger of fractures and even breaking-off in rough weather. The maximum size of scanners in the horizontal plane, fitted in merchant ships, is about 3·60 m. This means therefore that the beam-width in the horizontal plane, using 10 cm. wavelength will be wider than when using a 3 cm. radar with the normal 1·80 m. horizontal width.

(ii) Resulting from (i), it can therefore be stated that the bearing discrimination for a 10 cm. installation is less than the bearing discrimination for a 3 cm. set. Echoes are therefore less well defined on 10 cm. marine radars than on 3 cm. marine radars.

(iii) Another result of diffraction is that the dark sectors on a 10 cm. radar display, caused by blind and shadow sectors, will be far less, angularly, than those—caused by the same objects—on a 3 cm. display.

III. **Back Scatter**

(*a*) The *radar cross-section* of many targets decreases with the wavelength (see Table on page 92). Specular reflection will be less for the longer wavelength and for a radar reflector to have the same radar cross-section for a 10 cm. installation as for a 3 cm. one, its diameter would have to be increased by $\sqrt{10/3}$, i.e. 1·8 times (page 92). It is one of the reasons why radar reflectors are more difficult to detect on a 10 cm. radar than on a 3 cm. radar set, and it forms one of the limitations of 10 cm. radar one has to keep in mind. For a ship and a sphere, however, the radar cross-sections for the 3 and 10 cm. wavelengths do not differ much.

(*b*) *Back scatter of water droplets.* The rain clutter varies inversely as the square of the wavelength and inversely as the square of the distance.

With the echo of a target in a rain clutter area, the signal/clutter ratio varies as the square of the wavelength. *Hence there is a great advantage of using the 10 cm. radar set for the detection of ships inside rain areas.*

And as the clutter signal itself has far less magnitude on a 10 cm. display, *use of 10 cm. radar may prevent computer saturation for ARPA displays.*

(*c*) *Clutter signals from sea waves* and associated droplets cannot be described in terms of simple mathematical equations. Their responses depend on the signal wavelength, the type of polarization used, the angle of incidence, and the state of the sea. However, with certain exceptions at short ranges, it is found that the clutter density from sea waves is less on displays of 10 cm. radar than on displays of 3 cm. radar.

IV. Attenuation

Attenuation is due to absorption whereby the energy of the e.m. wave is converted into heat.

At centrimetric wavelengths the absorption, apart from some "humps" at about 0·15 and 1·25 cm., by *atmospheric gases* is small, but it can become significant over long paths. Attenuation by oxygen is of the order of 0·03 db/n.m. (two-way) in the 3–10 cm. wavelength range (standard pressure and temperature).

Absorption and resultant attenuation can be significant with *water vapour* and *snow*, and increases with a drop in temperature. However, it decreases quite rapidly as wavelength increases, and becomes practically negligible near the 10 cm. range.

The same is true for *rain drops* where the absorption decreases as the wavelength increases, firstly as a direct function of the wavelength, and secondly as a result of the changing value of the dielectric constant, which also is a function of the wavelength. The total effect is a variation inversely proportional to the wavelength squared.

Here also, it can be seen that the 10 cm. installation offers great advantages over the 3 cm. set when a vessel is in rain or has to sail through a region of rain squalls.

Another great advantage of 10 cm. radar in this connection is that the attentuation in a waveguide is far less (approximately inversely proportional to the wavelength used) than for 3 cm. waveguides.

V. Polarization

The reflection coefficient with vertical polarization (often the mode for 10 cm. wavelength) is less than with horizontal polarization and cancellation between direct waves and ground reflected waves will not be so complete with vertical polarization. Thus, in *good* weather, targets near the sea surface (where the first minimum is) will be illuminated more strongly with vertically polarized than with horizontally polarized waves. The echo signal will likewise be greater with vertical polarization. With rough weather, and the resulting destruction of the lobe pattern, the situation might reverse.

There is an increasing tendency to employ horizontal polarization also for 10 cm. radar sets.

CHAPTER 10

USE OF RADAR FOR NAVIGATION (2)

Presentation of Display

Relative Motion Display, Unstabilized, Head-Up

If there is no gyro-compass on board, or the radar display cannot be connected to a gyro repeater or a magnetic transmitting compass, the presentation of the display should be " Head-Up ".

During a course alteration, for example to port, the heading marker remains at zero, i.e. upwards, while the picture turns round inside the tube in a clockwise direction. This will cause considerable blurring due to the retainment of the afterglow. While the alteration takes place, and for some moments afterwards, little use can be made of the picture on the screen. This is a great disadvantage for blind pilotage during fog and in narrow waters, rivers and channels where frequent alterations of course take place. See Fig. 10.2.

Turn the gain down during a course alteration. This will prevent blurring.

Bearings are relative and when course is altered, the echo swings over the screen and the track of the echo is broken up. This makes it impossible to assess the relative motion of a ship's target during and shortly after an alteration. It also causes false "tadpole tails" when own ship is yawing. More will be said about this in Chapter 16 : Use of Radar for Anti-Collision.

Relative Motion Display, Stabilized Outer Azimuth (Bearing) Ring, Course-Up, Unstabilized

In this case the outer bearing ring is stabilized by the gyro-compass transmitter, but the picture itself is not stabilized so that blurring of echoes takes place when the ship's course is altered. The true course is read by the heading marker and true bearings can be read on the outer bearing scale, irrespective of any yawing of own ship. See Fig. 10.1. There is also an inner stationary bearing scale (not shown in Fig. 10.1) which has the zero point opposite the heading marker so that relative bearings can be read.

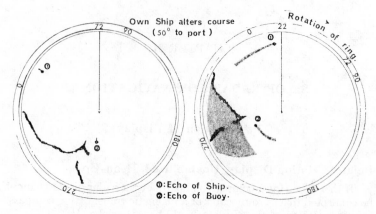

Blurring of echoes but bearings are true.

FIG. 10.1   RELATIVE MOTION DISPLAY, STABILIZED OUTER
AZIMUTH RING, COURSE-UP, UNSTABILIZED

Relative Motion Display, Stabilized, North-Up

The display is connected to a gyro compass or magnetic compass repeater system. The orientation is " North-Up " and the heading marker indicates the true course on the edge scale. Comparison with the chart is made easier.

During a course alteration, the heading marker swings round to the new course indication—acting like a gyro repeater—but the picture stays aligned in the same position and remains with a North-up presentation. No blurring takes place due to a course alteration. The presentation is as clear as when proceeding on a steady course. See Fig. 10.2.

Bearings are true and there is no discontinuity in the track of a ship's echo during an alteration. Observations can be continuous.

This presentation should always be used when the display can be stabilized.

It is true that this type of presentation is a bit difficult to get used to if one has already become accustomed to a head-up presentation. However, this presentation has far more desirable advantages than the head-up display. One may say that a navigator does not need to turn a chart up side down if a southerly course is laid-off, in order to see whether land will show up on starboard or port of the track, but this comparison is not correct. On a chart, the ship's position progresses along, but in this presentation the

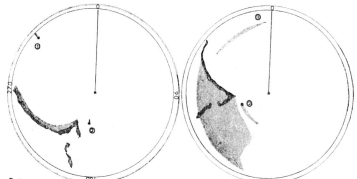

Echoes rotate and blurring occurs when ship alters course.

UNSTABILIZED RELATIVE MOTION DISPLAY

Ⓞ : Echo of Ship.
❷ : Echo of buoy.

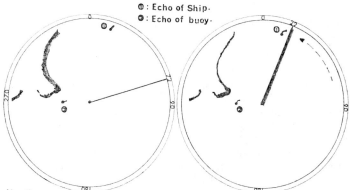

Headingmarker rotates when ship alters course. No blurring.

STABILIZED RELATIVE MOTION DISPLAY

FIG. 10.2   UNSTABILIZED AND STABILIZED RELATIVE
MOTION DISPLAYS

position of own ship is stationary while echoes move in the opposite direction to the course and automatically the navigator relates this to the visual view presented from the bridge where starboard is on the right hand and port on the left hand side. However, clear weather practice will help and experience will soon teach the observer to get used to it.

If a True Motion or Track Indicator Unit is provided, then the Relative Motion Stabilized Display must still be used for *landfall navigation*, as true motion is only introduced for the medium and shorter range scales where the tail of a moving echo can be observed.

Figure 10.2 shows a comparison between the Unstabilized and the Stabilized Relative Motion displays.

Relative Motion Display, Single Stabilized, Head-up

This is simply a Relative Motion Stabilized Display with the heading marker on zero of the relative bearing scale. Yawing will make the heading marker move and after a course alteration the heading marker *has to be reset to the "Up" position*. This is generally done by manual means (Picture-Rotate Control), but is sometimes achieved automatically (in which case it becomes a Relative Motion Head-up Double Stabilized Display). Sometimes a North-reference electronic marker is shown near the periphery of the display. This type of display presentation (or a Relative Motion Course-up Stabilized Display) is now an IMO requirement for ARPAs.

Relative Motion Display, Unstabilized, Head-Up, Off-Centred

In some types of radar sets, especially designed for *river navigation*, the centre can be off-set. A greater forward range is obtained, while using the same range scale. Bearing discrimination is better in some areas of the screen than the same areas on the next scale up. Echoes near the edge of the screen possess a greater relative motion than the same echoes displayed on the next range scale up. The ordinary bearing marker scale cannot be used for observation of bearings unless a Parallel Index is provided, see Fig. 10.8 (an electronic bearing marker would give more accurate bearing observations). The view abaft the beam is limited.

Relative Motion Display, Stabilized, North-Up, Off-Centred

This display can be presented when True Motion is provided. By means of the Zero Speed Switch the displacement of the electronic centre can be halted (see Chapter 4: Controls). The presentation can be useful for applying Parallel Index Techniques and for a quick estimation of the nearest approach for a ship at a range greater than that represented by the radius of the PPI at the range scale in use, because the rate of relative motion is not so clearly indicated on a longer range scale.

It provides an excellent display for *coastal navigation*.

True Motion or Track Indication (Chart) Display

The principle of True Motion has already been discussed in Chapter 4. Moving targets and " own ship ", if moving, are shown as moving spots on the face of the PPI, while stationary targets (no current is assumed) yield echoes which have no afterglow trail. The

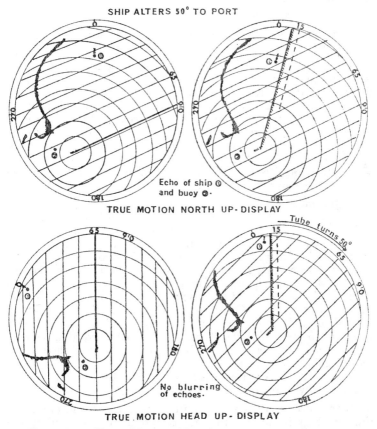

FIG. 10.3   TRUE MOTION NORTH-UP AND TRUE MOTION HEAD-UP DISPLAYS

display is stabilized. Fewer difficulties are encountered here in the interpretation of the picture—in contrast with Relative Motion Stabilized Display, North-Up—because it is similar to Chart Presentation. See Fig. 10.3.

(a) Sea-stabilized

Correct heading and speed through water are fed in, but no tidal information. It is an excellent display for *coastal navigation, pilotage* and anti-collision (see later Chapter 16). Echo trails of ships' targets and trail of " own ship " give an indication of *the heading of the vessels, irrespective of whether there is current or not.*

(b) Ground-stabilized

Correct heading and log speed of own ship plus direction and rate of tide are fed in.

When *Set and Rate Controls* are provided, then observe the movement, if any, of the echo of a known land-fixed object such as a lightvessel with the Rate Control on zero and the correct course and speed of own ship through the water fed into the radar display. If there is a current of any consequence the echo will now appear to move in a direction opposite to that of the set of the current so the Set Control can be adjusted accordingly; for example, if the echo appears to move S.W. then the Set Control is adjusted to read N.E. The Rate Control is then turned slowly from zero until the echo ceases to move over the display.

If the radar set is the type which has a " *Course Made Good Correction* " *Control,* then the quickest method of allowing for the current is probably to construct a velocity triangle (velocity of own ship through water, velocity of the current and velocity over the ground) on the chart. This triangle should be drawn anyway, even if one was not using radar ! The *length* of the resultant velocity vector will give an indication how the speed input should be re-adjusted. The *direction* of the resultant velocity vector will inform the operator if the " *Course Made Good Correction* " should be applied to starboard or to port of the steered course as shown on the gyro repeater. With the "Course Made Good Correction" control on zero, an increased (decreased) speed input would allow for a following (opposite) tidal current. This is useful when proceeding along a twisting river where the direction of the current changes continually with the direction of the river.

If the current is not known initially it can be found in a short time by observing and plotting the echo of a land-fixed object (see Fig. 13.5). If there is no land-fixed object available, then a sea-stabilized display is generally the best presentation to use.

A diagram showing the Sea-Stabilized True Motion Display versus the Ground-Stabilized True Motion Display is shown in Fig. 10.4. To understand the diagram, the reader should have some knowledge of the plotting procedure, as described in Chapter 13. See also Fig. 13.5.

The Ground-Stabilized True Motion Display can be very useful for *pilotage*, when the pilot wants to know his and other vessels' courses made good over the ground in relation to buoys, beacons or landmarks.

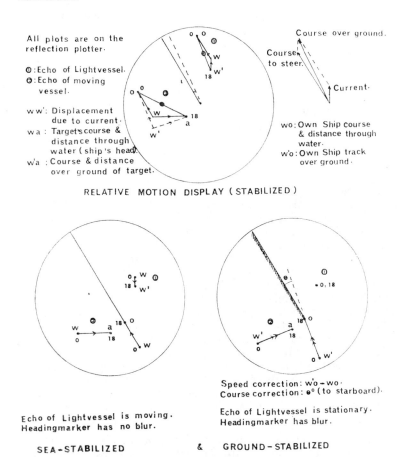

All plots are on the reflection plotter.

⊙ :Echo of Lightvessel.
ⵙ :Echo of moving vessel.

w w': Displacement due to current.
w a : Target's course & distance through water (ship's head).
w'a : Course & distance over ground of target.

Course over ground.
Course to steer.
Current.

wo: Own Ship course & distance through water.
w'o: Own Ship track over ground.

RELATIVE MOTION DISPLAY (STABILIZED)

Speed correction: w'o – wo .
Course correction: ⵙ° (to starboard).

Echo of Lightvessel is moving.
Headingmarker has no blur.

Echo of Lightvessel is stationary.
Headingmarker has blur.

SEA–STABILIZED   &   GROUND–STABILIZED

TRUE MOTION DISPLAY

FIG. 10.4   SEA-STABILIZED AND GROUND-STABILIZED TRUE
MOTION DISPLAY

If a ground-referenced two-axes Doppler Log is available, then its output can be fed directly to the display unit

Relative Motion Ground-Stabilized, True Motion Vector Display

This display is only possible where a computer is available which calculates the ground velocity of the vessel by tracking a stationary object. On ARPA displays, a control known as "Echo-Reference" or "Auto-Track" will instruct the computer to do this task. All true motion vectors will then display true ground velocity of own ship and targets.

True Motion Head-Up Display

This type of display is a True Motion Presentation, but double stabilization is applied, i.e. both the picture presentation (picture alignment) as well as the Cathode Ray Tube (and bearing ring) are coupled to the gyro-compass. This means that tube, bearing ring, parallel index and reflection plotter, all as one unit, rotate (mechanically or electronically) within the display unit in correspondence with every change of course. To understand its operation, it might be best to suppose that the ship is steering 000° T. The heading marker is "up". If the ship now alters to starboard then the action of the picture alignment is to swing the heading marker clockwise round (as would happen in the conventional stabilized display), but at the same time and at the same rate, tube and bearing ring swing in an anti-clockwise direction. The result is that the heading marker remains in the "upright" position, indicating the new course against the bearing ring while at the same time the echo formations on the tube have turned through an angle in an anti-clockwise direction equal to the alteration of course, but without causing any picture smear. This double stabilization clearly eliminates one of the main disadvantages of the head-up display, i.e. the constant blurring of echoes whenever alterations of course take place or when the ship is subject to yaw. It also gives the operator a much better check on bearings of other vessels as the bearings will always be true and their accuracy does not depend on the ship's head when the bearing was taken.

A diagram of the True Motion Head-Up Display is shown in Fig. 10.3. It illustrates the presentation before and after an alteration to *port* by own ship.

With radars carrying a double-stabilized tube arrangement, a Relative Motion, Stabilized Head-up Display can be selected by the operator instead of True Motion. This, nowadays, when centred, is known as a **Course-up Stabilized Display** because the heading marker shows the course steered.

Techniques

Use of Parallel Index for Navigation

The great advantage of the Parallel Index Technique is that no ranges and bearings need be taken and plotted, and an uninterrupted watch can be kept on the radar display of the movement of own ship over the ground. In other words, this technique is used to yield a *continuous fix*. Especially on the shorter range scales this technique is very sensitive in detecting deviations from the desired track over the ground. It needs pre-computing, but there is little work involved.

The underlying principle is that while using a Relative Motion Stabilized Display (Centred or Off-Centred) stationary echoes move on a course line about anti-parallel—depending on wind, current and tidal streams—to own ship's head, but while using a True Motion Display (Ground-Stabilized), echoes of land, buoys, light-vessels, etc. remain stationary, but the range marker circle, and the electronic bearing cursor move along with the spot representing own ship.

On a relative motion display whenever the reference echo moves off the reference line alter course to port or starboard to return the reference echo to the reference line. To help decide the direction of alteration of course needed, it should be remembered that the reference echo moves actually or approximately in a direction parallel and opposite to the direction of the heading marker.

Three examples will be given :

(a) *Leaving a fairway to drop anchor.*

A Relative Motion Stabilized Display is used. See Fig. 10.5.

Tracks and point of anchorage are marked on the chart, *A-B-C-D*, and a reference point *R* is chosen. The bearings and ranges of the lines *AR*, *BR*, *CR* and *DR* are transferred by means of the parallel index (or electronic bearing curzor) and range marker to a suitable scale on the PPI (shown as 1, 2, 3 and 4 respectively). Their terminal points *A'*, *B'*, *C'* and *D'* are then connected by means of chinagraph pencil, and a small circle, about 2 cables in radius, is drawn around point *D'*, showing the limits of anchoring space.

By careful manoeuvring, the echo-point of the reference buoy is then made to follow the lines *A'-B'-C'-D'*, and ending up within the small circle.

The use of *two* parallel index lines is advised when navigating in narrow fairways. These lines can be drawn on the display with reference to the same echo of a good radar reflecting object, both at the *same* side of the spot representing Own Ship. The cross-indices,

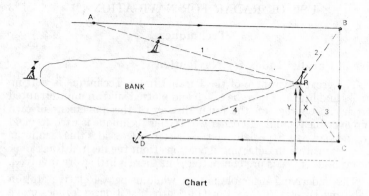

Chart

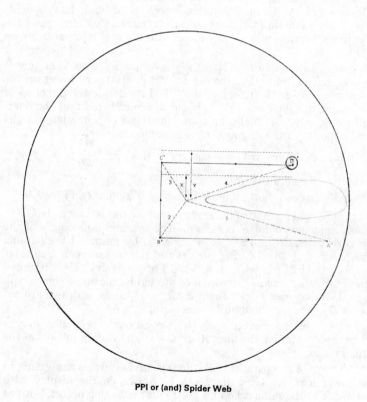

PPI or (and) Spider Web

FIG. 10.5 PARALLEL INDEX TECHNIQUE

taken from the chart (*x* and *y* in Fig. 10.5), should have magnitudes such that if the echo moves *between* the two lines, the master can be assured that his vessel is navigating safely in the selected lane.

A further refinement would be to plot an image of the bank on the PPI, in the same way as the track was plotted by means of the reference point. Under no circumstances should the echo of the reference point ever enter the plotted bank area.

(*b*) *Anchoring at a pre-selected place with no ranging mark ahead.*

Mark the anchorage (*A*) on the chart and draw through it the intended course of approach (track to make good). Transfer this line parallel through a point which represents an object possessing good radar reflecting properties (in diagram 10.6 this point represents a pier *P*). Drop a perpendicular from *A* on the latter line, the intersection being called *B*. *AB* is known as the *Cross Index Range*, and *PB* as the *Dead Range*.

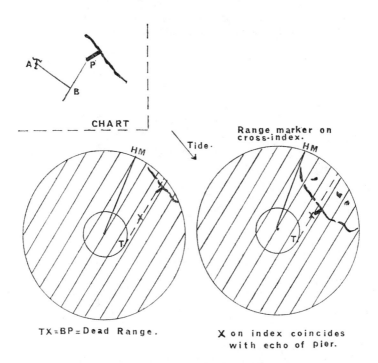

FIG. 10.6   USE OF PARALLEL INDEX ON RELATIVE
MOTION DISPLAY (STABILIZED)

Now go to the PPI, select a short range scale, turn the Parallel Index to the course to make good (indicated on the chart) and set the Range Marker on the Cross Index Range. Draw a tangent line in chinagraph pencil to the Range Marker circle parallel to the Index lines. Mark off the Dead Range from this tangent point T along the tangent line in the direction of the course to make good and call this point X.

Con the vessel in such a way that the echo of the pier comes under TX produced; then keep the ship conned so that the echo of the pier moves along the tangent line drawn on the Parallel Index. Drop anchor when X coincides with the echo of the pier (Fig. 10.6).

(c) *Passing a given distance from a navigational mark. Course alterations.*

Figure 10.7 shows the Relative Motion Stabilized Display (*a*) and the True Motion Display (*b*) of a ship which has to pass a distance of $\frac{3}{4}$ M from a buoy. The vessel needs an approach course of 090° T. with the current running towards the N.N.E. When the buoy is 1·5 mile abaft the beam, course has to be changed to 020° T.

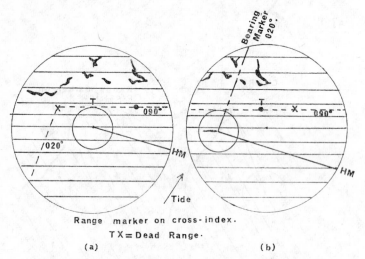

Range marker on cross-index.

TX = Dead Range.

(a)   (b)

FIG. 10.7   USE OF PARALLEL INDEX ON STABILIZED
RELATIVE MOTION DISPLAY AND ON TRUE MOTION DISPLAY

Using the Relative Motion Stabilized Display, the range marker s set at the cross-index range ($\frac{3}{4}$ M) and the Parallel Index is turned to 090°–270°. A chinagraph line is then drawn on the reflection plotter in direction 090°, touching the range marker at 'T'. Next the index or dead range TX is marked off (—1·5 M) and through X a line is drawn through the direction of 020°.

The ship is conned in such a way that the echo of the buoy trails along the TX line and when point X coincides with the echo of the buoy, course is altered to 020°.

Using the Parallel Index Technique on a True Motion Display (Fig. 10.7 (b)) which is ground-stabilized, align the Parallel Index as discussed above but lay off the index or dead range TX from the echo of the buoy (+ 1·5 M). Set the electronic bearing marker on 020°.

The ship is conned in such a way that the range marker circle remains tangential to the line drawn in chinagraph pencil through the echo of the buoy. Course is altered just before the curzor goes through point X on the Parallel Index and the vessel is steadied when heading marker and curzor coincide.

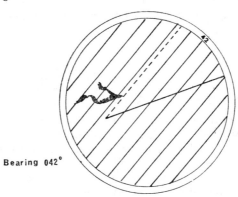

Bearing 042°

FIG. 10.8   USE OF PARALLEL INDEX FOR READING A
BEARING WHEN THE DISPLAY IS OFF-CENTRED

Some General Remarks

Unless a Parallel Index Technique is simple, and the navigator is familiar with the area, it has to be prepared *beforehand*. From the chart plan, the transfer procedure, instead of having applied directly to the PPI as described, is carried out on a plotting chart (spider web or manoeuvring board) with 'Own Ship' in the centre of the range circles. After all relevant points are properly marked and bearings and distances are indicated by representative figures, then having recorded the chart number on the spider web, it can be stored away. When required later on, it will then be a simple task to copy the plot from the plotting chart onto the PPI by means of chinagraph pencils.

The size of instrumental errors can only be guessed but its maximum is determined by the range and bearing error limits stated in the D.o.T. Marine Radar Performance Specification ($1\frac{1}{2}\%$ of maximum range of scale in use, and $\pm$ one degree either side of the main beam). In the worst case it would be (Maximum range error^2 + Maximum bearing arc errors^2)$^{\frac{1}{2}}$ and would amount to :—

> 256 metres (0·14 mile) on the 6 mile range scale,
> 128 metres (0·07 mile) on the 3 mile range scale,
> 85 metres (0·05 mile) on the $1\frac{1}{2}$ mile range scale,
> 74 metres (0·04 mile) on the $\frac{3}{4}$ mile range scale.

Another useful application of the Parallel Index, although we cannot strictly call it a navigational technique, is for reading a bearing when the display is off-centred. This is illustrated in Fig. 10.8 where the Parallel Index is aligned to a line going through the spot representing own ship and the echo of a landmark. The centre-line of the Parallel Index indicates the bearing on the bearing scale.

Similarly a bearing can be obtained between any two objects shown as echoes on the display.

Nav-Lines

These Navigation Lines are electronic lines displayed on the screen and they have already been discussed in chapter 4 under ARPA controls. Broadly speaking they come under three categories :

(1) Ground-stabilized digital lines in the correct registration on a sea-stabilized relative motion display (Sperry Approach).

(2) Ground-stabilized lines on a ground-stabilized PPI (Decca and Raytheon Approach).

(3) Electronically generated and stored lines displayed on a Relative Motion display (Iotron Approach).

The latter (3) are associated with Digiplot Displays and consist of a *pair* of dashed electronic lines drawn by the operator in any true direction and at any fixed range from the spot representing Own Ship. They form part of Own Ship and remain stationary on a Relative Motion display presentation—while echoes of fixed targets move "through" them, but move with Own Ship on a True Motion display. The Nav-Lines, therefore, differ from Track Lines, which can be displayed on some types of PPIs, for these are positioned geographically, i.e. they are fixed features with respect to land echoes.

To display and control the Nav-Lines an extra, optional, control box is fitted near the PPI (see Fig. 10.9) where their direction (100°) and distances from 'Own Ship' (2 cables to starboard, 8 cables to port) can be dialled in. Ten pair of lines (0–9) can be programmed and retained in the Digiplot memory, using the *left-hand* thumb-wheel switch of the line selector and turning direction and range controls for each pair of lines. However, only *two* pair of lines can be displayed at any one time, their numbers shown on the thumb-wheels of the line selector, but only the specific pair of lines showing its number on the *left-hand* thumbwheel can be controlled, and has its direction and ranges shown in the digital read-outs.

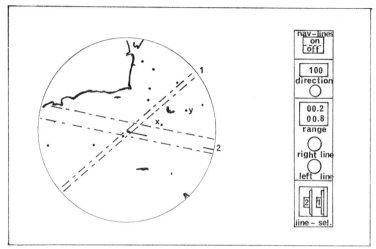

FIG. 10.9 NAV-LINES

For example, in Fig. 10.9, only Nav-Lines 2 can be controlled in direction and range. If the operator wants to control Nav-Lines 1, he should change the line selector setting from 2–1 to 1–2. With the setting on 1–1, No. 2 lines disappear and on 2–2 the No. 1 lines are removed. If only one line of a pair is required the operator can set the range of the unwanted line to any range greater than the range scale, for example 20 miles.

Nav-Lines are used for Parallel Index Techniques as described in the previous sections. When the echoes of the buoys, indicated in Fig. 10.9 by x and y come "under" the lines, the ship's heading can be set to the new course.

The advantages of Nav-Lines over the "Chinagraph pencil" method are:—

1. Lanes do not have to be re-drawn when the range scale is changed.

2. Pre-programming can be done without cramming the display with lines—and confusing the operator—as only two pairs can be

displayed at once, while future lines can be brought on the PPI by simply flicking the thumbwheel switch.

N.B.   When sailing in a buoyed channel and having an ARPA display available, instead of drawing lines with chinagraph pencils or generating Nav. lines, one could put 6-minute *relative* vectors on the echoes of the buoys. These vectors would indicate the ground track. With buoys on either side, and using the range marker, there will be an easy check to see if the vessel remains safely positioned within the channel.

CHAPTER 11

USE OF RADAR FOR NAVIGATION (3)

On the following pages, some types of radar aids for the use of the navigator will be studied.

They can be divided into two classes:

A. Passive aids such as corner or radar reflectors.

B. Active aids. Under this category come Ramarks, Racons, and Transponders.

A. Radar Reflectors

Radar reflectors can be compared with " Cat's eyes " in the road or with the prismatic buoys in the Suez Canal. A simple geometrical proof will show that when a light or radar beams falls upon three mutually perpendicular planes, the direction of the incident radiation is exactly the opposite to that of the reflected radiation. Hence the radiation is returned to the scanner. See Fig. 11.1.

During the reflecting process part of the radiation is lost, because some of it will " miss " a reflecting surface.

Radar reflectors are used to boost up the echo strength of targets which otherwise would return weak echoes as do buoys, wooden boats, etc. The three light-metal planes which make up the reflector are each shaped as a right-angled triangle and not as a square. This makes the reflector more accessible for the incoming radiation from different directions. The height *h* in Fig. 11.1 defines the size of the reflector. We speak of a 12-inch (30 cm. reflector or) 16-inch (40 cm.) reflector, etc. A 24-inch (60 cm.) radar reflector can give the same echo strength, if fitted high enough, as a medium-sized merchant ship. It is assumed in this section that 3 cm. radar is used.

For all-round coverage in azimuth, the corner reflectors are arranged in clusters. There are two main types:

1. *Octahedral clusters.*

2. *Pentagonal clusters.* See Fig. 11.1.

Octahedral clusters are formed by the intersection at right angles of three diamond-shaped plates, so making eight corner reflectors.

161

Pentagonal clusters consist of five corner reflectors arranged in a ring. For complete coverage, another pentagonal cluster should be fitted upside down in between the open spaces of the original one. This is sometimes done, but care should be taken not to make the buoy top heavy. However, new buoys can be designed in such a way, making use of a counterweight, to allow for the extra weight. Generally pentagonal clusters are fitted on buoys while octahedral clusters are used in small craft. See Fig. 11.1.

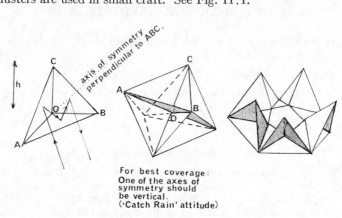

For best coverage:
One of the axes of
symmetry should
be vertical.
('Catch Rain' attitude)

FIG. 11.1　RADAR REFLECTORS AND GROUPING IN
OCTAHEDRAL AND PENTAGONAL CLUSTERS

An important concept related to the corner reflector is the "axis of symmetry". It is an imaginery line dropped from the corner perpendicularly on the opposing "fourth" plane (the mouth of the corner reflector). It makes an angle of 55° with each of the three straight sides and 35° with each of the three reflecting planes.

The polar diagram for reception in the horizontal plane received from the octahedral cluster depends very much on its orientation. If the line connecting the two apices is vertical (i.e. each of the axes of symmetry makes an angle of 35° with the horizontal), the polar diagram contains 4 main lobes, spaced 90° apart, plus 4 subsidiary, much thinner ones. A much better polar diagram can be produced when *one of the axes of symmetry is vertical*. In this case the polar diagram contains 6 main lobes, spaced 60° apart, plus 6 subsidiary lobes. This orientation of the cluster is known as the "*catch rain*" attitude and it is the ideal attitude (if attachment fittings permit it) for octahedral clusters in small craft.

Another attitude of the cluster, often advised for medium sailing yachts which have sufficient reflecting power for beam-on aspect, is to have *one of the axes of symmetry horizontal* and aligned parallel to the fore and aft plane. This orientation produces a polar diagram in the horizontal plane of two powerful lobes in the fore and aft plane, spaced 180° apart.

It has also been found that if quadrantal sectors are used, instead of triangular sides in the reflector, its radar cross-section will be much improved. This arrangement gives the cluster a ball-like or spherical appearance.

Referring back to the case—one but last paragraph—where the cluster has an orientation with one of axes of symmetry horizontal, thus producing a polar diagram with two strong diametrically opposed main lobes, the best solution for excellent all-round coverage is to have a series of clusters packed vertically into a stack, all with one of their symmetry axes horizontal and each one slightly displaced, angularly, from the previous one in the stack. In other words, arranged with an angular distribution of the axes in a horizontal plane. Another advantage of this arrangement is that the coverage diagram for reception in the vertical plane is filled in and no minima occur.

These stacked clusters, are, in fact, on the market. Their number is limited by the weight, which still is surprisingly low. They are encased in a cylindrical weather-tight container. They cause little windage problems and can be hoisted at the backstay of a yacht, preferably at least 4 m. above sea level. This arrangement would avoid interrupting the airflow and makes the cluster fully effective, all-round, over the required minimum radius of five miles.

Dihedral reflectors (two vertical plates at right angles) placed in a ring on top of a lighthouse which is a poor radar target, will improve the echo strength considerably (3–17 M).

The clusters and the reflectors individually must be of rigid construction. A slight displacement will make the reflection less effective. Remember in this connection that a displacement of one degree of the index mirror of a sextant displaces the reflected ray by two degrees.

A buoy fitted with a cluster of radar reflectors is a powerful target and possesses a good deal of " reserve " echo strength when displayed as an echo on the screen. Therefore by reducing the gain, it can be picked out among strong sea clutter, which is not always the case with the echo of an ordinary buoy except when it is very close. Alternatively, anti-clutter can be employed on short range scales.

Loss of the echo due to its " twinkling " character (Chapter 6) in calm sea conditions must be reckoned with. Maxima and minima could be avoided by fitting two clusters of corner reflectors, suitably

G

displaced in azimuth, one at half the height of the other. Wave motion may also affect the echo strength.

A 30 cm. pentagonal cluster increases the detection range of a third-class buoy from 1·5 to 3·5 miles. For can and pillar buoys it amounts to an increase from 3·5 to 7 miles. Spherical buoys fitted with radar reflectors can be easily detected at a range of 5 miles or more, depending on the size of buoy and reflector. Reflectors have been placed in a crown on lighthouses in order to increase their radar " visibility ".

One of the great uses of radar reflectors is to boost up the echo of a small craft or wooden boat. The echoes of such craft are very difficult to detect on the radar screen, so giving rise to danger to navigation. Their reflecting surface is poor or very small. They can also become easily obscured by shadow sectors, even at close ranges, as their echo strength is not powerful enough to emerge from the area of reduced intensity in such a sector. Several collisions have occurred between radar-equipped vessels and wooden vessels or small craft which have escaped the notice of the observer.

A 30 cm. reflector hoisted at a suitable height is known to have increased the detection range of a wooden fishing vessel from 2 to 6 miles. A lifeboat equipped with a 40 cm. radar reflector showed an increase in detection range from 3 to 7 miles.

Radar reflectors are also used in the whaling service. After the whale has been killed, the catchers place some radar reflectors in the carcass so that it can be easily detected by the factory vessel.

A special type of radar reflector is the Lens Reflector, or **Lensref** for short. Based on the invention of Luneburg (1948), a scientist in the field of optics, it consists of a spherical lens which has a variable index of refraction. It has the property that a plane wave incident on the sphere is focussed onto a point on the surface at the diametrically opposite side. Likewise, a transmitting point source on the surface of the sphere is converted to a plane wave on passing through the lens. Because of its spherical symmetry, the focussing property does not depend upon the direction of the incident wave.

For micro-wave application the lens—about 30 cm. in diameter—consists of several concentric hollow spheres made of different dielectric material. At a height of two metres above the sea surface, it may increase the detection range of a poorly-reflecting target on a calm sea to about 10 miles, assuming a scanner height of about 15 metres. The radar cross-section of the lens is over 30 sq. metres and it also reflects upwards, thus making it very effective for detection from aircraft.

IMO made a recommendation that all vessels under 100 g.r.t. should be fitted with a radar reflector of a determined minimum performance. In this connection, the Department of Trade has issued a "Marine Radar Reflector Performance Specification, 1977", so that manufacturers, making general purpose radar reflectors for small vessels, can obtain Certificates of Type-Testing for their products.

Finally a remark about *Scotchlite tape*, which is a retro-reflective tape fitted to survival craft and survival equipment to aid the detection of such craft at night. By illuminating the material with a spot light it will give off a bright return. It was found that the material not only reflects light but also radar waves. Some initial evaluation tests were conducted by the Department of Trade. The findings of this evaluation are that Scotchlite tape cannot be described accurately as giving the equivalent reflecting properties of metal, but it can be said to have considerable value as a radar echo enhancer when used on normally poor radar reflecting targets made of rubber, wood, glass, etc. It was emphasised that the material should not be called a Radar Reflector.

B.   Radar Beacons and Transponders
Ramark

The Ramark is a continuously transmitting beacon, sometimes compared with a lighthouse, and referred to as a "radar lighthouse". Its action is *independent* of own ship's radar. Its signal is detected by ships' radars, causing a bright radial line or a narrow sector of dots and (or) dashes to appear on the PPI showing the true or relative direction of the beacon. Its use is only for direction-finding.

The beacon either uses a *variable* frequency which sweeps through the marine frequency band thus making the signal available to all ships' radars working in that band, or it operates on a *fixed* frequency *outside* the marine radar frequency band. In the latter case the radar set has either to be re-tuned to receive the signal, or a separate receiver is incorporated in the radar set.

Experiments with Ramark transmitters have been carried out in Britain in the past and these have shown that they have a number of operational disadvantages (masking of the picture by many false-indirect and side-echoes). Development in this country was therefore not continued, though beacons of this type are established in Japan, mainly for the use of fishermen.

Racon (Radar Beacon)

This beacon, in the maritime radio-navigation service, is defined as a receiver-transmitter device which, when triggered by a surface search radar, automatically returns a distinctive signal which can appear on the display of the triggering radar, and thereby provides range and bearing information.

It is often called *"Secondary Radar"*. See Fig. 11.2.

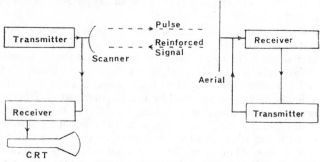

FIG. 11.2   SHIP'S RADAR AND RACON BEACON

The Racon consists of a receiver, signal processor and a transmitter, and should be capable of operating in conjunction with any radar equipment working in the marine radar frequency band. The 3 cm. band, in which Racons mostly operate, covers the frequency range 9320–9500 MHz, a bandwidth of 180 MHz.

The beacon responds to the pulse transmitted by a ship's radar and it is this pulse which triggers it off. The Racon then transmits a strong signal back which appears on the PPI as a bright radial flash, a series of dots or dashes, or a single morse letter (to provide identity-coding), starting from the Racon echo and making outwards towards the edge of the display. Hence a Racon echo provides range as well as bearing of the Racon installation from the ship.

FIG. 11.3   RACON SIGNAL AT LONG AND SHORT RANGE

The flash or dots do not start exactly at the range of the echo of the Racon station itself because there is a slight delay in the processor unit of the Racon installation. At short ranges this does not matter because the range can be measured to the Racon station itself, while at long ranges, where the Racon station itself is generally not recorded, the range can be measured from the electronic centre to the inner edge of the flash and the small error caused by the slight delay, can be ignored. (Fig. 11.3).

Types of Racons available or under development

Each type of Racon has its own advantages and drawbacks and this should be kept in mind when planning for different applications.

I. *Slow Frequency Racon.* This Racon, developed in the U.K. sweeps through the 180 MHz marine frequency band in a sweep time of 90 to 120 seconds. On a short range scale with a short pulse and the relatively wide radar bandwidth of 10–20 MHz, this presents no difficulties, and the signal can be received during a time period of 6–12 seconds, i.e. during 2–4 consecutive rotations of the scanner.

Using a long range scale with a long pulse and narrow radar band (2–5 MHz), this will be different, and the signal during a Racon sweep period may only be recorded during one scanner revolution or, sometimes, the scanner will miss picking it up altogether. There may be difficulties, therefore, in detecting this type of Racon at an extreme range.

An efficient and well-sited shipborne radar may detect a good Racon signal when the Racon beacon is at a distance of 18–30 nautical miles. The response time with this beacon is 48 micro-seconds, representing 4 nautical miles. The latest Racons have their maximum range and response time reduced respectively to approximately 18 n.m., and 20–24 micro-seconds.

II. *Stepped Sweep Racon.* The marine radar frequency band is divided into four equal sub-bands. The Racon sweeps through *one* of these in 12 seconds. When the Racon is triggered by a radar pulse, it will respond to the frequency to which it is then tuned and immediately jump to the corresponding time-point in the next sub-band, where it will respond to the next trigger. This process is repeated for each triggering pulse received from the radar. Transmissions therefore will take place in each of the four sub-bands for each scanner revolution. In this way the whole marine radar band is covered in 12 seconds, so that any radar set will receive Racon responses every 12 seconds. See Fig. 11.4.

The number of possible Racon returns per time-base is reduced by a factor of 4. This is of little consequence as this type of Racon is only used for short ranges.

III. *Fast Sweep Racon.* In this type the frequency of the transmitter sweeps over the marine radar frequency band in a period of about 12 micro-seconds. Due to this very fast sweep, which is repeated a number of times, the signal appears on the PPI as a series of small dots. This type of Racon is extensively used in France, and 10 cm. Racon beacons have also been designed and tested while operating the fast sweep.

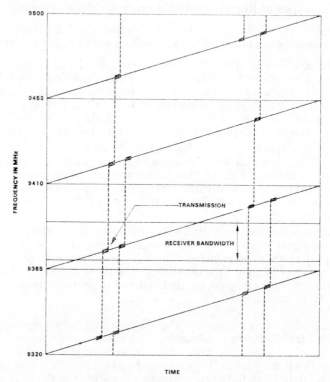

FIG. 11.4   STEPPED SWEEP RACON REPRESENTATION

IV. *Fixed Frequency or Edge-Band Racons.* In this Racon a wideband receiver is used, but its transmission takes place on a frequency at the edge of the marine radar frequency band. The band allocated to them ranges from 9300–9320 MHz, and generally they operate at 9310 MHz.

Because a different frequency of transmission is used from the one the radar is tuned to, special receiving facilities have to be provided. By switching between two pre-set voltages, the local oscillator in the mixer can be tuned to receive either the radar signals (radar mode) or the Racon signals (Racon mode).

With the radar in the Racon mode, the Racon signal will be presented in *isolation*. There can be no interference from the Racon signal on the radar picture, for example the masking of small echoes ; neither can the Racon response be masked by sea clutter or spurious echoes. This absence of mutual interference, especially at short ranges, where the Racon response will appear over a wider sector, is a distinct advantage of Fixed Frequency Racons over Swept Frequency Racons.

Another great advantage of the Fixed Frequency Racon over the Swept Frequency one, is that while in Racon mode, the Racon response is *available* on every scanner rotation and there is *no delay* in acquiring the signal once it is within operational range. With Swept Frequency Racons there will be a delay of about 100 seconds between successive sweeps which may cause inconvenience to the navigator.

V. *Frequency Agile Racons.* In this Racon the interrogating frequency is sampled, and its value stored. After the necessary signal processing the transmitter is switched on and its frequency compared with the stored information. Rapid retuning is then carried out, if necessary, so that the transmitted frequency over the majority of response time is at, or very close to the input frequency.

The accuracy of the transmitted frequency is dependent on the accuracy of the frequency-comparison process in the signal processor, and on the accuracy of the input frequency measurement which, again, is dependent on the length of the interrogating pulse ; the longer the pulse, the more accurate the measurement will be.

A Frequency Agile Racon provides a signal which will be displayed on every scanner rotation period, similar to the Fixed Frequency Racon. Since it operates within the radar band, it does not require additional receiving facilities in the radar set.

It does suffer, however, from the interference problems as does the Swept Frequency Racon, and since it is available on every scanner rotation, this interference could be even more severe than for the Swept Frequency Racon.

The circuitry in this type of Racon is considerably more complex than in other types and this may reduce the reliability.

Interference

There are two types of interference related to Racon operation. The first one is the reception of responses which have not been initiated by the interrogating radar. Fortunately this type of interference is not common and not very serious.

The second type of interference comes from unwanted signals initiated by the interrogating radar, and can be a major source of trouble. As the range between the radar and the Racon is reduced, the angle over which the aerial will interrogate the Racon, due to increased signal strength, will increase. This increases the response sector on the PPI, especially on the shorter range scales. This means that echoes of close targets can be obscured. Eventually triggering by the side-lobes, and/or by reflections off the ship's superstructure can occur. The result will be a series of bright sectors on the PPI in different directions (Fig. 11.5).

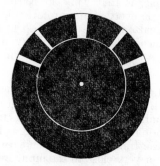

FIG. 11.5   MASKING BY RACON INTERFERENCE CLUTTER

This interference problem is even made more difficult on the shorter range scales because short pulses are used which require a wide band-width. As a result the Racon signal—and its spurious echoes—will be seen on several successive scanner rotations, thus reducing the time when there is no Racon signal when one can watch for echoes of small targets near the centre of the screen. *Temporary* reduction of receiver gain or application of F.T.C.—better not to use the fully automatic 'Clearscan'—will remove the indirect echoes caused by reflections from the ship's superstructure, and by the side-lobe echoes.

Polarization. Racons working in the X-band are designed to operate with marine radars using horizontal polarization, and do not respond to radars using vertical polarization.

With the increase of S-band radars in ships—and the use of vertically polarized e.m. waves—one may soon expect the appearance of Racons working in the X- *and* S-band. In fact the first was demonstrated recently (the AGA-Ericon Racon) which, besides operating in the X-band (with automatic suppression of side-echoes), also operates in the S-band employing the frequency range of 2900–3100 MHz.

Echo Enhancers. Owing to the application of integrated circuits (see Appendix) the weight and physical size of radar beacons can be made very small. They can also operate from primary or secondary batteries. Sometimes, therefore, they are placed on small craft or on certain types of buoys, so that their echoes can be detected at a good distance and cannot be lost in sea clutter. This type of beacon is known as an *Echo Enhancer* and they re-transmit only *one* short echo pulse, generally not longer than the received pulse.

Uses of Racons

(1) Racons are used to provide *identification of important navigational marks*, either because of poor reflecting characteristics of the latter or because they are surrounded by several objects yielding echoes on the radar display of approximately the same strength. For example, the installation of a Racon on a lightvessel situated at the entrance of a fairway or at the mouth of an estuary, can be extremely helpful for a navigator. Without a Racon, it can sometimes be very difficult to identify such a light vessel from its echo on the display due to the number of similar looking echoes of radar reflector buoys and of ships nearby.

(2) Racons can provide marks for *identification of a coastline* when the coastal features are masked by ice, as is the case in Sweden during the winter.

(3) A Racon can warn the navigator of a *new danger*, for example a new wreck not shown on the chart. IALA Buoyage Scheme A makes provision for this case so that a Racon can be fitted at the outside of the primary danger mark (using a second buoy, for example), showing a code signal on the radar screen—1 n.m. in length—to indicate the type of hazard, for example the morse signal *W* for Wreck.

As a navigator, in most cases, would not be aware of such a new danger, the Racon must be either the 'Swept Frequency' type or the 'Frequency Agile' type. It should never be a 'Fixed Frequency' type where the navigator himself has to take certain steps to bring the Racon response on to the radar screen, steps he would not be likely to take if there were no Racons marked on the chart.

To reduce the difficulty of interference, it has been suggested in Trinity House circles that permanent Racons, whose position has been recorded on charts and in official publications, should be fixed frequency edge-band Racons, while temporary Racons, placed, for example, near hazards, should use a sweeping frequency through the complete frequency bands of marine radar.

(4) It has been recommended that Racons should be placed on some *drilling* or *production platforms* (for use by helicopters or service craft), and also on buoys or platforms used for measuring Ocean data (Ocean Data Acquisition Systems).

(5) A Racon can provide *identification of supply ships* for deep sea fishing fleets (for homing purposes).

(6) Proposals have been made for the use of Racons in *Search and Rescue* applications. The fitting of Racons to survival craft could greatly assist rescue operations.

(7) Racon can be used to provide information about the *exact position of bridge supports*, centre of free spans etc. When approaching a bridge, it is often not possible to see from the radar screen what position the ship has to take up and what course she has to shape before going under the bridge. All that can be seen on the display is a very strong echo-band across the waterway and beyond this echo and many shadow areas. Many serious accidents have taken place while navigating under these circumstances, either due to direct collision with part of the bridge itself, or with ships approaching from the other side of it.

Transponder

A Transponder differs from a Racon in that, besides echo enhancement, and information about bearing and range, it must also provide *positive identification, and,* sometimes, *other relevant data* when properly interrogated. IMO has defined a Transponder as being a receiver-transmitter device which transmits automatically when it receives the proper interrogation, or when a transmission is initiated by a local command. The transmission may include a coded identification signal and/or data. The response may be displayed on a radar PPI, or on a display separate from any radar, or on both. Such transponders could be installed on board vessels for interrogation by other ships and/or by shore stations so as to improve collision avoidance techniques, and to facilitate marine traffic control.

If marine Transponders are implemented it would seem therefore that the transponder frequencies should be on the edge of the radar band but not coincident with racon frequencies.

Two types of Transponders are being envisaged :—

(*a*) Those which establish positive identification and recognition of a target, but further communication is continued on other channels such as V.H.F./R.T.

(b) As above, but further communication (data transfer) takes place via the Transponder itself.

Whatever the type, owing to selective calling, Transponders are more complicated than Racons. Firstly, on board the interrogated ship the transponder must be able to receive omnidirectionally since the direction from which the interrogating pulse arrives, is unknown. Furthermore, the receiver of the transponder needs a wide band-width in order to cope with the digital code of the interrogating pulses.

Secondly, on board the interrogating vessel, there must be a special receiver which will accept the enhancement signal *and* also decode the signals from the Transponder. Presentation of this data —echo + decoded signal—may be on a conventional PPI and/or on a peripheral display.

Selective azimuth recognition will be important and interrogating pulses are sent out via the conventionally rotating radar aerial.

Operationally it is envisaged that a radar operator would select the target he wishes to identify by means of a strobe on the primary radar. The target causing this echo would then be interrogated and the information relevant to it would appear on the radar display or the auxiliary display.

To prevent mutual interference dual frequency transmission is used and besides the X- or S-band, the lightly loaded C-band (5 GHz or 6 cm.) is adopted.

Presently, the type of Transponder described above has been introduced in Hovercraft Ferries. Radar displays in conventional ships would see the presence of a fast moving Hovercraft as a long echo tail displayed on their PPI, but viewed on the radar of another Hovercraft of similar speed, there is no scan-to-scan overlap of the echoes on its radar. This reduces the intensity of the PPI afterglow so that the afterglow tails of Hovercraft and conventional ships have approximately the same length and cannot be distinguished from one another. This could prove extremely dangerous in reduced visibility as appreciation of close quarters would be too late.

Fig. 11.6 shows a block diagram of such a Transponder fitted in a Hovercraft Ferry. Interrogation takes place on *two* separate frequencies, one in the X/S-band using the conventional ship's radar and one in the C-band. Both transmissions are synchronized and directional. In the interrogated vessel there is an X/S-band receiver with an omnidirectional aerial and a C-band receiver equipped with a directional aerial (scanner). When the two pulses—from X/S-band and C-band—are received at exactly the same moment, and AND-gate is enabled and a response is transmitted on the C-band in the *correct direction* of the interrogator. The response is coded such that its reception causes a synthetic video signal of appropriate identifying structure and length to be displayed on the radar screen. Owing to the use of the two frequencies and the directional responder aerial, *only the interrogating ship* will receive the signal on her PPI and no interference can be produced on other radars.

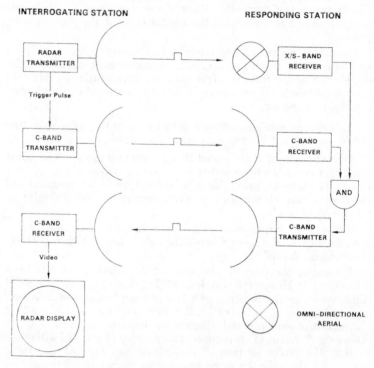

FIG. 11.6 TRANSPONDER USING TWO FREQUENCIES

If the Transponder is used for further data transfer, for example, the transmission of course (direct from gyro-compass) and speed (direct from log), then the device is often called a *Maritime Radar Interrogator Transponder* or *MRIT*.

In the field of Collision Avoidance a MRIT would be of great benefit. It has been shown that the values of CPA and TCPA computed *directly* from course and speed of a target (besides instantaneous range and bearing) are more accurate than when extracted from the echo *motion* (i.e. historical development) on a radar display.

The question of carrying Transponders on board ships would have to be agreed internationally. It is already introduced in some local waters where the pilots take the Transponders on board to be interrogated by shore stations, but where it concerns larger areas or world-wide use, it would be necessary to make the carriage *mandatory*.

Besides in the field of Collision Avoidance, Transponders can be used for the following purposes :—

1. For identification of ships to assist shore surveillance in harbour approaches and restricted sea channels in national and international waters. Indirectly, the Channel Navigation Information Service employs light aircraft carrying Transponders. These aircraft are used to identify vessels in the channel and in order that the identification is positive, it is necessary for CNIS to follow the proper track of the aircraft on the radar screen.

2. For identification and approach information to or from a specific point or into a channel or harbour. Transponders could be taken on board by the pilot. As each transponder would be individually coded, the identity of a vessel would be established.

3. To assist in Search and Rescue operations and to identify uncharted navigation hazards. These items have already been discussed under the Racon section. It should be noted that, in these cases, "Swept Frequency" Transponders must be used.

Radaflare

The Radaflare or Radar Lifesaving Rocket is a radar aid, consisting of a rocket which can be fired from a pistol. After firing it emits a very strong light at a height of about 400 m. but also ejects a cloud of tuned di-poles which strongly reflect 3 cm. radar waves. The patch of echoes displayed on the PPI lasts for about 15 minutes and the di-poles have a maximum detection range of 12 nautical miles. It can be used as a distress signal and can assist in a quicker discovery by radar of small craft, lifeboats and liferafts.

CHAPTER 12

RADAR LOG

A great deal of administrative work is already carried out on board ships, but a Master should make it his duty to enlist one of his officers to organise the upkeep of a radar log. Insertions in the log should also be made by other officers so that they, too, may better understand the capabilities and limitations of the set. Such a log would have a two-fold purpose.

1. It helps in checking the performance of a radar set;

2. It informs the observers about the responses of certain targets and about their responses under different meteorological conditions. A radar set may be quite efficient, yet still perform only a limited duty.

The log's first function is of value for the owner to assess the benefits of the radar installation, and for the manufacturer to guide him in the processes of research and development. It should contain the following information:

(a) Date of installation;

(b) Heights of scanner for various mean draughts of the ship;

(c) Periods of use—
    (i) Duration of single period;
    (ii) Total hours run (in many cases there is a clock in the transceiver to record this);

(d) Where used;

(e) The reasons for use;

(f) Benefits and limitations experienced;

(g) Time saved;

(h) Reading of the performance monitor, together with the maximum reading;

(i) Repairs carried out and modifications made.

The second function of the radar log is to assist the navigator himself. It is also of importance for general information and publication about the interpretation of the display (*Navigational Journals*). In this connection, the following items can be usefully recorded:

(*a*) State of weather and sea. The effect of precipitation, fog, sandstorms, dust and smoke. Comparison of targets inside and beyond an area of precipitation. The deterioration of echoes due to snowfall. Detection ranges of ice. Super- and sub-refraction. Multiple-trace returns. Ghost echoes.

(*b*) The range at which a target is first detected. The echo at this range should paint well and not be too flimsy.

(*c*) The average detection range. This can only be compiled after the observer has become familiar with the first detection ranges of specific targets.

(*d*) The range at which a target becomes recognisable, either from the chart, visual deduction or local knowledge. (Note the difference between (*b*) and (*d*)).

(*e*) The actual recording of the identification following from (*d*). Of ships, record their approximate size, type and aspect, for example 10,000-ton tanker, loaded, beam-on. Land targets should be indicated by their name or position, bearing from ship and approximate height, for example Worle Bury Hill, 147 degrees true, 100 metres. Recordings of buoys should include the name or position of the buoy, what shape it is and whether it carries a radar reflector.

(*f*) Radar conspicuous objects.

(*g*) Performances of Racons and Ramarks at long and at close ranges.

(*h*) Interference phenomena due to other ships using radar.

(*i*) Maxima and minima effects of echoes of radar reflector buoys.

(*j*) Maximum reading of the performance monitor and the reading when operating the radar.

(*k*) A statement whether the radar has been used for anti-collision or navigational purposes. If used for navigational purposes, the type of navigation should be mentioned (e.g. landfall, coasting or pilotage).

(*l*) The behaviour of ARPA displays, in particular their acquisition and tracking capabilities for 3 and 10 cm. radar sets under different weather conditions.

Additional diagrams of shadow sectors and indirect echoes are very useful. A diagram showing the extent of shadow sectors should be displayed prominently near the PPI. By covering the PPI with a piece of tracing paper, one could record echoes of coastal features, especially when they become first detectable. These first detectable echoes depend on the aspect of the targets and hence ranges and bearings should also be recorded. By the time a coast becomes recognisable, a fix can be obtained. The first detectable echoes, however, may have disappeared or merged with other echoes, but it will then be possible to " work back " to the original position and mark the first detectable echoes on the chart, though it will be difficult, though not impossible, to find the cause of these echoes.

Coasting round the British Isles, one can employ the services of a photographer of the manufacturer. He will board the ship with special equipment and take photographs of the PPI when making a landfall or approaching harbour entrances, etc. The additional expense might be worthwhile.

That part of the log which concerns the performance of a radar set is often referred to as the " operational log ". The other part, which gives a record of suitable or unsuitable external conditions, is known as the " log of targets ". The manufacturers sometimes send an operational log on board.

It is already realised that the benefits of a radar log are many. Some of them are:

(i) The log will give a picture of the particular characteristics and the limitations of a certain radar set and will teach the observer how to be critical in interpreting the display.

(ii) Standards of radar performances are drawn up under certain conditions and can be compared with future performances under similar conditions.

(iii) The log will greatly help in identifying particular features of coastlines from previous experience. It will assist the navigator while making a landfall, and is of particular importance for ships which sail along regular routes.

(iv) The log is of great value when new officers or relief-officers come on board. It provides the history of the radar set and its behaviour under different conditions and circumstances, and gives details which cannot be discussed in a short time.

CHAPTER 13

PLOTTING

Plotting has two purposes:

(a) It can show us whether danger of collision exists, how close we will pass off the target (nearest approach or distance of the closest point of approach from own ship) and how much time there is left before this will take place.

(b) The approximate determination of the course and the speed of the other vessel from previous observations, so that sensible avoiding action can be taken when needed.

The second purpose is connected with one of the limitations of cm. radar which does not show up the aspect or leading edge of an isolated small (in relation to the horizontal beam-width) target except at very close range.

Plotting does *not* reveal to us the *shape* of a target and hence not the present heading. It will inform us, however, about the *motion* of the target during the plotting interval.

Reporting and Recognition of Collision Hazards

To provide the Master with information about collision hazards, about the possibility of planning avoiding action and about the taking of avoiding action, a good method to be adopted by the radar observer is to report according to a standard pattern. Such a report would consist of two main parts:

1. (a) Last bearing, drawing forward or aft (passing ahead or astern respectively);

    (b) Last range, decreasing or increasing;

    (c) Nearest approach (distance of closest point of approach from own ship) as forecast (CPA) ;

    (d) Time interval to the nearest approach (closest point of approach) from the last observation (TCPA).

2. (a) True course or relative course or aspect of target;

    (b) Speed of target.

First consider part 1. Whenever an echo is observed on the screen, the Master, especially during reduced visibility, is naturally anxious to know whether there is an appreciable change in the

179

bearing and if the range is increasing or decreasing. If little change in the bearing is observed, and the range is decreasing, at once the question arises how far off will the target pass if both ships maintain their course and speed and how much time is there left before this will occur. Even if there is an appreciable change in the bearing, the Master is likely to want to learn whether the target is passing ahead or astern.

If it is apparent that avoiding action has to be taken, then part 2 must be completed. In that case the Master *must* know the approximate motion and speed of the target. This is the same as in the visual case where one can only plan avoiding action properly if the other ship's course and speed can be estimated. Alterations of course without establishing the target's direction and speed are irresponsible actions.

The *aspect* is defined as the *relative bearing of own vessel taken from the target*. A starboard or port bearing is indicated as *Green* or *Red* respectively. For example, an aspect of *Red* 90° means that the target's port side is observed to be beam-on to own ship; a target head-on has zero aspect, stern-on 180° aspect.

Strictly speaking, as we have seen, the aspect cannot be deduced from a plot, but we will assume that *the most probable aspect can be deduced from the motion of the target during the plotting interval*.

As seen from Fig. 13.1 the direction of movement of the target can be expressed in terms of either her aspect, or relative course or true course and it is entirely up to the Master which he prefers. Often a glance at the plot will suffice.

Aspect appeals to a lot of sailors because it is so closely related to the visual conception when they sight a ship. They automatically estimate the angle between her bearing and her course. It also gives us insight as to whether the target may see us on her starboard or port side or whether own ship is in the overtaking position. Such considerations may help us to form an idea about the possible reactions of the target and they are of special importance when sailing through fogbanks and the Steering and Sailing Rules 11 to 18 must be applied when the fog lifts suddenly.

The report has to be enlarged if avoiding action is going to be taken. Here we must assume initially that the other ship maintains her course and speed. We cannot predict her actions with certainty.

After the alteration of course or reduction in speed, the observer must watch the plot closely and re-estimate the nearest approach and the time when this will take place. If the nearest approach remains dangerous, then the best practice is generally to either reduce speed substantially or to stop own ship or to alter course to put the target right astern. The other ship in such a case, probably, has taken avoiding action at about the same time which has cancelled own ship's action.

If, however, the avoiding action is successful and the nearest estimated approach is safe, then the radar observer can *continue his report by informing the Master when the original course or speed can be resumed with safety.*

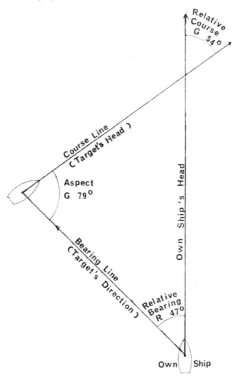

FIG. 13.1 TARGET'S DIRECTION, TARGET'S HEAD AND ASPECT

Two remarks must be made about the nearest approach:

(a) In clear weather, when at close quarters, one can almost immediately see what the other ship is doing and one can act accordingly, if danger of collision is involved. Radar, on the other hand, is not suited for close quarter situations in this connection. Because cm. radar does not possess enough discrimination, it is very *slow* and sometimes unable to tell us what the other vessel is doing.

The target, if she has radar on board, is placed in the same predicament and there is insufficient appreciation on the part of each vessel of the other's movements. If the target has no radar on board or cannot use her radar, then, in thick fog, she will be completely unaware of our position and movements. Hence in fog, when there is a great loss of information, even when both vessels are using their radar, a wide margin of safety has to be introduced.

(b) There are many factors which easily cause errors in plotting (Chapter 14) and the nearest approach as obtained from the plot may differ considerably from the actual value.

Therefore taking these two considerations into account, the new nearest approach selected should not be too small. Give the target a wide berth. The situation is not the same as when in clear weather.

It is good practice generally, in the open sea, for the Master to base his plan for taking avoiding action on a bold alteration of course and/or speed initially so that the other ship, if she is using radar, will be able to detect own ship's action as quickly as possible. After taking avoiding action careful plotting should be continued to see if the other vessel is keeping her course and speed. If she does, then a prediction should be made from the plot when it will be safe for own vessel to resume her original course and/or speed. The following three factors should then be taken into consideration:

(i) The closest distance which one considers it safe to pass the other vessel under the existing circumstances (three miles is generally accepted as a safe minimum distance for average types of merchant ships in the open sea).

(ii) The time factor. In case the other vessel is using radar she should be allowed to have sufficient time to detect own ship's alteration in course and/or speed (a minimum of about twelve minutes is generally considered necessary).

(iii) If own ship took action by altering course to bring the echo across from the starboard to the port side or vice versa, then, when the original course is resumed, care should be taken to avoid, if at all possible, bringing the echo back to the opposite side again (if this was done a misunderstanding of the situation might arise if both vessels suddenly came into sight of one another).

The report can be computed either by plotting on a sheet of paper or directly on a screen covering the PPI, or by mechanical or electronic plotting devices, some of which will be discussed in Chapter 15.

There are two main types of plots:

(a) Relative Motion Plot.

(b) True Motion Plot.

Relative Motion Plot

The motion of the echo is plotted relative to own ship, which is considered as a fixed reference point. In other words, the motion is plotted as it appears on a Relative Motion Radar Display. The centre-point of the plot represents the electronic centre of the radar screen, i.e. own ship. The heading marker, representing the fore- and aft-line of own vessel, is drawn on the plot and indicates the direction of own course.

All the plots shown in the following diagrams, are referred to as *Compass Datum Relative Plots*. This means that the compass bearing scale is fixed. When course is altered, the heading marker swings round in the same way as it does on a stabilized display, and the movement of the echo is not broken up as it would be on a Head-Up Relative Motion Display. The relative bearing scale is not used, though one can, if one wishes, lay-off relative bearings from the heading marker wherever this is positioned.

In nearly all the diagrams it is assumed that North is " up ". One may, of course, if this is preferred, always start off with the heading marker upwards, provided one turns the heading marker to the new direction when course is altered. By turning the plot bodily around one could then bring the new heading marker to the upward position again.

After putting in the heading marker, the different bearings (relative to the heading marker or true) and ranges are plotted from the centre-point according to a suitable scale which should not be too small (about one inch to represent one mile is suitable). A time interval can be chosen which is related to own ship's speed, for example five minutes for a ship of twelve knots, so that the distance moved by own ship during that time is one mile. Or one can take intervals of six minutes during which the ship covers a distance of one-tenth of the speed. Anyhow, this can be best left to the observer as *it depends on the rate of approach of the other vessel*. If the rate of approach is fast, a three-minute interval between successive observations is advisable, but the plot should not be completed before *at least three observations* of the target have been made. If the log is in operation, it should be read when observations are taken as this will indicate the distance own ship has travelled through the water. This is of special importance after reduction of speed when one is not quite sure about the average speed of own ship, and we will see later that inaccuracies in the distance covered by own ship during the plotting interval will introduce errors in the estimated course and speed of the target.

In Fig. 13.2 three bearings and ranges of each of two echoes have been taken which are laid off from the centre. The first bearings and ranges were at 0000 hrs., the last ones at 0012 hrs.

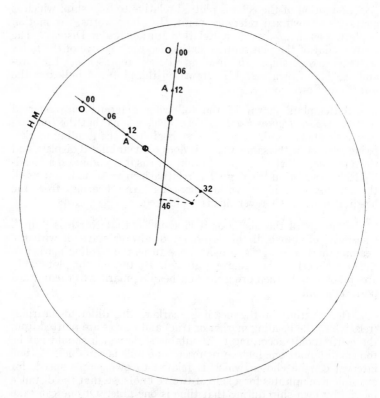

O A —⊖— Echo track or Apparent Motion of targets.
Predicted Nearest Approaches of Ships occur at
0032 & 0046 respectively.

FIG. 13.2 RELATIVE MOTION LINE AND PREDICTED
NEAREST APPROACH

The plotted point which is laid-off first (0000) of each echo is called O (for " Origin "), that one which is laid-off last (0012) is called A. OA represents the movement of the echo in 12 minutes as seen on the screen of a relative motion display. The nearest distance of approach is the length of the perpendicular dropped from the centre on to OA produced. The unit of time is OA, representing 12 minutes. The arrival times of the estimated closest points of approach of the two targets, whose echoes are depicted on the plot, will take place at 0032 hrs. and at 0046 hrs. (foot of perpendicular from centre) if all ships concerned maintain their courses and speeds.

Items 1 (*a*), (*b*), (*c*) and (*d*) of the report are now known.

OA is the relative motion of the echo and its length and direction are determined by the course and speed of own ship and the course and speed of the target vessel. Therefore, one may expect that if both vessels maintain their course and speed, and bearings and ranges are correct, the three points will lie on a straight line with the 0006 point halfway between the 0000 and 0012 points. In practice, however, one may find that the respective points are staggered even when the target follows a steady course and speed. This is so because the bearing accuracy, especially of the unstabilized display is not very high. Also, own ship may yaw and this will affect the relative motion.

If the plotted points are situated nearly on a straight line, and if the distances between them are roughly proportional to the time intervals between the respective observations, one may draw a mean line through the plotted points and assume that the other vessel has kept her course and speed.

We thus see that the information derived from the relative motion line is the nearest approach and the time it takes to the closest point of approach. It can also tell us if the other vessel maintained her course and speed.

The relative motion plot is compact; the image of own ship is fixed and generally echoes are only plotted of targets whose ranges are *decreasing*.

Determination of Course and Speed of Target

The determination of the course and the speed of the target is sometimes called *Completing the Plot*. See Fig. 13.3.

If the target has been stationary (lightvessel or buoy), then at 0012 hrs. its plotted echo would have reached *W* (Zero Speed point) where *WO* is parallel to the heading marker and its length corresponds to the distance travelled by own ship in 12 minutes, i.e. a distance of 12/60 × own speed. However, at the end of the plotting interval, the echo is not at *W*, but at *A* (0012 hrs.). This can only mean that *WA* must represent the true motion of the target during these 12 minutes. Measure *WA*, multiply by 60/12 and the speed of the target is obtained (in practice, one compares the length of *WA* to the length of *WO* and estimates the speed relative to own ship's speed). Measure angle *OWA* and the relative course of the target is known. The aspect can also be read off. In the diagram, the aspect is roughly Red 70°.

Items 2 (*a*) and 2 (*b*) of the report are now established.

If the *WO* component is plotted during the time interval, then nearest approach, course and speed of the target are obtained *simultaneously*.

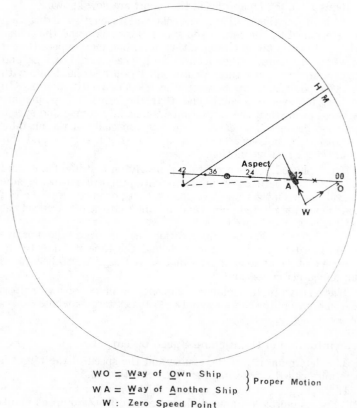

WO = Way of Own Ship } Proper Motion
W A = Way of Another Ship
W : Zero Speed Point
Nearest Approach at 0042.

FIG. 13.3 DETERMINATION OF COURSE AND SPEED
OF TARGET

An advantage of practising the complete relative motion plot is that it enables an officer to acquire a better understanding of the *relative motion display*—it helps him in interpreting the true meaning of the motion and changes in motion of echoes on the display. This can be useful on those occasions when lack of time precludes plotting or, when there are many echoes, it helps the observer to be selective in his plotting.

When plotting on a reflection plotter, the direction and length of *WO* can be obtained by means of the parallel index, first aligning it parallel to the heading marker for direction and then swinging it through 90° for marking the distance (using chinagraph pencils).

Target alters Course or (and) Speed

If the plotted points are lying on a straight or nearly straight line, but *the distances between them are not proportional to the time intervals between the observations*, then the other ship has altered course, or altered speed or has done both (Ship *Q*, Fig. 13.4).

The same is true when the plotted points do not lie on a straight or nearly straight line (ship *P* and ship *R*, Fig. 13.4). In all these cases it is accepted that own ship has maintained course and speed.

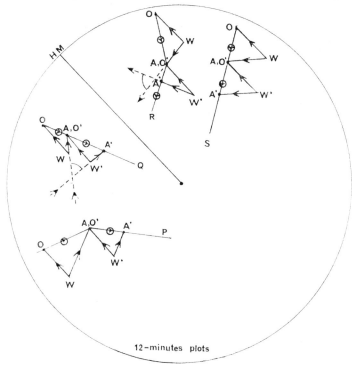

P reduced speed ; Q altered course to starboard.
R altered course to starboard and increased speed.
S maintained course and speed.

FIG. 13.4 EFFECT ON RELATIVE MOTION LINE WHEN
THE TARGET ALTERS COURSE AND/OR SPEED

Irregularities in the *OA* line tell us that manoeuvring action has been taken by the target, but it does *not* inform us of *what* type this is. This can only be ascertained by making a new plot *when the new OA line is well established.*

One can never know the exact time when the target took manoeuvring action. Hence close observation of the echo on the screen is advised.

Ship *S* in Fig. 13.4 has maintained course and speed—in practice the second velocity triangle (*W'O'A'*) need not be completed to verify this fact, because it can be seen at a glance that *O'A'* is equal in direction and magnitude to *OA* and because own ship has maintained course and speed *W'O'* is the same as *WO*; thus *W'A'* must be the same as *WA* vectorially.

Determination of Set and Rate of Currents and Tidal Streams

If it is desired to find the set and rate of a current, the echoes of a stationary target, for example, a lightvessel, should be plotted. The diagram (Fig. 13.5) shows an apparent motion of the lightvessel

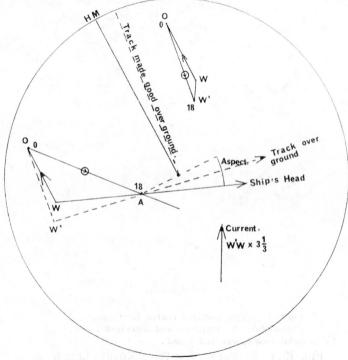

FIG. 13.5　DETERMINATION OF SET AND RATE OF CURRENT

on the starboard bow from W to W'. This cannot be the case. As the lightvessel cannot move towards own ship, own ship must have moved towards the lightvessel, and the set is represented by the direction from W' to W. The drift and the rate can be deduced from the length of $W'W$.

Also illustrated in Fig. 13.5 is the construction to determine the course and speed made good over the ground of another vessel at 0018 hrs. Instead of using WO for one side of the triangle, $W'O$ (parallel to the track made good) is employed.

It should be noted that this construction is seldom used at sea, where for collision avoidance, the observer is interested in the other ship's *heading* with respect to own ship's head. But it is worthwhile to study the construction in order to grasp the " Ground Stabilized True Motion Display " (Chapter 10).

N.B. The nearest approach which can be deduced from the direction of OA is not altered by the presence of current or a steady wind.

True Motion Plot

A True Motion Plot is shown in Fig. 13.6. 'Own Ship' is near the bottom left-hand corner.

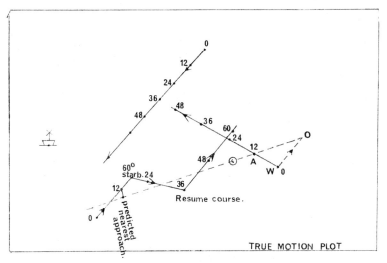

FIG. 13.6   TRUE MOTION PLOT

Bearings and ranges are taken of the targets at 0000 hrs., 0006 hrs. (not in diagram) and 0012 hrs. The positions of own ship at these times are plotted along a line representing the course line.

From these positions bearings and ranges are laid-off. A protractor can be used for laying-off the bearings, a suitable rule should be employed for the ranges; or a chart, parallel rules and dividers can be employed.

Course, speed and aspect of the target can be determined directly and there is a constant check that the other vessel maintains her course and speed. As the position of own ship constantly moves along, one needs a large sheet or plotting board.

There are some officers who prefer the True Motion Plot to the Relative Motion Plot, because it gives them a better realisation of what happens. It is advised in any case to get thoroughly acquainted with the relative motion plot and the OWA triangle first. Once this is mastered it will not be difficult to determine the nearest approach and time from the true motion plot. This is illustrated in Fig. 13.6, where in chain lines the OWA triangle, the relative approach line and the nearest approach are shown. Note that the perpendicular, indicating the predicted nearest approach is dropped from the 0012 position of own ship, the time *terminating* the plotting interval for which the triangle WOA is constructed.

This type of plot is sometimes called the *Complete True Motion Plot* as it yields the distance and time of the estimated nearest approach besides the aspect and speed of the target.

Alternatively, one does not complete the True Motion Plot but uses the Table printed at the back of the book which gives the nearest approach directly from two successive ranges and bearings (provided the target maintains course and speed).

It has been observed, however, that because the True Motion Plot does not provide the observer *directly* with the time and distance of the estimated nearest approach, selective plotting becomes more difficult and there is a tendency to try to plot too many ships at once, thereby losing sight of the crucial encounters.

When the True Motion Plot is carried out in the vicinity of land approaches, plotting on a large scale chart can be extremely useful.

Avoiding Action of Own Ship and its Effect on the Relative Motion Line

Suppose that own ship takes avoiding action. The questions arising are:

Will the action be effective and when is it safe to finish it?

Own Ship Reduces Speed

The diagram (Fig. 13.7) shows a plot carried out for two targets (one of them a lightvessel) for a 12-minute interval. First bearing and range is at 0000 hrs., last bearing and range at 0012 hrs. for each of the targets.

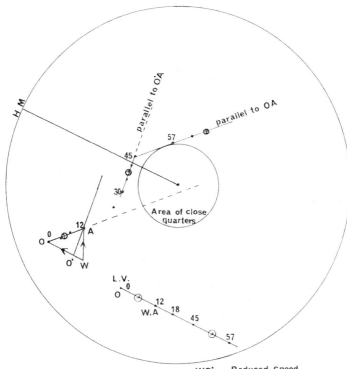

Dead Slow at 18 minutes past. $\dfrac{WO'}{WO} = \dfrac{\text{Reduced Speed}}{\text{Original Speed}}$.
Back to Original Speed at 45 minutes past.

FIG. 13.7   EFFECT ON RELATIVE MOTION LINE WHEN
OWN SHIP CHANGES HER SPEED

At 0018 hrs. own ship reduces speed to one quarter of her original speed. How do we find the new relative motion assuming for the time being that the other vessel maintains her course and speed ?

A new velocity triangle ($WO'A$) must be constructed in which WA represents the target's present velocity, WO' represents our own ship's *new* velocity and $O'A$ represents the new relative motion.

As, at this stage of plotting, the WA line (for 12 minutes) is already available on the plot, this vector is best utilised to commence the construction of the $WO'A$ velocity triangle. From W lay off WO' representing own course and distance made good in 12 minutes; then $O'A$ gives the new relative motion for 12 minutes.

As soon as own ship has settled down to her new speed—and this might take about ten minutes for the larger class of vessel—a new range and bearing should be taken of the other vessel's position and transferred to the plot (the 0030 position in Fig. 13.7). A line should then be drawn through this position parallel to the $O'A$ line and this line from now on will be known as the *Prediction Line* or *Predicted Track*.

If the echo does not follow the predicted track at the predicted rate on the plot, it must be taken that WA has changed and the target has carried out an alteration of course, a reduction of speed or both, and a new velocity triangle might be required.

The question now is: At what time can the original speed be safely resumed, assuming the echo is following the prediction line? Once back to the original speed, the *direction* of the apparent motion of the target, and its rate, will be the same as before the speed reduction, provided the target maintains course and speed.

The construction (assuming instantaneous new speed) is as follows:

Around the centre, draw a circle of radius representing the intended minimum distance of the nearest approach. Draw a tangent to this circle parallel to the original OA. In diagram 13.7, this tangent intersects the prediction line at the 0045 point and when the echo is at this point speed can be resumed.

Keep observing the echo closely after the resumption of speed. Report if any deviation occurs from the predicted track. The time unit after 0045 is, of course, the length of the original OA (which represents 12 minutes in this example).

Own Ship alters Course

Suppose that own ship is going to alter course 90 degrees to starboard at 0015. The new prediction line has to be constructed assuming for the time being that the other vessel maintains her course and speed. See Fig. 13.8.

Again, as in the previous example, part of the original velocity triangle is used for the construction as the WA vector is already available. There is no alteration in speed of own ship and the length of the WO vector for 12 minutes will remain unchanged, although its direction will change. Therefore, circle WO round W through an angle of 90 degrees so that the new vector WO' points in the direction of the proposed course reading on the bearing scale. $O'A$ gives the predicted relative motion for 12 minutes after the course alteration is carried out.

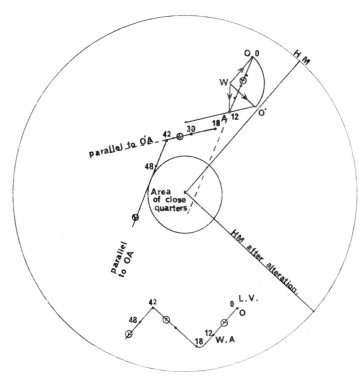

90° to Starboard at 15 minutes past. $WO'=WO$. $\angle OWO'=90°$.
Back to Original Course at 42 minutes past.

FIG. 13.8   EFFECT ON RELATIVE MOTION LINE WHEN
OWN SHIP ALTERS COURSE

As soon as own ship has steadied up on her new course, a new range and bearing should be taken of the other vessel's position and transferred to the plot (the 0018 position in Fig. 13.8). A line should then be drawn through this position parallel to the $O'A$ line and this line represents the prediction line.

If the echo does not follow the predicted track at the predicted rate on the plot, then it must be taken that the target has carried out an alteration of course, a reduction of speed or both and a new velocity triangle might have to be constructed.

At what time can the original course be safely resumed?

After all that has been said before, the construction is obvious:

Draw a circle about the centre with the radius representing the intended minimum safe nearest approach. In diagram 13.8, the tangent to this circle parallel to the original *OA* line intersects the apparent motion line at the 0042 point. When the echo has followed the prediction line and has reached this point, the original course can be resumed.

Keep observing the echo closely after the resumption of the original course. Maintain this until the target is abaft the beam and report till then any irregularities in the echo's track. The time unit after 0042 is, of course, the length of the original *OA* (representing 12 minutes in this example).

Passing a given Distance off a Lightvessel

Fig. 13.9 shows an echo of a lightvessel which is on the starboard bow. In the case of Fig. 13.9 (*a*), own ship is not affected by current, but in the case of Fig. 13.9 (*b*) the current sets easterly.

Take Fig. 13·9 (*a*) first. It is the intention to alter course at 0022 (mean time of the manoeuvre) in order to pass the lightvessel

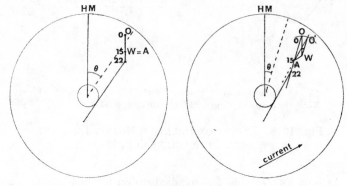

θ = Course alteration

FIG. 13.9   ALTERING COURSE TO PASS A GIVEN DISTANCE OFF A LIGHTVESSEL

at two miles distance (say) on the starboard side. What alteration of course is required?

Draw a circle, representing 2 miles, round the centre of the plot, then lay off a tangent to the circle from 0022.

At 0022 make an alteration of course equal to the difference between the original heading and the direction of the tangent towards the point marked 0022.

If current is running (Fig. 13·9 (b)) plot the lightvessel as if it were a moving vessel, using the OAW triangle.

Plot the 0022 position—altering course the same time as before—and draw the tangent. Transfer the direction of the tangent back through point A. Then circle WO round clockwise until it intersects the transferred tangent in O'.

The angle OWO' represents the required alteration of course for 0022.

Note

It can happen when plotting on a non-rotatable reflection plotter using an unstabilized Head-Up display, that the observer is compelled to produce a Head-Up Datum Relative Motion Plot, which might lead to some confusion after an alteration of course by own ship.

The best thing to do in this case is to start a fresh plot after the course alteration, but *leaving* the old plot on the plotter. Comparison with the old plot will soon reveal if the other vessel has maintained its course and speed (if the other vessel has maintained her course the directions of the WAs differ by the amount of own ship's course alteration).

Put the range marker on the intended minimum safe nearest approach distance and, using the Parallel Index, rotate the original OA-line through the angle of the course alteration (anti-clockwise if the alteration was to starboard). Then draw a tangent, parallel to the Index, touching the range marker. The point of intersection of this tangent line and the present echo track, represents the point where the echo should arrive before a resumption of course is contemplated —on the assumption that the other vessel maintains its course and speed.

Recapitulation

Assume that it is required to produce a 12-minute plot on a plotting sheet of a ship's echo which has appeared on the radar display. Three sets of ranges and bearings are taken with an interval of six minutes. Only a Relative Motion Plot will be considered. (Consult Figs. 13.3, 13.7 and 13.8).

H

1. Draw a line representing the heading marker.

2. Lay-off the first position of the target, marking it "O" and indicate time.

3. Draw a line through "O", parallel to, but in opposite direction to, the heading marker. Terminate this line at "W", where WO represents the distance travelled by own ship during the plotting interval (12 minutes). Mark in the direction from W to O.

4. Plot the 6-minute and 12-minute position of the target and label the latter with "A". Mark the time. Check that, for all practical purposes, the produced echo motion is uniform in its direction and rate.

5. Draw a line through "O" and "A", producing it well past the centre of the plot. This line represents the apparent motion of the target and its symbol is an encircled arrow.

6. Drop a perpendicular from the centre of the plot onto the OA-line produced. This yields the Closest Point of Approach (CPA).

Using the OA-distance as a time unit (12 minutes) determine the time at which the target is expected to arrive at CPA (TCPA).

If the range of the object is increasing it has already passed the CPA.

7. Connect W to A with a straight line. This line, from W to A, represents the true motion of the target during the plotting interval.

If the target is at anchor W and A should coincide, provided there is no current. If there is current there will be a displacement between W and A where the line, A to W, indicates the direction and drift of the current (or tidal stream) during the plotting interval.

8. If the aspect is required, measure the angle between WA (produced) and the line connecting the two ships (line of sight), adding "Green" or "Red", when own ship is to "starboard" or to "port" of the other ship, respectively.

That completes the basic plot. Suppose now that the Master decides to take avoiding action, either by an alteration of 90° to starboard or by a reduction in speed from "Full" to "Slow" (approximately $\frac{1}{4}$ Full Speed). Execution of the manoeuvre assumed to be about 18 minutes after starting the basic plot.

Alteration of Course	*Reduction in Speed*
9. Draw in new heading marker. Turn WO, round W, through an angle of 90°. Label the new vector WO'.	9. From W, in the direction of O, measure off WO', where $WO' = \dfrac{WO}{4}$

10. Join O' and A, producing the line well beyond A. The line O' to A represents the relative motion for 12 minutes for the new heading or the new speed, respectively.

11. After own ship has steadied on her new heading or settled down at her new speed, take a range and bearing of the target and plot the position of the target on the plotting sheet. Mark the time.

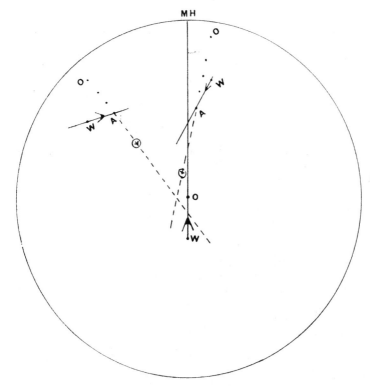

FIG. 13.10 VISUAL ESTIMATION OF COURSE AND SPEED OF TARGETS

12. Draw a line through this new fix parallel to the $O'A$ line, already plotted above.

13. Check that the direction and rate of movement of the echo along this line is in accordance with the $O'A$ vector. If this is not the case, the target must have taken some action and it is best to start a new complete plot.

The "middle" part of the plot has now been completed and, on the assumption that the target has maintained its course and speed (see 13), the Master has to consider when he can order own ship to go back to the original course, or when he can put the telegraph on "Full" again.

14. Around the centre of the plot, draw a circle of radius representing the minimum safe CPA distance.

15. Draw a tangent to this circle parallel to the original OA-line. Mark the point where this tangent intersects the present relative motion track of the target.

16. When the echo has reached the above point, it will be safe to resume the original course or speed.

17. Having resumed the original course or speed, keep the echo under observation until the target is completely clear.

If in (4) the echo motion is not uniform in direction (developing a kink) and/or in rate (unequal progress during each of the 3-minute time intervals), then one simply has to wait until the motion becomes uniform, and start the plot from then.

Recommendation

With several echoes on the display, it is advisable to draw a small line on the Parallel Index or Reflection Plotter from the start of the time-base in the opposite direction of the heading marker. The length of this line should correspond to the distance progressed by own ship during the plotting interval (WO). See Fig. 13.10.

On the relative motion display it is not too difficult now to make an estimation of the courses and speeds of the targets by visualising WO transferred to the various OA's (which, of course, must be marked by spots on the display).

A further refinement can be carried out on the reflection plotter by drawing, from the start of the time-base, the velocity vector OO' which is the difference between the new and the original velocity vector (a 60 degrees alteration to starboard is contemplated in the example shown in Fig. 13.11). Again, applying this vector (OO') by eye to the relative motion tracks will give the prediction of the relative motions after avoiding action is taken, provided the other vessels maintain their courses and speeds.

More Advanced Methods of Plotting

When faced with a multi-target situation, avoiding action taken for one ship might result in dangerous approach courses involving other vessels, which were considered perfectly safe before action took place.

In order to get a good overall assessment of the situation before avoiding action is taken and to decide upon an optimal solution, if this is possible, special plotting procedures are available. Two of these will be discussed.

A. The construction of PPCs and PADs (Possible Points of Collision and Predicted Areas of Danger).

B. The construction of SODs and SOPs (Sectors of Danger and Sectors of Preference).

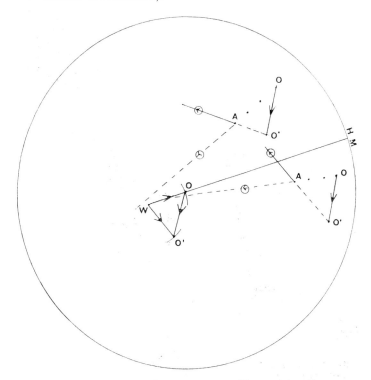

FIG. 13.11   VISUAL ESTIMATION OF PREDICTION LINES

A. The Construction of PPCs and PADs

The PPC is that point on the display which, if brought dead ahead of own ship whilst maintaining speed, would result in interception with the target (if it also maintained its speed). In other words it is a point the navigator must avoid, at all costs, coming and remaining on, or near to, the heading marker of a radar display.

The PAD is a bound area shown on the display in which the PPC is situated. Its perimeter is based on a minimum safe distance the navigator wants to keep from the other ship. In other words, it depicts an area which the navigator should avoid his vessel entering.

Fig. 13.12 shows the construction of the PPC, and its movement over the plot (display).

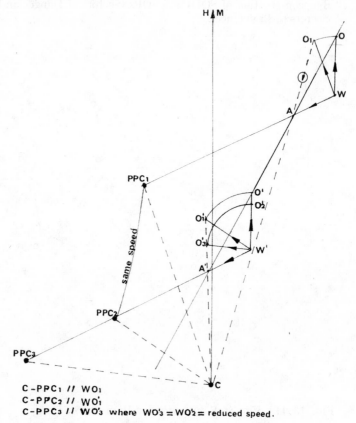

C -PPC₁ // WO₁
C -PPC₂ // WO'₁
C -PPC₃ // WO'₃　where　WO'₃ = WO'₂ = reduced speed.

FIG. 13.12　CONSTRUCTION AND MOVEMENT OF PPC

1. Construct the usual basic triangle *OAW* for an echo coming towards the centre of the plotting sheet.

2. Connect the centre of the plot, *C*, to *A*, thus forming a line along which the echo would come down if there was a pure collision case.

3. Rotate WO round W and the point where the arc of O intersects the line CA produced, is marked O_1.

Angle OWO_1 represents the course alteration to be carried out by own ship (when the echo is at A) in order to intercept the target.

4. Draw a line from the centre of the plot parallel to WO_1. This line intersects WA produced at the PPC (PPC$_1$ in Fig.).

This construction could be carried out for various other echoes on the plot, although for some targets—depending on their speed and aspect—PPCs are non-existent. Such a series of PPCs on the plot will give the navigator a sound idea about the type of actions he should *not* take.

5. To study the movement of the PPC over the plot, while maintaining speed, repeat the construction when the echo is at A', using an identical triangle. In this case, the line connecting the centre of the plot to PPC$_2$ is parallel to WO_1'

6. It can be seen that during the time the echo moved from A to A', the PPC moved along a line nearly parallel to the heading marker, but going slightly away from it.

Suppose that the navigator, having plotted the first triangle OAW and PPC$_1$, feels that he may get into a close-quarters situation and decides to alter course to starboard. This action will move his ship away from the PPC.

Alternatively, when the echo is at A, he could, say, reduce speed from "Full" to "Half". In this case a further construction is necessary, shown in Fig. 13.12.

7. In the second triangle $O'A'W'$, measure $W'O_2'$ to be $\frac{3}{4}W'O'$.

8. Rotate $W'O_2'$ round W' until the arc of O_2' intersects CA' produced at O_3'.

9. Draw a line from C parallel to $W'O_3'$. Where this line intersects $W'A'$ produced, is the new PPC (indicated in diagram by PPC$_3$).

Own ship has moved away from the collision point, and the final result, although taking place at a slower rate due to the speed reduction, is similar to an alteration to starboard.

The construction of a PAD is shown in Fig. 13.13.

1. Construct triangle OAW.

2. Draw a circle round the centre of the plot with a radius equal to the desired or required minimum CPA.

3. From A draw three lines :—

 AC (collision line),
 Tangent line AL (target passing astern line),
 Tangent line AM (target passing ahead line),

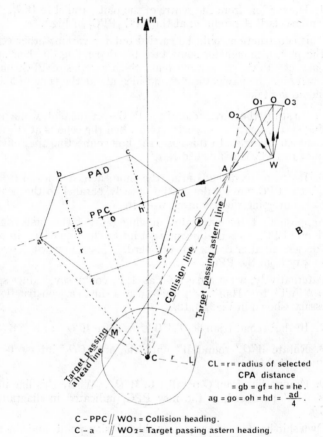

CL = r = radius of selected CPA distance
= gb = gf = hc = he.
ag = go = oh = hd = $\dfrac{ad}{4}$.

C – PPC // WO₁ = Collision heading.
C – a // WO₂ = Target passing astern heading.
C – d // WO₃ = Target passing ahead heading.

FIG. 13.13   CONSTRUCTION OF PPC AND PAD

4. Rotate WO round W. The arc so obtained intersects
    CA produced at O_1,
    LA produced at O_2,
    MA produced at O_3.

5. From the centre draw three lines, respectively
   parallel to WO_1, intersecting WA produced at PPC ;
   parallel to WO_2, intersecting WA produced at "a", the cross-ahead point for own ship ;
   parallel to WO_3, intersecting WA produced at "d", the cross-astern point for own ship.

6. Divide ad into four equal parts and at "g" and "h" erect perpendiculars gb, gf, hc and he, each equalling the desired minimum required CPA distance.

7. The figure $abcdefa$ forms a hexagonal PAD. Instead of a hexagon, an ellipse could be constructed (a little more difficult) with major and minor axes respectively equal to ad and bf (or ce). (The PAD of a *stationary* target would be an equal-sided hexagon or a circle).

With several of these PADs constructed for different targets and for a given time, the navigator can decide upon the best manoeuvre available, generally keeping clear of the PADs while keeping in mind and applying the Collision Regulations so as to avoid cancelling actions in particular. It must be remembered that when vessels are in sight that own ship might be the stand-on vessel in which case it is the duty of the give-way vessel to take action to avoid own ship, and this should result in the movement of the relevant PAD away from the heading marker.

Notes

(i) It can be seen from the construction of the PAD that the PPC does not necessarily coincide with the centre of the hexagon.

(ii) PPCs *forward* of the beam move in time in a direction approximately opposite to the direction of the heading marker. With the PPC on the heading marker the PAD will follow exactly the line of the heading marker.

(iii) The existence of PPCs, cross-ahead and cross-astern points for own ship and the shape of the PAD depends on the aspect of the target and the speed ratio of own ship and target. Let us consider the case of Fig. 13.13 again. In Fig. 13.14 part of the AWO triangle is drawn again, but WO is variable, both in magnitude and direction, the point "O" moving along the various arcs.

Case 1 : $WO \geqslant WA$.

One PPC and one single PAD is produced. If speed is reduced by own ship, the WO' lines in this case rotate anti-clockwise and PPC and PAD move along WA produced, further away from "A". (Fig. 13.13).

Case 2: In *all* the next cases $WO < WA$ and the situation becomes more complex.

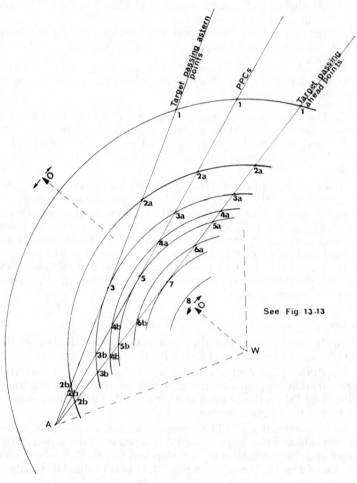

FIG. 13.14 EFFECT OF SPEED RATIO ON THE PRODUCTION OF PPCs AND PADs

As soon as WO becomes smaller than WA, besides the original PPC and PAD (see points 2a in Fig. 13.14), a second PPC and PAD appear from infinity (points 2b) which are closing in towards the original PAD the more own ship's speed is reduced.

Gradually the following cases will develop as the speed ratio own ship speed/target speed becomes smaller. See Fig. 13.14 (Note that target passing-ahead points are equivalent to own ship passing-astern points, and vice versa).

Case 3 : Two PPCs, two target passing-ahead points, one target passing-astern point.

Case 4 : Two PPCs and two target passing-ahead points.

Case 5 : One PPC and two target passing-ahead points.

Case 6 : Two target passing-ahead points.

Case 7 : One target passing-ahead point.

Case 8 : Collision is not possible.

(iv) In cases with end-on or overtaking encounters with aspects around zero or 180 degrees, it is often, construction-wise, not possible to obtain a passing-ahead and/or passing-astern point. This is logical because the passing will be to port or to starboard. The hexagonal PAD in such a case can be drawn symmetrically around the PPC with its "length" and "width" equal to twice the minimum required CPA distance.

B. The Construction of SODs and SOPs

The construction is based on the following problem : Determine the alteration of course and/or the change in speed at a particular time to pass a given minimum distance from a target having constant velocity.

The solution is shown in Fig. 13.15. Suppose that action is taking place when the echo is at P. Then from P draw two tangents to the circle with a radius equal to the given distance. Transfer these two tangent lines parallel through A (see inset), then rotate WO round W till the arc of O intersects one of the lines at O'.

For the echo to remain on one of the tangent lines through P, one can either alter the ship's head from WO to WO', or reduce speed from WO to WO'', or increase speed from WO to WO''' if this were possible (see note Fig. 13.15). In fact, one can combine any change of course with a change of speed providing the new O-point is on one of the lines. If the O-point falls outside the sector, the passing distance will be larger than the given one. The sector, originating at A and bounded by the transferred tangent lines and the arc of the maximum speed circle (radius WO_m) is the *Sector of Danger* or *SOD*.

In practice, the SODs are constructed from the *centre* of the plot. The sequence is as follows :—

1. Draw *OAW* triangle, say for 12 minutes.
2. Plot the circle with radius equal to minimum CPA distance.

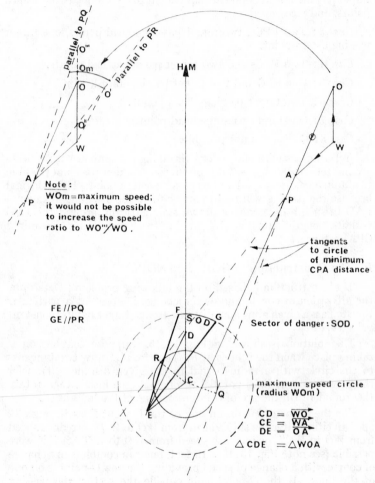

Note:
WOm=maximum speed;
it would not be possible
to increase the speed
ratio to WO'''/WO .

FE //PQ
GE //PR

tangents
to circle
of minimum
CPA distance

Sector of danger : SOD.

maximum speed circle
(radius WOm)

CD = $\overrightarrow{WO}$
CE = $\overrightarrow{WA}$
DE = $\overrightarrow{OA}$
△ CDE = △ WOA

FIG. 13.15 SECTOR OF DANGER

3. Plot the point *P*, where the echo will be when action is being considered.

4. From *P* draw in the two tangent lines.

5. Measure off *CD* (equal to *WO*) along the heading marker.

6. Lay off vector *DE* (parallel and equal to *OA*).

7. Draw two lines through *E*, opposite and parallel to *PQ* and *PR*, the tangent lines.

8. Circle the maximum speed round from the centre *C*, its arc intersecting the two lines through *E*, at *F* and *G*. Sector *FEG* is the SOD.

9. Assuming that own-ship is not the stand-on ship, alter course and/or speed, such that the upper point of own ship's velocity vector (point *D* in diagram 13.15) comes outside the sector and in accordance with the "Rule of the Road". Note that the word 'sector' does not correspond to the geometrical definition.

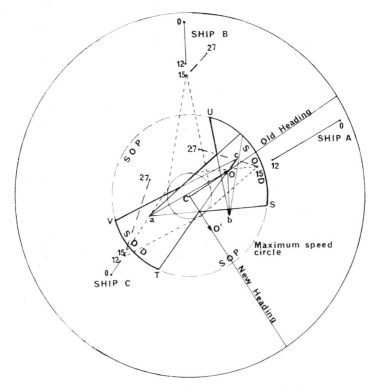

— — — New predicted relative motion after 00–15 .

For explanation see text and Fig. 13.15 .

FIG. 13.16 ALTERATION OF 90° TO STARBOARD CLEARS SODs

Fig. 13.16 shows the various SODs with three echoes of ships on the plot (display). All the plots are drawn for 12 minutes and only the OA lines are shown for the three ships' echoes A, B and C. Action is going to be taken at 00-15.

Measure off WO (12-min. length) from the centre of the plot, add the vector OA for each of the three vessels, yielding "a" for ship A, "b" for ship B and "c" for ship C. The SODs can now be completed by drawing two lines from each of "a", and "b" and "c" parallel but in *opposite* direction to the tangent lines drawn from 00–15 for echoes A, B and C; the third boundary lines for each sector is the arc of the maximum speed circle.

It can be seen at once that a speed reduction will continue to involve own ship in a close quarters situation. With the *present speed*, however, the sectors bounded by arcs ST and VU are clear. These clear sectors are known as *Sectors of Preference* or *SOPs*. Provided that it is permitted by the Collision Regulations, one can make therefore a bold alteration either to starboard or to port. In the diagram a 90° alteration to starboard is carried out. A close check should be kept on ship C in case she alters to starboard.

It can be easily deduced that the relative motion after 00–15 for the three ships A, B and C is represented by the 12-minute vector $O'a$, $O'b$ and $O'c$ respectively, CO' being the own ship's velocity vector after 00–15 (Fig. 13.16).

A similar construction could be carried out for a stationary target (a light vessel, for example), with or without current being in existence. Suppose that it is required to pass a light vessel between 1 and 3 miles. The result on the plot will be three sectors with one common point. The two outer sectors will be SODs and the one wedged between will be the SOP.

Note that in these plots the PPCs, PADs, SODs and SOPs are constructed for one particular time. For an assessment at *any* time, one would need the assistance of a computer.

CHAPTER 14

COMMENTS ON PLOTTING

Errors in Plotting

Errors in plotting can be due to:

(i) Errors in bearings;

(ii) Errors in ranges;

(iii) Wrong estimation of the course and distance followed by own ship during the plotting interval, i.e. an error in own ship's velocity;

(iv) Errors in the time of the plotting interval.

Errors in bearings taken on the Relative Motion Non-Stabilized Display can be out $\pm 2°$. The chance in errors in bearings taken on a Stabilized Display is less and the greatest bearing accuracy is obtained on a Stabilized Display with an electronic curzor, although errors in the time interval of the plot may occur with some electronic curzors which are difficult to align quickly.

Errors in ranges can attain a maximum value of $2\frac{1}{2}\%$ of the maximum range of the range scale in use on modern sets and 5% on older sets.

In diagrams 14.1, 14.2 and 14.3 *two* observations are plotted. It is logical that errors due to wrong bearings and ranges can occur in *each* observation but the discussion will be easier to understand if we assume that there is *no error* in the *first observation* while the *total* errors of both is concentrated round the *second observation*. Such assumption does not affect the final conclusion. In some of the diagrams " *A* " is depicted as a circle. This means that on account of bearings and range errors—which to certain limits are unknown — " *A* " *can be anywhere within the circle.*

Let us now have a closer look at the diagrams.

Figure 14.1 shows the effect on the plotting triangle owing to *errors in bearings and ranges.* All the three ships, P, Q and R are on near-collision courses.

Ship P's speed is slow compared with own ship's speed but the speeds of ships Q and R are nearly the same as the speed of own ship. Large percentage errors may be made in the estimated course and/or

speed of the very slow ship P and much smaller errors may occur
in the estimated speed (ignoring the course error) of the other vessel
coming from a direction *fine on the bow* (Q) or in the estimated course
(ignoring speed error) of the other vessel, coming from a direction
broad on the bow (R).

Figure 14.2 shows the effect on the plotting triangle owing to
errors in the estimated speed of the observer's ship. All the three
ships, X, Y and Z are on near-collision courses. Large errors may
occur in the estimated course of the other ship if her speed is slow
compared with own ship's speed (X). Much smaller errors may be
made in the estimation of course and speed where the speed of the
other vessel is nearly the same as the speed of own ship (Y and Z).

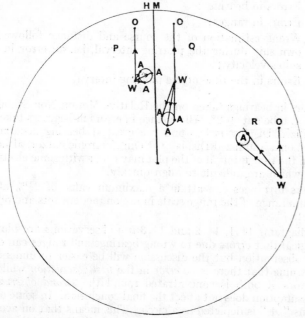

FIG. 14.1   EFFECT ON PLOTTING TRIANGLE OWING TO
ERRORS IN BEARINGS AND RANGES

For ships, in these cases, coming from a direction *fine on the bow*,
the error is confined mainly to wrong estimation in speed (Y) while
for ships coming from a direction *broad on the bow*, the error applies
mainly to the course (Z).

A similar diagram for estimated course errors—not so likely as
speed errors—with WW practically horizontal, would reveal large

errors in speed for ship X, errors mainly confined to the estimated course for ship Y and mainly confined to the estimated speed for ship Z.

Figure 14.3 shows that owing to errors in *bearings* and *ranges* a greater plotting interval between two observations will decrease the uncertainty in the nearest approach.

The conclusion can be summed up as follows:

(*a*) Errors in ranges and bearings and errors in own estimated speed may give rise to large estimated course and speed errors of another vessel on a near-collision course if her speed is slow compared with that of own ship's speed. These errors become smaller for ships on near-collision courses which have speeds similar to that of the speed of own ship. They may be confined mainly to errors in estimated speed or estimated course, depending on the relative direction of approach of the other vessel.

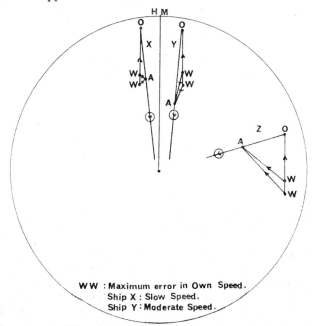

FIG. 14.2   EFFECT ON PLOTTING TRIANGLE OWING TO
ERRORS IN ESTIMATED SPEED OF THE OBSERVER'S SHIP

(*b*) Slow speed of a target vessel makes the plot very unreliable. It is generally impossible to find out from a plot whether another vessel is steaming slowly, or is stopped. When a target vessel is stopped, it is, of course, impossible to deduce aspect or heading.

The unfortunate consequence must be faced that during fog when most ships slow down, the plots tend to become less accurate.

(c) Doubling the plotting interval will halve the chance of errors in nearest approach.

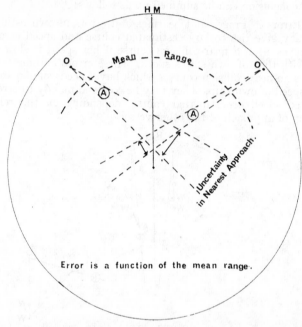

Fig. 14.3. A GREATER PLOTTING INTERVAL DECREASES THE UNCERTAINTY IN THE PREDICTED NEAREST APPROACH

Results obtained from plotting are often not very accurate. However, the navigator can take comfort from the fact that when the results are inaccurate the situation is generally not very dangerous and conversely, when the situation is potentially dangerous the plot is generally sufficiently accurate to help him in planning satisfactory action. For example, when the OA line is short, the accuracy of the predicted nearest approach is poor but as the relative motion is small, there is usually plenty of time for more observations of the echo to be plotted so that a more accurate estimate of the time and distance of the predicted nearest approach can be found. Again, when the target is moving slowly, the course of the target cannot be accurately obtained—but, obviously when a target is moving slowly it is not likely to present much of a hazard and a knowledge of its precise course is not so essential.

General Remarks about Plotting in Reduced Visibility

It is impossible to lay down rules about how much plotting should be carried out and what type of plot should be used. Each ship will present her own problem and each Master should make his own arrangements for observation and plotting drill most suitable to his ship, depending on the traffic density and the bridge manning during fog, the number of displays, the type of presentation, plotting facilities, illumination problems, chartroom-wheelhouse arrangement, etc. There are, for example, still many ships which carry only one unstabilized display; on the other hand, there are several ships nowadays which possess two displays with reflection plotters so that relative motion presentation can be offered simultaneously with true motion presentation; some ships possess additional plotters of various kinds and still others more sophisticated plotting devices as ARPA, for example.

With *one officer on the bridge* (besides the Master), continuity of radar observance is essential and one officer should be detailed for radar duties only. But, of course, he can be relieved from time to time as the study of the radar screen when there are several echoes present, can be a tiring duty. On a relative motion display the direction of the " tadpole " tails should be watched and danger of collision exists if their movement is towards the electronic centre of the screen. The use of the Parallel Index, in this connection, is advised. An unstabilized display, especially when the ship is yawing, yields unreliable tadpole tails and closely-spaced echoes are easily mixed up. If a feeling of uncertainty or impending risks is arising, it is often best to stop the vessel and try to sort out the echoes. In areas where many ships and especially small craft can be expected, the plotting on a construction sheet or plotting board is not advised as it will break the continuity of close observance. If no reflection plotter is provided in this case then the observer should single out the targets whose bearings are not changing appreciably and whose ranges are decreasing and *compile a record* in an orderly manner (that is, besides the time, bearing and range of a target, the heading and speed of own vessel should be written down with each observation).

The study of these record sheets needs a certain amount of fundamental understanding. See Fig. 14.4. When another ship passes clear of own ship, it is not only the bearing which should open out, but—until the nearest approach—the *rate of change of bearing should increase*. In Fig. 14.4 the straight echo track on the right shows the plotted positions of another ship at equal time intervals with the bearings and the change in bearings printed in the right-hand side quadrant. It can be seen that in this case the change in the bearings is increasing. The second fully drawn line in this figure shows an echo track which is straight in the beginning, but then starts to curve towards the centre. The bearing of this target is opening out (17°–19°–23°–29°–34°–38°–51°) but the rate of change is not always increasing (2°–4°–6°–5°–4°–13°) and it is said that the bearing is *not opening out sufficiently*. It is therefore important that the observer should also compare the difference between successive bearings taken at equal time intervals. If this difference remains the same or is decreasing while at the same time the range is decreasing he should be warned that something has gone wrong (the other ship might as in the case, shown in Fig. 14.4, have taken a series of small avoiding actions causing a *curved* apparent motion line to appear on own ship's radar display) and he might have to stop the engines or even stop the ship.

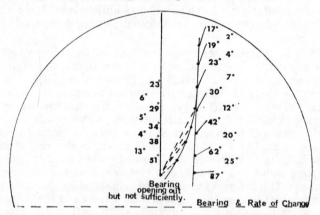

FIG. 14.4   CHANGE OF BEARING AND RATE OF CHANGE
OF BEARING

If a radar display is fitted with a reflection plotter one officer is generally capable of keeping an efficient radar watch. If the display is on Relative Motion, he will be able to construct a complete relative motion plot by means of chinagraph pencils and a plastic ruler for the ships he considers to be dangerous, while if the display is on True Motion he can construct a true motion plot of the targets and for those he considers to be a risk he can determine the predicted nearest approach as shown in Fig. 13.6. Or he might prefer to use the Zero Speed Switch (Chapter 4) for some minutes during which the relative motion of the targets will be shown on the display until the predicted nearest approach is established by the Parallel Index. As the true motion plot in the latter case is interrupted by the introduction of the relative motion, kinks and irregularities develop in the plot on the reflection plotter, but this should not worry the observer as he knows the reason for it. But he should offer an explanation to other persons (Master, Pilot etc.) who, at times, also watch the radar display.

Care should be taken when changing range scales while working a true motion plot on the reflection plotter. For example, if the observer, while making a true motion plot using the six mile range scale decides for a particular reason to switch down to the three mile range scale for some time and then later switches back to the six mile range scale, a displacement of the electronic centre is introduced in the direction of the heading marker on the longer scale and this applies for the spot representing own ship *and also for all other echoes*. This is because the spot representing own ship moves with twice the speed on the three mile range scale as it does on the six mile range scale. This is illustrated in Fig. 14.5. At 0009 the operator switches down from the six mile to the three mile range scale and then switches back again to the six mile range scale at 0018 *and continues the original plot*. The result is that all the echoes are displaced a distance $ab - cd$ in the direction of the heading marker and the echo of the lightvessel to port has developed a drift while, apparently, the ship on port bow has altered to port and the ship on starboard bow has reduced her speed.

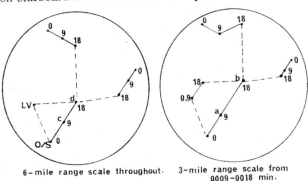

6 – mile range scale throughout.   3 – mile range scale from
0009 – 0018 min.

FIG. 14.5 DISTORTION IN PLOT OWING TO CHANGING RANGE SCALES WHILE USING A TRUE MOTION DISPLAY AND REFLECTION PLOTTER

After the picture on a True Motion Display has been reset, it will help for reference purposes not to wipe the old plot off immediately but to wait until the new plot of echoes and own ship has been properly re-started.

In all the cases described above, plotting on a *transparency* can also be of great assistance if one man has to deal with watching the display and extracting the proper information from the many echoes he might observe on the display (see next chapter).

With *two officers on the bridge* (besides the Master) and proper plotting facilities available an uncompleted true motion plot (showing aspect and speed only) on a construction sheet can complement an uncompleted relative motion plot (showing the *OA* line only) on the reflection plotter. Similarly, an uncompleted true motion plot on the reflection plotter can be supplemented by an uncompleted relative motion plot on a construction sheet or plotting board. If no reflection plotter is available and plotting facilities are cramped, a completed relative motion plot (triangle) of ships involving a risk, carried out on a construction sheet (spider web) placed on a small desk and done by the second man, is the practical solution. Or, without reflection plotter, a complete true motion plot (showing also the nearest approach of ships which are considered a danger) may be carried out on a plotting table with track plotter. Here again, it is always advisable to draw up a record sheet stating times of observations, own ship's head and log speed, and the ranges and bearings of targets. The plotter can then study the sheet, make a selection and plot the ships in order of urgency, paying particular attention to the range, the rate of approach and the rate of change in bearing. Such a sheet also might become a valuable document in case one becomes involved in a collision through no fault of the Master and officers of own ship.

Where available, mechanical plotting devices such as the *R.A.S.* plotter or any other rotatable plotter should be employed for relative motion plots as they can speed up the action of plotting considerably.

The type of plot which will develop—true motion or relative motion—depends for a great deal what the display presentation is offering. With one watchkeeping officer on the bridge and a reflection plotter available, a relative motion plot will be carried out on a relative motion display and a true motion plot on a true motion display. The same applies if the observer is using the transparency method. If he uses neither, then he should keep a recording sheet.

With two or more officers on the bridge besides the Master and there is a reflection plotter fitted, duplication has not much sense and again the complementary plot on the construction sheet is determined by the display presentation. If there is no reflection plotter, but there is a small plotting table, then with two officers, obviously one carries out a relative motion plot on a " spider web ", a *R.A.S.* plotter or any other rotatable device. If there is ample space for plotting, the second man may prefer a true motion plot with the track plotter, if available.

In general terms one can say that a true motion plot involves some more work, but requires less training, while the relative motion plot can mean less work but the observer should be thoroughly trained. The relative motion plot also affords better selective plotting while there is sometimes a tendency for true motion plotters to try to plot too many ships.

There is, of course, a limit to the number of ships which can be plotted, or even recorded and in such a case the Master has to sacrifice his speed. The engines should be put to slow or dead slow and the echoes observed whose ranges are decreasing within the six mile range. When a vessel is closing in, the telegraph should be put on " stop " or the engines reversed depending on the circumstances. At this slow speed, the display presentation can be considered to be practically on " True Motion " and only the very nearest targets are of concern and need be marked on the tube, if necessary, then disregarded when they are cleared and one can then deal with the next one. In this way the ship is inched along and navigated with extreme caution.

Plotting helps to eliminate one of the limitations of cm. radar where the present aspect cannot be deduced unless the vessel is very close by. We can only deduce the aspect from *past* motion and when the target is *stopped no* heading is indicated.

Plotting should be done *quickly*. If there are not enough or no trained officers on the bridge for plotting, then Master and owner should realise that the value of a radar set for collision warning is greatly reduced. No alteration of course must be attempted if no plotting information is collected about the approximate aspect and speed of the target vessel. The Master should never act less cautiously in such a case than he would have done if there was no radar on board. On the contrary, he can still use his radar set and increase the safety of own vessel by treating echoes, forward of the beam, within a distance of three miles as sound signals, and he should, except in the case where he is certain that a risk of collision does not exist, reduce the ship's speed to the minimum at which she can be kept on her course. V.H.F. R/T contact can also be tried (see Chapter 16).

A clockwork Radar Plotter Timer is extremely helpful. It enables the officers to concentrate on the radar more than the clock.

The Department of Trade Notice *M* 983 emphasises amongst other things the *need for clear weather practice*. Thus it can be expected that in a Court of Inquiry it would be unfavourable for the

Master and officers of a ship involved in a collision if they can give
no evidence of having practised plotting in clear weather.

Useful Rules

It can be useful for the observer if he is able to *predict*, without
actual plotting, the approximate relative motions of ships' echoes
on the PPI *before* he makes a manoeuvre or takes avoiding action,
in fact, to estimate mentally a trial manoeuvre.

First, when a target is on a collision or near-collision course, the
point of possible collision must be on or near the heading marker.
If the rate of approach is fast, the PPC will be reached soon and on
the PPI it cannot be very far from the point representing own ship.
This must also mean that the aspect of the target will be small.
Conversely, with a target on a slow approach on a collision course,
its aspect must be large. Furthermore ships on collision courses have
green aspects when in own ship's port-hand semi-circle and red
aspects when in own-ship's starboard-hand semi-circle. See Fig. 14.6.

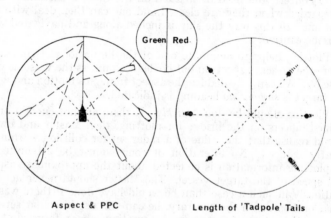

Aspect & PPC   Length of 'Tadpole' Tails

FIG. 14.6   RATE OF APPROACH AND ASPECT FOR SHIPS
ON COLLISION COURSES

Secondly, there exist a number of Rules applicable to ships'
echoes for any bearing on a *stabilized relative* motion display,
irrespective whether the ships are on collision courses or not.

1. For echoes moving in the same direction and parallel to the
heading marker, an alteration to starboard or port by own-ship
changes their motion in an anti-clockwise or clockwise direction
respectively.

2. For echoes moving in a direction parallel and opposite to the heading marker, an alteration to starboard or port changes their motion in a clockwise or anti-clockwise direction respectively.

3. For echoes moving at right angles to the heading marker an alteration of course by own ship has, initially, little effect on the direction of motion of the echoes.

4. If own ship reduces or increases her speed, the relative motion lines will change in such a way (direction and/or speed) that their points are displaced in a direction the same as or opposite to, the heading marker respectively.

5. With a fast or slow relative motion rate, the effect of manoeuvring action by own ship is relatively small or substantial respectively.

Of course, these Rules are logical when thinking about it, but to observe and confirm the effect each time a manoeuvre is made, will help the observer a great deal in coming to understand relative motion displays.

CHAPTER 15

PLOTTING DEVICES

Since radar was introduced in ships during and after World War II, plotting devices have been designed, first of a simple construction —still widely used—but as time progressed much more elaborate and sophisticated designs came on the market. Their main object is to reduce the burden on the radar observer, but they also aim to increase the accuracy.

Broadly speaking, they can be divided into three classes :—

1. Simple plotting devices.

2. Electronic plotting devices which are not computer-based.

3. Plotting devices which make use of digital computers for data collection, calculation and assessment of multi-target situations.

The latter are now known as *Automatic Radar Plotting Aids* or ARPAs. During the last 15 years a great variety have been produced by different manufacturers, and others are still being developed. They form a very important group because, gradually over the coming years, their carriage on board will become compulsory for ships of 10,000 g.r.t. or more.

It would not be possible in this book to discuss the many types, but IMO on 15 November 1979 adopted a resolution stating the Performance Standards for ARPA. This document has brought some uniformity in ARPA facilities and control, thus making discussion easier. However, some important features *not* mentioned in the resolution, are incorporated in some ARPAs and special attention will be paid to them.

I. Simple Plotting Devices

Plotting Charts

Plotting charts are especially designed for plotting. They are a representation of the screen with a graduated bearing scale and concentric rings which are spaced at equal intervals. Some of these " spider webs ", produced in the United States, can be used at night with the aid of special illumination (ultra-violet light), which is not a distraction in the wheelhouse.

Spider webs are of great assistance for quick plotting. All that is needed are a small ruler and a pencil. Furthermore they can be kept and stored for some time for record purposes.

Rotatable Plotting Devices

(a) The plotting surface is fixed and transparent. Under it a graduated disc with a uniform grid can be rotated and affords quick and easy laying-off of bearings and ranges.

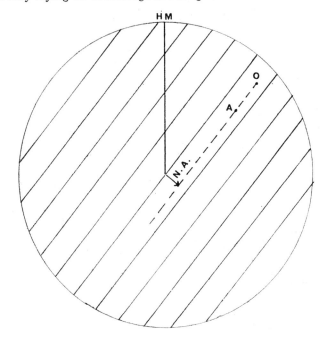

FIG. 15.1   USE OF PARALLEL INDEX FOR ESTIMATING
THE PREDICTED NEAREST APPROACH

(b) The plotting surface can be rotated and is engraved with the compass bearing scale. It is mounted on a base on which the heading marker and relative bearing scale are engraved. A ruler, with slots for moveability, is attached to the pivot.

Plotting on these devices is done in exactly the same way as described in Chapter 13. For plots concerned with alterations of course by own ship, the plotting surface is moved around an amount corresponding to the course alteration. We could have done the same thing with our plots on a sheet of paper. By rotating the paper—and the plot—we can bring the new heading marker again in the upward position.

The *R.A.S.* (*Radio Advisory Service*) Plotter is an example of (*b*).

Parallel Index

The Parallel Index consists of equidistantly spaced lines engraved on a transparent screen. When fitted on the PPI it can be made to rotate and by aligning the parallel lines with the echo trail, the nearest approach can be estimated directly. In order to increase the accuracy, intermediate lines can be drawn in chinagraph pencil. See Fig. 15.1. The Relative Motion Display is used in this case.

Similarly, it can be used on a True Motion Display so that the course of the target can be measured.

Reflection Plotter

A cross-section of this plotter is illustrated in Fig. 15.2.

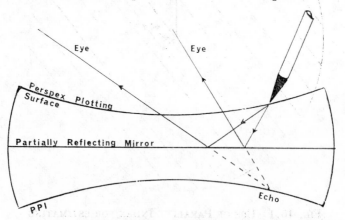

FIG. 15.2 CROSS-SECTION OF REFLECTION PLOTTER

The plot is carried out on a transparent screen, covering the PPI. Between the plotting screen and the PPI there is placed a flat partially reflecting mirror, which makes the avoidance of parallax errors possible. When the pencil point is directly above the radar echo, then the image of its mark on the plotting surface covers the echo on the PPI from whatever direction one may look. The marks can be removed easily and when a plotting light is switched off, the marks disappear and the screen can be observed normally. Different colour wax pencils can be employed to trace the echoes of different ships.

On a Relative Motion Stabilized Display the usual Relative Motion Plot is produced; on a True Motion or Track Indicator Display a True Motion Plot is constructed.

Although the scale of the plot is small the results it yields are often more accurate than those obtained from a larger scale transferred plot. The reason for this is that blunders can be more easily made when observing bearings and ranges of several targets at once and in laying them off on the plotting sheet, whereas with a reflection plotter the positions of the echoes are not so likely to be marked incorrectly on the screen.

Note that care should be taken when using an Unstabilized Display. In this case, for the sake of accuracy, it is essential to make sure that the ship is right on course at the moment the positions of the echoes are being marked on the reflection plotter. When the vessel is yawing, it is recommended to ask someone to sing out when the ship is on course.

Focussing should be checked occasionally (see "Maladjustments", Chapter 4).

Plotting Board and Track Plotter

The Plotting Board, made by radar manufacturers, consists of a flat sheet of glass which can be marked with wax pencils. Concealed edge illumination with controllable brilliance allows plotting to be carried out in darkness. A time/speed distance table, illuminated in the same way, is incorporated in the plotting surface. Bearing and distance can be quickly laid-off by means of a *Track Plotter*. This Track Plotter consists of a bearing and range scale, suspended by means of a parallel linkage system thus providing translational but no rotational movements.

The whole assembly is mounted on a $\frac{3}{4}$ in. baseboard for securing to an existing table, or it can be supplied in a console with provision for a built-in illuminated clock, timer, log repeater and distance counter.

The Plotting Board and Track Plotter are extremely suitable for carrying out a True Motion Plot. Near land, for this purpose, a navigational chart can be inserted under the plotting surface.

Plotting by means of Transparencies

The use of transparencies affords a very quick method of making a complete plot. They are shaped in such a way that they can slide over the reflection plotter or over the tube mask of the PPI. A line

drawn on the transparencies represents the heading marker and
allows for easy alignment with the picture. They can be used with
the Relative Motion Displays and with the True Motion Displays.
In both cases put the transparency over the PPI (reflection plotter
or tube mask), align by means of the heading marker so that the
point representing the spot (and starting point of the line indicating
the heading marker) of the transparency is superimposed on the spot
representing own ship on the PPI. Mark the echoes with chinagraph
pencil both on the transparency and on the PPI.

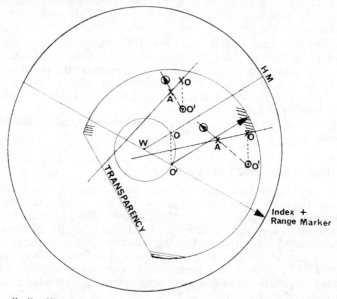

X X   Marks on PPI
⊙   Marks on Transparency

FIG. 15.3   ESTIMATION OF PREDICTION LINE BY MEANS OF
A TRANSPARENCY

When the presentation on the PPI is " Relative Motion " set the
range marker to the distance representing *WO* (distance travelled
by own ship during the plotting interval) and move the transparency
this distance *against* the direction of the heading marker—downwards with a Head-Up Display—and at the end of the plotting
interval, with centre of transparency now on range marker, mark the
echoes again on the transparency and the PPI. The line joining the
echoes on the PPI indicates the directions and rates of the relative

motion of the ships while the line joining the marks on the transparency represents the directions and rates of the true motion of the ships. This will be understood by studying Fig. 13.10.

When the PPI presentation is " True Motion " the transparency should be moved *in the direction* of the heading marker at the same rate as the movement of the spot representing own ship over the PPI. The plot on the transparency now indicates the relative motion of the other ships, while the plot on the PPI, naturally, represents the true motion of own ship and targets.

A prediction line can also be produced by the same method. *WO* is marked off on the line representing the heading marker on the transparency, *W* at the starting point of the line. The range marker is set on the same distance and the transparency moved *laterally*, by keeping the heading lines parallel, until *O* has been displaced along the range marker circle through an arc equal to and in the direction of the proposed course alteration (set Parallel Index on the new heading). The direction of the *O'A* line is given by the lines joining the points marked at the *beginning* of the plotting interval (now the *O'* points) on the transparency and the points marking the *end* of the plotting interval (*A* points) on the PPI. See Figs. 15.3.

If the prediction line is required for a given reduction in speed then the points *O'* on the PPI are equivalent to the points marked at the *beginning* of the plotting interval on the transparency *after* the transparency has moved against the direction of the heading marker equal to a distance from the centre of *WO* − *WO'*, where *WO'* represents the distance moved by own ship during the time of the plotting interval with her reduced speed.

These transparencies are best used in conjunction with a reflection plotter otherwise great care has to be exercised in order to avoid parallax errors.

II. Electronic Plotting Devices which are not Computer-Based

" Anti-Collision Radar " Display

It is the intention with this particular system, produced by Decca Radar Ltd, that the display presentation should be "True Motion ". See Fig. 15.4.

The echoes illustrated are those of a lightvessel and of a moving ship. The same echoes are used again in Figs. 15.5 and 15.8.

To assess collision danger, interscan lines (a total of five and known as " electronic relative motion markers ") can be brought on the screen. These markers, often called " matchstick " markers, are electronic lines, 25 mm. long, which are painted on the PPI by interscan technique so that the brilliance of the markers is independent of the position of the rotating trace. Each marker has a bright spot at one end while the other end of the marker remains pointing towards the spot representing " Own Ship ". Hence, on a True Motion Display, and with Own Ship moving *the marker moves across the radar screen* with the bright spot at a constant range and bearing.

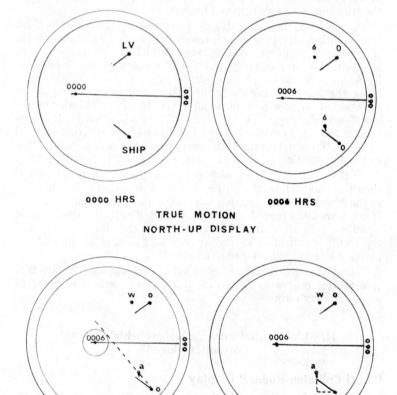

TRUE MOTION
NORTH-UP DISPLAY

DETERMINATION NEAREST APPROACH, COURSE & SPEED

Fig. 15.4   A–C Radar Display

Each marker can be brought separately on the display and can move independently of the others. The marker controls are grouped together in a compact control unit mounted on the true motion pedestal. The controls and indicators comprise five illuminated switches, range, bearing and brilliance control and an amber window which is illuminated when the 3-, 6- or 12-mile range scale is in use (these are the range scales on which the markers can be operated).

To introduce and position a marker, a key switch is pressed down and the range and bearing control used to align the marker with the selected echo by superimposing the bright spot of the marker on the echo under observation. The marker switch is then centred and the marker will now remain pointing towards the spot representing own ship. See the two upper diagrams in Fig. 15.4 representing the situation at 0000 hours and 0006 hours.

A collision risk exists if the echo remains on the marker or very near to the marker. The line connecting the bright spot of the marker to the actual position of the echo portrays the relative motion line (OA) so that the predicted nearest approach can be estimated. The tadpole tail of the echo gives an indication of the true motion line (WA). All this information can now easily be collected and completed by means of a reflection plotter.

With five markers, up to five echoes can be kept under observation.

Note:—Often the display brilliance has to be increased slightly to make the markers visible.

The True Motion/Electronic Plotting System

With this type of display one can simply do all the plotting by means of electronic interscan lines and an incorporated clock. The display unit is equipped with an electronic interscan curzor which is geared to the mechanical curzor. By means of a joy-stick one can position the origin of the interscan line on any selected echo and the line can then be rotated by means of the mechanical curzor crank. In this way it is possible to bring an electronic plotting line on the display showing the relative or true motion of an echo. Adjustable marker pips at equal intervals can be superimposed on the interscan line so that a time-base is formed for calculations involving the time of the nearest approach or the speed of another vessel. It is also possible to make the variable range marker run out along the interscan line instead of having its zero point at the own ship position.

A 12-minute clock is incorporated in the display unit and by means of a switch the origin of the interscan line can be connected to the true motion computer so that it behaves as the echo of a stationary target. These arrangements make it possible for the operator to carry out a plot " electronically " on the display.

I

For example, suppose that the North-Up Relative Motion presentation is used and an echo needs observation. Position the origin of the interscan line on the echo and start the clock. The origin of the interscan line now moves in a direction opposite to that indicated by the heading marker and at a velocity corresponding to the distance travelled by own ship per time interval. At any time later, for example six minutes, rotate the interscan line to intersect the echo and move the variable range marker along the interscan line until it has reached the new position of the echo. The direction of the interscan line now indicates the true course of the target and the reading of the variable range marker, multiplied by ten, will give the target's speed.

With the display presentation on " True Motion, North-Up " a similar manipulation can be carried out, but, in this case, the origin of the interscan moves forward in the direction of the heading marker corresponding to the distance travelled by own ship during the time interval. After, for example, six minutes, the interscan can be rotated to intersect the echo and the direction of the interscan will indicate the direction of the OA-line. By means of the marker pips the time of the predicted nearest approach position can be easily estimated.

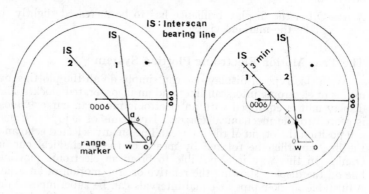

RELATIVE MOTION NORTH-UP & TRUE MOTION NORTH-UP
DISPLAY

FIG. 15.5 THE TM/EP DISPLAY

These two cases are illustrated in Fig. 15.5. If, in both cases, the initial position of the echo is plotted on the reflection plotter, then, in the first case the interscan can be aligned with the OA-line after the WA alignment, while in the second case the interscan can be aligned with the WA-line after the OA alignment. The interscan lines do *not* appear msiultaneously on the display as one might think looking at Fig. 15.5. There is only one interscan line and it appears in succession on the display in different positions, the succession being denoted in Fig. 15.5 by the numbers 1 and 2.

The TM / EP display is produced by Raytheon Ltd.

Situation Display

In a way one could say that the Situation Display has evolved from the photoplot system (now obsolete, but was discussed in previous editions), but what was done by optical means in the latter is achieved by electronic means in the former. In the photoplot system the radar picture appearing on a small ($3\frac{1}{2}$-in. diameter) PPI is retained on a film which is then transmitted by means of light rays to a flat horizontal screen. Transmission is by intervals so that (say) every 6 minutes the projected picture is renewed. In the Situation Display system the blue light forming the picture, emitted by the short-persistence CRT (same diameter

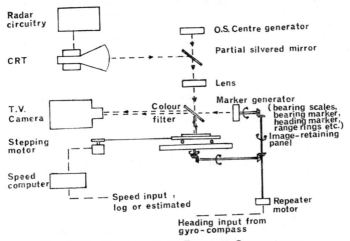

FIG. 15.6   SCHEMATIC OF DISPLAY COMPONENTS
OF SITUATION DISPLAY

as in the photoplot) is projected on an image-retaining panel. This panel, as the name implies, retains the picture as long as an electrical charge is applied to it. The panel re-emits the picture, but now in orange light, which is televised by a T.V. camera and transmitted to several television screens (one main display and up to four slave displays). At the end of a plotting interval the electrical charge to the retaining panel is switched off, the tails of echoes disappear and the cycle can be re-started. A plotting interval is normally 6 minutes, but can be 3 minutes (on the shorter range scales) or 12 minutes, using an override control.

It can, therefore, be seen that there are a lot of similarities between the photoplot system and the Situation Display. In both cases the observer is presented with a bright image and no hood and/or visor is required. In the Situation Display the picture can be chosen as " radar negative " (black echoes on a bright green background) or " radar positive " (bright green echoes on a black background) which used to be also a facility of the earlier photo-plot systems. As in 'Photoplot', more clutter suppression is required than is needed for a conventional display.

From the diagram it can be seen that the retaining plate is mechanically connected to the transmitter of the master gyro and changes in the ship's heading will make the plate rotate. With the display on relative motion, the plate can only rotate, but no translatory motion is imparted to the plate and the display will show a *head-up stabilized relative motion* presentation with all echoes moving relatively and producing solid relative echo trails. These trails are far more pronounced than the tadpole tails associated with PPI displays and their direction and length may be measured directly from the screen to give reasonably accurate assessment of the distance and the time of the closest point of approach.

With the display on " true motion track ", the retaining plane is also connected to a motor which inputs the ship's speed (log or estimated) resulting in the plate moving *backwards* in proportion to the ship's speed. What will be seen now on the display is that the echoes move relatively, but the strong trails which move *bodily* with the echoes having a direction showing the *true motion* of the target. In other words, for ships on collision courses other than from right ahead or astern, the direction of the movement of the echoes differs from the direction of the trail. (To obtain a fuller understanding, read the previous section, " Plotting by Means of Transparencies ", where it is shown that a true motion track can be produced by moving a transparency over a relative motion display presentation *against* the direction of the heading marker.) At the end of a cycle when the electrical charge to the retaining panel is switched off, the plate is reset before once more applying the electrical charge.

Other information, such as the heading marker, bearing marker, range rings, range scale indicator, etc., are generated separately and viewed by the T.V. camera through a colour filter at the same time as the reflection of the radar picture is projected on the retaining panel.

Fig. 15.7 shows the main display on True Track presentation.

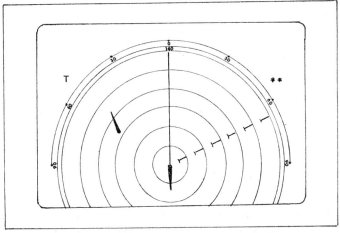

True track mode.End of plot. Bearing cursor on 208° true.

FIG. 15.7   MAIN DISPLAY (Situation)

The tail of the ship's echo on the port bow indicates the true motion of the other vessel, but by plotting the echo on the screen the relative motion is displayed.   Own ship also develops a tail in this case (not, of course, when the display is on " Relative ") and by comparing the length of the other vessel's echo-tail with that of our own ship, a good idea about the other vessel's speed can be obtained provided its echo was on the screen when the cycle started.   If the echo of a fast ship appears on the display, for example, in the middle of a cycle and the observer wants to assess the speed of the other vessel as soon as possible, then he should erase the existing picture (control provided) so that a new plot cycle is started.

Towards the end of a 6-minute plot cycle an asterisk appears in the upper right-hand corner of the screen and the plotting can be extended for a further six minutes by means of an override control.   If, however, a double asterisk appears it will be too late to override and the picture will be erased.   The picture then disappears, and builds up again within 10 seconds—about three consecutive sweeps of the trace.   In this way the solid tails are prevented from obliterating the picture, as could easily take place in coastal waters.

It will be noticed from the figure that a third of the picture, normally seen on a PPI astern of our own vessel, is cut off.   Bearings of targets astern are read by setting the bearing cursor over the echo of the target and reading the *reciprocal* bearing from the edge scale. No variable range marker is provided, but the electronic bearing cursor is in the form of a broken line, which facilitates easy interpolation between the range rings.

Slave displays are for viewing only, and not for taking measurements. A North-up stabilized display, either relative or true motion, is not available. However, a conventional main display with relative motion facilities only, can be incorporated or interswitching arrangements between radars and between 3 and 10 cm. radar sets are available. The Situation Display is produced by Kelvin Hughes Ltd.

The "Predictor" Display System

This system, produced by Marconi Marine, incorporates a completely automatic electronic plotting procedure, the principle of which depends on time-expansion, i.e. the slowing down of the rate of arrival of echoes after each transmission pulse. The degree of time-expansion varies from $\times 2$ on the longest range scale (48 miles) to $\times 256$ on the shortest range scale ($\frac{3}{8}$ mile). The result is that on the display the velocity of the CRT scanning spot on each radial trace is much slower than as it is for a conventional CRT. Thus the spot action of this new display corresponds to the velocity of the spot in a conventional CRT with a range scale of 96 miles. Moreover, the spot velocity is the same on all range scales and therefore, during each trace, spends longer time on the screen than it would do on an equivalent conventional display. Hence, the screen displays a stronger intensity which will become progressively brighter as one switches down the range scales.

The time-expansion reduces the frequency of the video bandwidth to a relatively low frequency which can be recorded on a magnetic tape. The loop makes one revolution and passes the record/play head every two minutes which corresponds to 48 revolutions of the aerial and PPI trace. A radar picture obtained once per aerial scan, is recorded every 10 seconds and remains stored on the tape during three revolutions of the two-minute loop before being erased. Hence a series of 36 radar pictures covering the last six minutes is continuously kept in store and updated every 10 seconds. The use of the two-minute loop ensures that every recorded picture presents itself at the tape head for replay when it is two minutes old, four minutes old and six minutes old. It is erased when it is eight minutes old and the vacant place on the tape can be used to record the current radar scene so that the tape is continuously re-employed.

Fig. 15.8 shows the four display presentations which are available. The echoes illustrated are those of a lightvessel and of a moving ship.

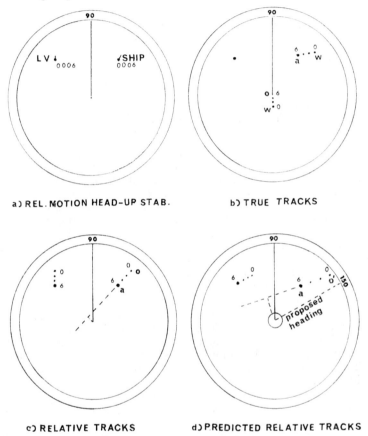

a) REL. MOTION HEAD-UP STAB.   b) TRUE   TRACKS

c) RELATIVE TRACKS   d) PREDICTED RELATIVE TRACKS

FIG. 15.8   PREDICTOR DISPLAY

Fig. 15.8(a) represents the *Relative Motion Head-Up Stabilized Display.* Double stabilization is applied, i.e. the tube as well as the picture on the tube are stabilized. Time : 0006.

Fig. 15.8(c) represents what is known as the *Relative Tracks Mode.* Four echo pips (for each target) can be seen showing the echo where it was at 0000, 0002, 0004 and, of course, where it is now at 0006 hours. The four paints of the echo show the relative movement made by the target during the last six minutes and any collision risk can be assessed. The cycle then goes on repeating, painting a similar scene but each successive painting is ten seconds younger in time.

Fig. 15.8(*b*) represents what is known as the *True Tracks Mode*. Again echoes are shown at 0000, 0002, 0004 and 0006 hours, but the WO-shift is applied to the *picture* for six, four and two minutes time interval respectively for the echoes showing their position at 0000, 0002 and 0004 hours. This brings the start of the time-base each 10 seconds back to the geometrical centre of the tube. Note that the true track of own ship can also be seen. The display therefore shows a True Track Head-Up Stabilized Display with own ship always in the centre of the tube after each 10-second period.

Fig. 15.8(*d*) represents what is known as the *Predicted Relative Tracks Mode*. In this case the effect of a *proposed* alteration of course and/or speed can be predicted and the display will show a picture *as if* the proposed course and/or speed had already been carried out. Besides the present echo at 0006 hours the echoes at 0000, 0002 and 0004 hours are displayed if it is assumed that our ship has been on the new course or at her new speed during the last six minutes. In fact, it shows the $O'A$ line and a shift of OO' for a time interval of six minutes, four minutes and two minutes has been applied respectively to the position of the echoes shown at 0000, 0002 and 0004 hours of the " Relative Tracks Mode " of Fig. 15.8(*c*).

In order to carry out the appropriate manipulations the Predictor Display is provided with two calibrated controls. One is calibrated in degrees, and performs a dual duty as electronic bearing marker and as a means for setting the proposed new course. The other control is for setting the proposed speed and is calibrated in knots.

It can be seen from Fig. 15.8(*d*) that an alteration of 60 degrees to starboard is contemplated and the display now yields the direction of the $O'A$ line and its length for six minutes time interval. The little circle round the centre which is drawn in Fig. 15·8(*d*) is also displayed on the actual screen. Its radius represents the advance of own ship during the turn and the predicted nearest approach should be measured from the outside of the circle rather than from the centre of the display (start of time-base). When helm is applied and the ship's head is coming round to the new course, the "advance" circle automatically decreases in size in accordance with the diminishing amount of turn still to be made.

Remark: The " True Tracks Mode " shows in fact a *Relative Motion True Track* display. That is, if one plots an echo successıvely *in time* on a reflection plotter, a relative motion line is produced by the plotting pencil. The true motion track, however, is generated electronically and is always present. By observing the direction of the " tadpole " tails *and* the direction of the true motion tracks, information can be obtained about possible danger of collision and the true motion of the target.

The " predictor " display has a number of advantages above the conventional display:—

(*a*) Automatic solving of velocity triangles for past and *future* occasions for a determined time interval.

(*b*) Bright echo tracks.

(*c*) No re-setting has to be employed when using a True Motion Display. This, indeed, eliminates the danger of the frequent occurrence of late re-setting and makes the display also eminently suitable for fast-moving vessels in clear weather.

III.   Automatic Radar Plotting Aids (ARPAs)

Introduction

It has already been mentioned that in the foreseeable future every ship of 10,000 g.r.t. or more shall have to be fitted with an Automatic Radar Plotting Aid (for details see ARPA Appendix I (a)). To appreciate the capabilities and limitations of an ARPA, it is worthwhile for the operator to study the IMO Performance Standards (ARPA Appendix I (b)). It starts by stating that ARPA should improve the standard of collision avoidance at sea by reducing the work-load on radar observers and by providing continuous, accurate and rapid evaluation of encounter situations. In order to comply with these requirements various ARPA features are considered and minimum standards are spelled out. To understand the definition and meaning of some of the terms used, one is referred to ARPA Appendix I. The account below is *not* a literal record.

1. *Detection.* If automatic detection facilities are supplied, then the performance should not be inferior to that obtained on the conventional radar display.

2. *Target acquisition.* This can be either manual only or manual and automatic. The manual acquisition facility must be provided— the observer may wish to acquire certain selected targets only, or he may be compelled to use the manual mode if, in automatic mode, the computer becomes overloaded ; manual *cancellation* facilities must also be available. If automatic acquisition is incorporated, facilities should be provided to suppress acquisition in certain areas (due to the presence on the display of coastal echoes, sea and rain clutter).

3. *Target tracking.* Processing, displaying and updating should be continuous and an ARPA should be able to track at least 20 targets if automatic acquisition is available, and at least 10 targets if manual acquisition only is provided (more targets in the automatic acquisition mode, because the computer is, in most cases, far less selective than the observer who will probably choose his targets carefully when in the manual acquisition mode).

If automatic acquisition is provided, the criteria for selection of targets should be known to the user—there is either guard ring or guard zone selection, or the observer should be acquainted with the priority programme for selection or "threat profile"; this is especially important when the computer becomes saturated and may "shed" certain targets.

A qualitative description of the effects of error sources on automatic tracking, and corresponding behaviour, should be provided to the user. This includes the effects of low signal-to-noise and low signal-to-clutter ratios caused by sea returns, rain, snow, low clouds and non-synchronous emissions. It should also include a description of "target swop"; if this takes place, then, if both targets were being tracked, information displayed about the targets will have exchanged and will be wrong for a period, or if only one target was being tracked, its data and tracking would become attached to the other, formerly untracked vessel. This data displayed on the latter vessel will correct itself after a short period but the result could be that the tracking has changed from a dangerous to a less dangerous vessel.

An ARPA should continue to track a target which is clearly distinguishable on the display for 5 out of 10 scans, and targets which are being tracked should be clearly indicated on the display.

Track *history* should also be provided on *request*, consisting of at least four equally spaced positions of the echo of the target being tracked over a period of at least eight minutes. (This information, which is based on "long" memory capacity, is more reliable than the vector presentation : it provides a check on the tracking process and gives clearer indication if changes in the target's course and/or speed have taken place).

4. *Display.* This may be a separate or an integral part of the ship's radar installation, but any malfunction of ARPA parts should not affect the integrity of the basic radar presentation. Its diameter should be at least 340 mm., and ARPA facilities should be available on at least two range scales (12 and 3, or 16 and 4 miles). Presentations should be stabilized with a choice of North-up, Head-up *or* Course-up. If true motion presentation is incorporated, true or relative motion vectors, showing the predicted motion of the target, should be available.

Course and speed information should be displayed in vector (time-adjustable or with a fixed time scale) form, or in graphic form. If this information is given in graphic form, vector information in relative or true motion should, *on request*, also be provided to the operator. If the information is given in vector form *only*, the observer should have an option of both true and relative motion vectors.

It should be possible to cancel unwanted ARPA data, and there should be independent brilliance controls for the ARPA data and the radar data, so that the former can be eliminated, if required.

The display should be a bright radar display so that it can be observed by more than one observer simultaneously in daylight conditions.

After acquisition, the ARPA should present in a period of not more than one minute an indication of the target's motion trend and display within three minutes the target's predicted motion in vectorial and graphical form. After changing range scales on which ARPA facilities are available or resetting the display, full plotting information should be displayed within a period of time not exceeding four scans.

(Note, by the way, that no mention is made of Points of Possible Collision (PPC) or Predicted Areas of Danger (PAD)).

5.1. *Operational warnings.* Audible and/or visible warning signals should be provided :—

(*a*) When any target closes to a range or crosses a zone selected by the observer. This is important in connection with the concept "keeping a proper look-out", and also, if manual acquisition only is provided, it warns the observer to start the tracking process in relation to (*b*) below.

(*b*) When any target is predicted to pass within a minimum range and time chosen by the observer.

An indication should also be given if any tracked target is lost other than by getting out of range. In all the cases mentioned above, warnings should also be clearly indicated on the display.

5.2. *Equipment warnings.* Suitable warnings of ARPA malfunction should be provided to enable the observer to monitor the proper operation of the system. Additionally test programmes should be available so that the overall performance of ARPA can be assessed periodically against a known solution (this can be done by the generation of artificial targets as on a radar simulator).

6. *Data requirements.* At the request of the observer the following

information should be immediately available in *alphanumeric* form in regard to any tracked target: Present range and bearing, predicted CPA and TCPA, calculated true course and speed.

7. *Trial Manoeuvres.* ARPA should be capable of simulating the effect on all tracked targets of an own ship manoeuvre without interrupting updating of target information. The simulation should be initiated by the depression either of a spring-loaded switch, or of a function key, and should give a positive indication on the display that simulation is taking place.

8. *Accuracy.* In this section two Tables are shown, each for four different ship-target cases ("scenarios" called here). The first Table shows the prescribed accuracy values for the relative motion (direction and speed) and CPA, within *one* minute of steady state tracking for the four scenarios; the second Table shows the prescribed accuracy values (95% probability values as in the 1st Table) of the relative motion (direction and speed), CPA, TCPA, true course and speed (no distinction is made between ground and sea velocity (course and speed) within *three* minutes of steady state tracking. The second Table, of course, should show higher accuracy values than the first one—the greater the number of observations, the more accurate the plot will be.

These accuracy values are based on the best possible *manual plotting performance* achieved under environmental conditions of plus and minus 10 degrees of roll. It is also assumed that the errors in the plot are due to the errors in the input sensors (scanner, gyro, log, etc.) and are not associated with ARPA design.

The four scenarios are shown in Table 15.1. Courses and bearings are in degrees (true), speeds in knots and ranges in nautical miles.

Scenario	Own Ship		Target Ship		Bearing	Range	Relative Mot.	
	Course	Speed	Course	Speed			Direction	Speed
1	000	10	180	12	000	8	180	22
2	000	10	045	14	000	1	090	10
3	000	5	237	16·8	045	8	225	20
4	000	25	307	17·8	045	8	225	20

TABLE 15.1

Note that there are three scenarios for medium range and one for short range, three for collision courses (no. 2 is the exception), two scenarios for medium speed of own ship, one for fast and one for slow speed. Two fast-speed target ships are also included.

The prescribed accuracy values are shown in Table 15.2 where the two Tables published in the IMO Specification are compressed into one. Courses are given in degree (true), speeds in knots, CPAs in nautical miles and TCPA in minutes.

Scenario	Within	Relative Motion		CPA	TCPA	True Motion	
		Direction	Speed			Course	Speed
1	1 min.	11	2·8	1·3		7·4	1·2
	3 min.	3	0·8	0·5	1·0		
2	1 min.	7	0·6	—		2·8	0·8
	3 min.	2·3	0·3	—	—		
3	1 min.	14	2·2	1·8		3·3	1·0
	3 min.	4·4	0·9	0·7	1·0		
4	1 min.	15	1·5	2·0		2·6	1·2
	3 min.	4·6	0·8	0·7	1·0		

TABLE 15.2

Note the following :—

(a) The close relationship existing between the accuracies of the relative motion and of the corresponding CPA value.

(b) The accuracy on the short range is much better than that for the longer ranges. This is because bearing errors have more influence on the direction of the relative motion line and CPA of targets being observed at the longer ranges.

(c) Quite large errors in the course of the target may occur if it is on an opposite course with medium (or slow) speed.

(d) The large errors in the relative motion direction in scenarios 3 and 4 are due to bearing errors caused by rolling and pitching and the resulting scanner movements. The distribution of these errors is quadrantal with maximum values on relative bearings 045°, 135°, 225°, 315°.

It is also stated in this section that when a target, or own ship, has completed a manoeuvre, the system should present, in a period of not more than one minute, an indication of the target's motion trend, and display within three minutes the target's predicted motion, in vector and/or graphic form in accordance with the data requirements (point 6) and the prescribed accuracy requirements given above in this section.

9. *Equipment used with ARPA.* The final section of this Perform-

ance Standard Document simply states that log and speed indicators providing inputs to ARPA equipment should be capable of providing the ship's speed through the water.

Simple Technical Description

Nearly all the components which make up an ARPA are incorporated in the Display unit. The display itself might be the conventional one, or the television-type video display unit, or a quantized display.

Conventional Display. In this case the interscan period would have to be used for the writing of symbols on the screen, drawing of vectors and tracks, etc.

Television-type Display. A good example is the

Situation Display, which has been discussed already in this Chapter. Light from a CRT, coupled to the main display, falls upon an electrostatic retaining panel (Fig. 15.6) and is then reflected via a mirror into a television camera which transmits the picture to the different monitors. This type of display, however, is not an ARPA display.

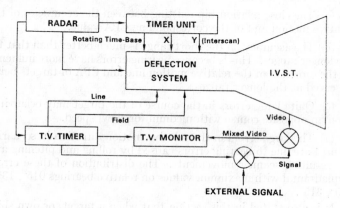

FIG. 15.9   SCAN CONVERSION

Another method, often used in port radar systems, is :—

Scan Conversion. Here an Indirect View Storage Tube or I.V.S.T. is used. Radar video information is stored (written in) on the I.V.S.T. in the usual radially scanned form and then read out as a standard raster scan signal. By careful adjustment of the threshold levels, echoes of targets can be made to stand out against low-level noise returns from sea waves or (and) rain. By inverting the video signal such that targets appear black on a white (or green) background, a high-brightness display can be achieved.

The signal generated at the I.V.S.T. can be fed directly to T.V. monitors. Signals from other sources, such as labels, can, because of the standard signal format, be shown simultaneously. A block diagram is shown in Fig. 15.9.

Quantized Display. The word "quantization" refers to a method of producing a set of discrete or quantized values that represents a continuous (analogue) quantity. The analogue quantity is "chopped up" into a finite number of sub-ranges. The technique is used whenever a set of discrete values is required to produce data for a digital computer, so that storing, sampling, comparing, etc., can take place. Another advantage of quantization is that the discrete values can easily be converted into binary numbers so that they can be used for "sorting" and calculations.

Write Cycle

Quantization in Range. One of the methods by which this can be achieved is shown in Fig. 15.10. An electronic (AND) gate is kept open during the time the enabling pulse (radar signal) is applied and the clock pulses are allowed to go through and stored one by one, successively, in a register (for more details see Appendix). The clock is synchronized with the transmission. The register which in a sense replaces the time-base, is where all the signals are loaded up and stored in real time. The successive elements in which the signals are stored, are known as "cells" or "bins".

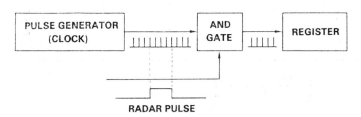

FIG. 15.10   QUANTIZATION OF ANALOGUE SIGNALS

For a 16-inch display the number of resolvable elements across a diameter is about 600 to 800, which would provide 300 to 400 range elements per sweep. An off-centre display displaces the origin to

70% of the range sweep opposite the heading marker, thus adding
70% more elements in the forward direction. Inexpensive registers
are now available with 512 (2^9) elements. Using *all* of these to store
data, the offset sweep would use all 512, and the centred sweep
would fill about 300. Sometimes, depending on the pulse-length
(range scale)—as will be seen in a moment—the cells are not fully
used, where in other cases there are more samples than cells. In the
latter case, two or four successive pulses are placed for storage in a
single cell.

It will be clear that the rate of sampling must depend on the
range discrimination of the raw radar display, i.e. the 'equivalent'
pulse-length ($\frac{1}{2}$ actual pulse-length). For a short pulse of 0·05 micro-
seconds the range discrimination would amount to $0·05 \times \dfrac{300}{2}$, i.e.

7·5 m. The sampling rate improves on this figure and is often chosen
as being equivalent to 1/400 mile, representing a range discrimination
of 4·63 metres. For longer pulse-lengths of 0·5 and 1 micro-seconds,
the respective sample length in range would amount to 1/50th and
1/25th of a mile respectively.

The Table below gives an illustration, and also indicates the clock
frequency required.

Scale n.m.	Pulse-length micro-sec.	Sample length n.m.	Clock freq. Mhz	Number cells
$\frac{1}{4}$	0·05	1/400		100
$\frac{1}{2}$	0·05	1/400		200
$\frac{3}{4}$	0·05	1/400	32·40* (basic)	300
$1\frac{1}{2}$	0·05	1/400		300×2
3	0·05	1/400		300×4
6	0·5	1/50	Basic $\div$ 8	300
12	0·5	1/50	Basic $\div$ 8	300×2
24	1	1/25	Basic $\div$ 16	300×2
48	1	1/25	Basic $\div$ 16	300×4

$$* \ 1 \div \left[\frac{1/400 \times 2 \times 1852}{300} \right]$$

TABLE 15.3

An echo-pulse, for example, at 1 mile range displayed on the 6-
mile range scale, would be stored in cell 50 and can be represented
by the binary number 110010 i.e. $1 \times 2^5 + 1 \times 2^4 + 0 \times 2^3 + 0 \times 2^2 + 1 \times 2^1 + 0 \times 2^0$.

For each transmitted pulse, the received signals are quantized in amplitude, sampled and stored, all in real time. This process is known as the *data write cycle*.

It is possible to store analogue pulses in special analogue registers for a limited time. This is useful for land echoes which do not require calculation or selection processes, but makes the quantized picture indistinguishable from raw radar presentation, yet possesses greater brightness due to a slower read-out (see later).

Quantization in Azimuth. One method uses a binary-coded disc, consisting of concentric rings, starting at the periphery and working inwards. The number of rings determines the number of digits of the binary read-out. With six rings, for example, 2^6 or 64 binary numbers will be available. In Fig. 15.11, to clarify the principle, a

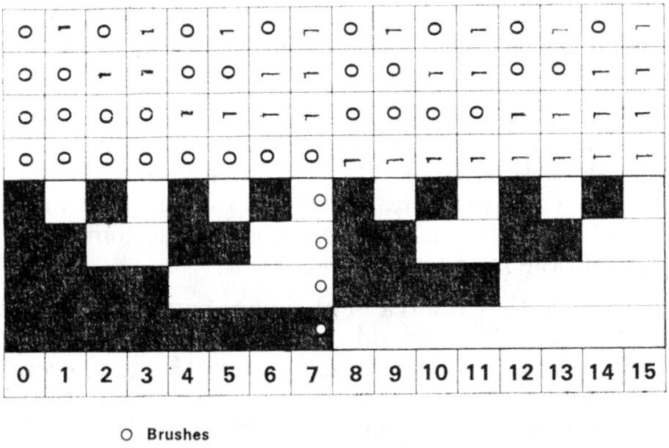

O Brushes

FIG. 15.11 SHAFT POSITION ENCODER
("straightened" portion of disc)

portion of the disc has been straightened out to show a four-digit binary progression. For optical read-out by means of photo cells, the dark areas are made opaque and the white areas are left clear, while for electrical read-out by means of brushes the dark areas are made of insulating material and the white areas of conducting material. Either the rings rotate and the brushes are stationary or vice versa. But, whatever the case, the rotating motion needs to be synchronized with the scanner rotation and phased in such a manner that the binary 0000 is read when the centre-line of the scanner passes through the fore and aft plane of the vessel. Note that in Fig. 15.11 all the dark (insulating) segments represent zeros and the white (conducting) segments represent ones. With nine rings, for example, a single discrete quantity would amount to $(360 \div 2^9)$ degrees, i.e. 360/512 or 0·7°.

The assembly of discs, rings and brushes is also known as a *Shaft Position Encoder*.

In another method of Azimuth quantization, each single trace or sweep is sampled and counted by means of a clock which is synchronized with the PRF and starts counting when the heading flash is triggered off. With a PRF of 900 for example, and an aerial rotation rate of 30, quantization would be down to an accuracy of 0·20°. A third method uses bearing resolvers which convert the bearing into analogue voltages, which then can be quantized.

It can therefore be seen that each echo-pulse can be defined by two binary numbers, one associated with range and the other with its azimuth. These numbers can be placed and stored in a matrix register with one ordinate being equivalent to range and the other perpendicular ordinate being equivalent to the azimuth.

Read Cycle

After completion of the write cycle, the read cycle is initiated and at the same time the CRT sweep is started. The read cycle timing and the CRT sweep rate need not to be in real time and are selected to fit into the time available between the completion of the write cycle and the next transmission cycle.

Let us assume that the read-out time is constant, amounting to 150 micro-seconds. This would correspond approximately to the time-base length of the 12-mile range scale (148 micro-seconds). On the 12-mile range scale the range selector switch would then include an additional position to give a choice between a raw analogue video display and a video processed display (this can be very useful in case something goes wrong with the video processor or if the operator wants to make sure that no important echoes are lost during the processing procedure).

Assume also that during the interscan period following the transmission, the information contained in the prime register is shifted bodily to a second register. Such arrangement keeps the stored echo-pulses in memory and leaves the prime register open for pulses during the next transmission cycle.

The consequences of these assumptions are that the CRT spot can move at a slower rate over the display for the shorter range scales, thus increasing the brightness, and secondly, affords an opportunity to compare the position of the echo-pulses during successive transmission cycles.

(a) *Bright Displays.* An illustration is shown in Fig. 15.12 for the ¼ mile range scale using a PRF of 3000 Hz.

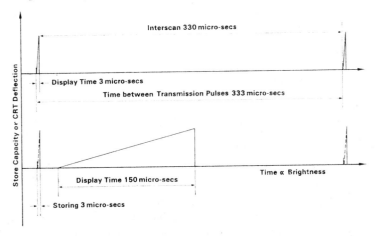

FIG: 15.12 BRIGHT DISPLAY TIMING

The spacing between pulses is $10^6/(3 \times 10^3)$ i.e. 333 micro-seconds. The duration of the time-base on a conventional radar display would be $2 \times \frac{1}{4} \times 1852/300$, i.e. approximately 3 micro-seconds. Using the bright processed display the reading out of the same echoes will take

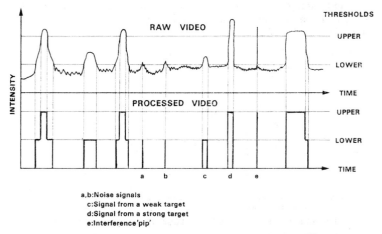

a,b:Noise signals
c:Signal from a weak target
d:Signal from a strong target
e:Interference 'pip'

FIG: 15.13 DIGITAL PROCESSING TWO-LEVEL VIDEO

150 micro-seconds, theoretically amounting to an increase in brightness of 50 times.

Memories, however, are digital devices and can only store the presence (one) or absence (zero) of the signal, not the strength of the signal. For this reason, *two threshold detectors* are used which categorize the incoming signals as either noise, echoes from weak targets and echoes from strong targets. See Fig. 15.13.

Echoes which pass the criteria as representing a weak target are stored in one memory while the strong echoes are stored in a second identical memory. During the display interval echoes of weak targets, sea clutter and rain clutter are painted at a lower brightness level, but echoes of strongly reflecting targets which cross the higher threshold are painted at a much higher brightness level.

Sometimes the threshold levels of *tracked targets* are *"adaptive"*, which means they are automatically adjusted depending on the intensity (amplitude) of the curve enveloping noise and clutter signals. Nevertheless the range between the levels, which is proportional to the dynamic range of the radar set, remains constant.

In this connection it should be mentioned that these modern displays in the analogue video signal stage use *logarithmic amplifiers* (for details see Appendix III) instead of linear ones. This is because the manual gain adjustment of linear receivers will affect all echoes. Logarithmic receivers do not require gain controls and therefore are simpler to operate under conditions when severe clutter is present. The reason is due to the shape of their characteristic input/output curve. The logarithmic receiver compresses the high level signals while enhancing weaker signals and does not saturate with normal signal levels as is the case with the linear receiver. An example is shown in Fig. 15.14.

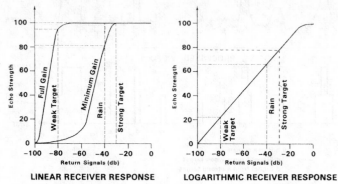

LINEAR RECEIVER RESPONSE   LOGARITHMIC RECEIVER RESPONSE

FIG: 15.14 LOGARITHMIC VERSUS LINEAR RECEIVER

The vertical scale of this diagram shows the relative output strength (volts) of the returning signal (echo); the units on the horizontal scale are in db (see Explanatory Note, Appendix IV). From left to right, although the scale is numerically decreasing, the input signal strength (voltage) is *increasing* from 10^{-5} db (20 log 10^{-5} = -100) to 1 (log 1=0). It simply means that the input signal strength is expressed in fractions of the full response range of the receiver, from minimum to maximum.

Suppose the signal strength of a rain squall at 5 to 6 miles range equals -40 db and a ship of -30 db signal is inside the squall. A small object nearby at -80 db is also present. At *full* gain with a linear receiver, the echo of the small object is clearly visible on the display, but the target inside the rain squall cannot be detected as both rain and target display the same brightness. At *minimum* gain the brightness of the echo of the target inside the rain—at saturation level—is slightly higher than the rain clutter which has now a brightness just below saturation level. Thus the ship's echo can be distinguished from the rain clutter. The echo of the small object, however, has disappeared.

With the logarithmic receiver, all three types of echoes would be clearly displayed on the screen at the same time.

With a standard read-out time of 150 micro-seconds (corresponding to the real time 12-mile range scale display), one might ask if the brilliance is now reduced on range scales longer than 12 miles. This would be the case but it can easily be overcome, either by stretching the echoes an additional number of micro-seconds, or by replacing the read procedure a second time during the interscan period. The latter arrangement is possible owing to the reduced PRF on the longer range scales and hence the far longer interscan period.

(b) *Time-base to time-base comparison.* This can be useful for *interference* and *clutter rejection.* Radar interference, noise and clutter are random in nature and will not often occur at the same time on successive time-bases. If the contents of the prime register, containing present echoes, is compared with the contents of a second register, containing echoes from the succeeding transmission, a comparison circuit can be arranged to accept only those signals which are present in both registers at the same range. Data which do not correlate are discarded leaving only echoes of bona fide targets (Fig. 15.15).

This scheme was already implemented in most conventional radar sets using delay line storage (similar to analogue shift registers), but the latter circuitry is quite complex compared with the system described above.

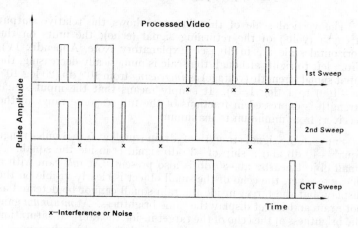

FIG. 15.15 INTERFERENCE REJECTION (SUPPRESSION)
PRINCIPLE

Fig. 15.16 shows a block diagram illustrating the write and read
cycle, including the two threshold levels.

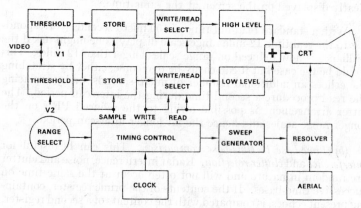

FIG. 15.16 VIDEO PROCESSED RADAR DISPLAY
BLOCK DIAGRAM

Tracker

We have already seen that in the quantized display, each target
is represented by a range and azimuth (relative bearing) number
(the Rho-Theta presentation). With gyro information and using
programme techniques (software) these numbers by transformation

can be converted into Y- and X-coordinates (northings and eastings). CPA and TCPA are *directly* derived from stabilized relative motion data in the tracker file. If the speed input of own ship is also taken into account, software inputs will be able to provide courses and speeds of targets. Activation of alarm signals for critical CPAs, TCPAs, lost echoes of targets, changes of targets tracks, over-loading of programme (and computer), etc., can also be programmed in : so can trial manoeuvres.

A very important part of the software is the *smoothing* or *filtering* procedure. This is similar to putting a mean line through a series of plotted positions on a paper radar plot. The traditional method is according to the $\alpha - \beta$ technique, using the following formulae in x- and y- coordinates.

$$G_n = F_n + a\,(P_n - F_n)$$

where G_n=smoothed position just after nth plot.

$$V_n = V_{n-1} + \beta/T\,(P_n - F_n)$$

F_n=forecast position just before nth plot.

$$F_{n+1} = G_n + V_n\,T$$

P_n=nth plotting position.
V_n=velocity estimate after nth plot.

$$\alpha = \frac{2\,(2n-1)}{n\,(n+1)} \ , \ \beta = \frac{6}{n\,(n+1)}$$

α, β=damping factors.
T=plotting interval.

Fig. 16.17 illustrates the procedure. (It is not the intention that students should follow the mathematics involved).

This approach is quite satisfactory, but sometimes it can be slow in detecting manoeuvres of targets. Kalman filtering would be better in this respect ; this is worked the same way as the $\alpha - \beta$ technique, but the derivations of α and β are a little more complicated. For example, $\alpha = \dfrac{\text{variance of } F_n s}{\text{variance of } P_n s + \text{variance } F_n s}$. In other words, the α and β factors are varied.
(for more details see ASWE Technical Report TR-72-14.)

Smoothing times are of the order of 10 to 20 scanner rotations depending on the range scale. For each of the targets tracked, a tracking "window" is opened at the position where the echo of the target is expected to be, based on past measurements of position (the F_{n+1} position). This simply means that a series of range measurements (from the clock counter) and bearing measurements (from the shaft encoder, for example) are recorded and passed to the tracker. In some ARPAs the joy-stick control operates two linear potentiometers. A small printed board mounted under the joy-stick control contains circuits to convert the analogue joy-stick signals into digital form. These signals are then routed to the matrix register(s). The window size is adjustable (included in the software package), starting large and shrinking down as smooth tracking proceeds. It depends on the distance between P_n and F_n, which gives an indication of the degree of confidence in the tracking procedure. If tracking becomes erratic or the target is lost, the window opens wide in an attempt to re-acquire the target.

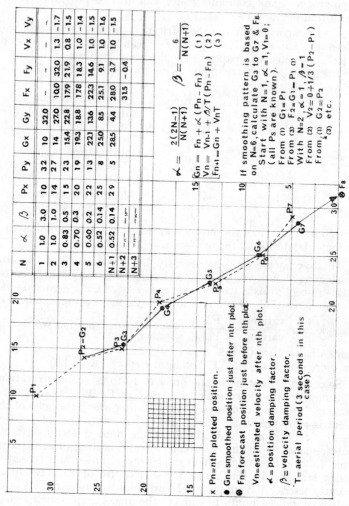

FIG. 15.17 SMOOTHING PROCEDURE

The distance and direction from P_n to F_{n+1} i.e. the vector $\overline{P_n F_{n+1}}$ will give an early indication if there is a change in the apparent motion of the target, and certain limiting conditions to this effect should be incorporated in the programme if Kalman filtering is not employed. The same vector will help "rate-aiding", i.e. forecasting ahead for a short period and then confirming that there is, in fact, an echo in or near the forecast position. This would reduce the likelihood of "target" swop and would assist in tracking a target whose response is intermittent due to the presence of rain or sea waves.

Target acquisition can be either manual or automatic. In the *manual acquisition* mode, the operator employs a joy-stick to position a tracking window over the echo of the target. To the operator this window is represented as a circle or ring whose diameter is about the same as the width of the tracking window which is approximately square. Tracking proceeds upon release of the joy-stick.

In the *automatic acquisition* mode, either the computer is aware of the position of most of the ship-sized echoes stored in the matrix (as many as 200 for a certain ARPA), or acquisition occurs when an echo with decreasing range, crosses the zone between two guard rings whose radial depth is the same as the depth of the tracking window. In the latter case, it is often found that the maximum acquisition and tracking capacity amounts to 20 targets (as per IMO Specification) and when this amount is exceeded some targets of lesser importance have to be dropped. A selection procedure or priority programme has then to be incorporated in the software package. A procedure, sometimes used, showing the various grades of priority, is illustrated in Fig. 4.2. For the Sperry CAS II display, the PAD distribution affords easy priority decisions by the operator himself.

Acquisition occurs when signals are detected on successive time-bases of transmission cycles, at which time a tracking window is placed where the signal was detected. This is achieved by transferring the contents of each register to the next one during successive interscan periods, leaving the prime register cleared for the next transmission. A comparison is then made of the contents of the same cell in N registers; if there are M or *more* than M "ones" in the same cell, a cell in the same position in another register, known as "hit" register is then set to "one". This arrangement prevents the acquisition of false echoes and also affords a chance for the weaker echoes to be recorded.

The micro-processor board responsible for this extraction process is known as the M.O.O.N. (M out of N) Board. For any value of N there exists an optimum value of M, but it appears that the latter is not dependent on the signal-to-noise ratio. For example with N equal to 10, M could be 6.

The M.O.O.N. Board is also responsible for placing the "eyebrows" over the echo, a confirmation that the target has been acquired.

A diagram of the tracking loop is shown in Fig. 15.18.

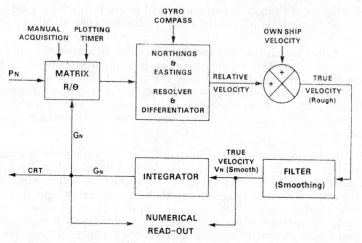

Pn , Gn , Vn : See text or Fig. 15.17

FIG. 15.18   TRACKING LOOP

PPCs

Point of Possible Collision or PPC. This point is defined as the point at which a collision could take place if the target (ship) maintains course and speed, and own ship maintains speed and has her course shaped in such a way that the PPC lies on the heading marker. The PPCs are points of great importance and the tracker compiles their positions and their movements on the PPI each radar scan. These movements can be quite complex, needing programmes which are much more complicated to perform than the relatively simple calculations of CPA, TCPA, course and speed, etc. The operator himself often becomes puzzled in the interpretation of these movements, so more attention will be paid to PPCs in this section.

In the special case in which the target is on a collision heading with own ship, the subsequent motion is entirely predictable. The PPC will move down the heading marker with the exact negative of own ship's motion, i.e., one heading marker segment every 6 minutes.

Before reading this section, it is worthwhile studying that part of Chapter 13 where the construction of PPCs is outlined, in particular Figs. 13.12, 13.13 and 13.14.

Some General Remarks about PPCs

PPC(s) are always positioned on the target's projected track.

With a target on a collision course, the PPC, coming closer, moves along the heading marker. If a secondary PPC is in existence, than its movement will be along a steady bearing line ($V_o/V_t < 1$).

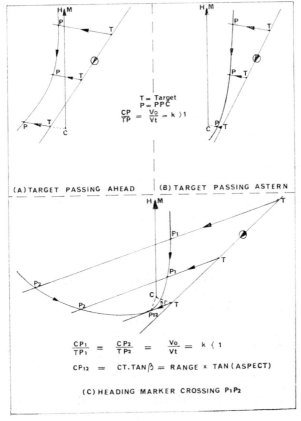

T = Target
P = PPC
$$\frac{CP}{TP} = \frac{V_o}{V_t} = k > 1$$

(A) TARGET PASSING AHEAD   (B) TARGET PASSING ASTERN

$$\frac{CP_1}{TP_1} = \frac{CP_2}{TP_2} = \frac{V_o}{V_t} = k < 1$$

$$CP_{12} = CT.TAN\beta = RANGE \times TAN (ASPECT)$$

(C) HEADING MARKER CROSSING P_1P_2

FIG. 15.19 (a), (b), (c)   MOTION OF PPC(s)

With the target *not* on a collision course, and maintaining its course and speed, the PPC(s) on own ship display will never cross the heading marker. The motion(s) of the PPC(s) on own ship display are illustrated in Figs. 15.19 (a), (b) and (c) for the following cases respectively : Own ship crosses astern of a target (target passes ahead); own ship crosses ahead of a target (target passes astern) ; own ship's heading marker crosses the line connecting two PPCs.

Note the passing-astern and passing-ahead sectors (Fig. 15.20) with two PPCs. At one PPC, if reached, own ship's head would be pointing *towards* the target, while at the other PPC, it would point *away* from the target.

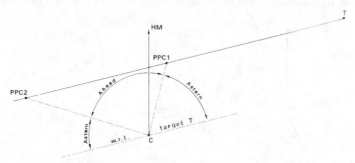

FIG. 15.20   PASSING-ASTERN AND PASSING-AHEAD SECTORS FOR OWN SHIP W.R.T. TARGET

The actual position(s) or non-existence of PPC(s) depends on (a) the speed ration k i.e. $\dfrac{V_o \text{ (Own ship)}}{V_t \text{ (Target)}}$ and (b) the aspect.

(a) *Speed ratio* (k)

(i) If own ship is assumed to have a constant speed, then for a stationary target, k would be infinite and the PPC is at the target's position (Fig. 15.21). With k>1 and the target on a constant course and speed, the PPC will have moved ahead of the target. When k=1 (equal speeds), a secondary PPC appears at infinity and with k<1 and decreasing, the secondary will move towards the primary more rapidly than the primary moves towards the secondary. With a further decreasing value of k the two PPCs will eventually merge at a single point on a line which is at right angles to the sight-line (the line connecting the two ships). If k drops down below this critical value, the single PPC will disappear. Collision is no longer possible because the speed of own ship is so slow compared with that of the target that she will never be able to reach it.

Similarly with k<1 the two PADs will merge, each end finally indicating a target cross-ahead (own ship passing astern) situation only.

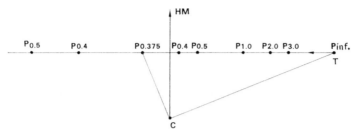

K — Vo/Vt
T — Position Target for different Ks
Pk — Position PPCs for different Ks
For K < 0.375 PPC will disappear ($\angle$ TCP = 90°)

Fig. 15.21   Effect of Speed Ratio on PPC(s) movement

(ii) The reverse process will take place for different speeds of own ship and constant course and speed of the target. With k=0 (own ship being stopped) there is, of course, no collision point in this case (Fig. 15.21). With increasing k a PPC will appear, splitting into two, with one moving at a relative fast speed away from the primary collision point, and the target, and finally disappearing at infinity.

Of course, with Own ship stopped and the target's aspect equal to zero, the PPC will be at the Own Ship's position. A PAD will be centred around 'Own Ship' if the aspect is less than Arcsin
$$\left[\frac{\text{(selected CPA)}.}{\text{range}} \right]$$

(b) *Aspect*

The PPC loci are shown in diagrams 15.22 (a), (b) and (c), for values of k>1, <1 and equal to 1.

The loci are circles with their centres situated on the line, or line produced, connecting own ship and target T, and with radii equal to Vo/(k–1) when k>1, and Vo/(1–k) when k<1. With k=1 (equal speeds) the radius becomes infinite and the *circles degenerate into a straight line bisecting the sight-line.*

(i) *With k>1*, the centre of the locus is situated beyond the target position at a distance kVo/(k–1) from own ship. The target is placed *inside* the circle and *one* PPC is possible (it also depends on the *value* of k as seen before) for aspects ranging from 000° to 360°. There exists, however, a *limiting heading* for own ship, making a relative bearing of arcsin (1/k) with the sight-line. With own ship

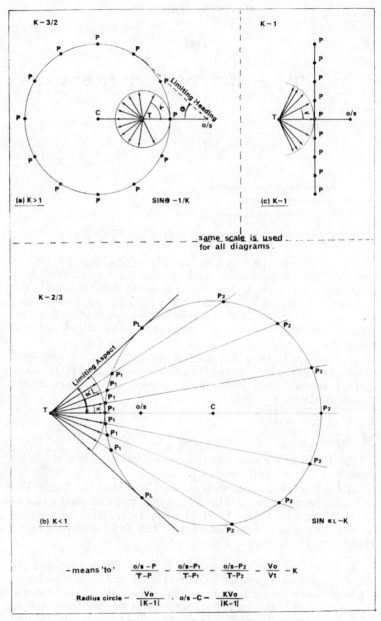

FIG. 15.22 (a), (b) and (c)
EFFECT OF ASPECT ON PPC(S) MOVEMENT FOR DIFFERENT SPEED-
RATIOS

heading with the target on a greater relative bearing, no collision is possible and the PPC remains on one side of the bow for all aspects of the target. If own ship has a heading inside the limiting angle, the PPC will move across the heading marker as the aspect of the target is changed. With increasing k (slower target's speed) this limiting relative bearing decreases.

(ii) With $k<1$, the centre of the locus is situated in a direction from own ship opposite to the target's direction at a distance $kVo/(1-k)$ from own ship. The target is now placed *outside* the circle and *two* PPCs are possible (for a suitable k value) when the target's track intersects the circle at two points. PPCs are possible for all headings of own ship, but there is a *limiting angle* for the *target's aspect*, namely arcsin (k) and with a greater aspect no PPC can exist. With k decreasing (faster target's speed), the limiting aspect angle will also decrease.

(iii) With $k=1$, the PPC will move along a straight line, bisecting the sight-line moving further away as target's aspect increases. Theoretically the limiting aspect would be 90° with the PPC at infinity.

Errors and Precautions

Errors can be divided into groups :—

1. Resulting from inaccuracy of the sensors.
2. Generated in ARPA itself due to imperfections in hardware or software. If the latter is the case, it might have something to do with (1).
3. Errors in interpretation of the actual display.

1. **Resulting from the Sensors.** These are already itemized in the IMO ARPA publications, Appendix 3 and will be briefly mentioned again. Their errors and standard deviations are relatively small.

(i) *Bearing Errors.* These are due to :—

(a) *Target glint.* It is not always known exactly which part of a target yields the strongest reflection. To a certain extent it depends on the aspect of the object.

(b) Some *backlash* in the *aerial* drive gear.

(c) *Rolling and pitching.* This gives rise to a quadrantal error, maximum on relative bearings of 045°, 135°, 225° and 315° with the minima in between. It is due to the angular tilting motion of the scanner. Superimposed on this quadrantal variation is a sinusoidal wave form caused by the lateral displacement of the scanner position.

(d) *Beam shape in the horizontal plane.*

(e) *Quantization in azimuth.*

(ii) *Range Measurement Errors.* These result from :—

(a) *Target glint.*

(b) *Rolling and pitching* causing lateral displacement of the scanner position.

(c) *Pulse-length, echo-shape and strength* (associated with pre-set threshold levels).

(d) *Quantization in range.*

(iii) *Course Input Errors.* These are caused by gyro-compass deviations and will affect tracking accuracy if their time-constants equal those of the tracker filters.

(iv) *Speed Input Errors.* These are caused by log errors and can become important. They affect course and speed calculations of the target and display true motion vector errors and predicted relative motion vector errors when using the 'Trial Manoeuvre' facility. Range, bearing, CPA and TCPA values are *not* affected.

2. Errors generated in ARPA itself

(i) *Smoothing Errors.* Especially, owing to rolling and pitching errors (a combined effect of scanner movement and gyro-compass errors) slight changes in vector quantities and digital read-outs are continuously taking place for all targets in rough weather. It should, however, be remembered that a targets' velocity vector, even under ideal conditions, is always subject to slight changes, depending on type of steering facilities employed, weather and ship's parameters.

When own ship or the target ship change their velocity vectors, smoothing will oppose the change and true velocity information of targets (vector and digital read-out) becomes unreliable. Some ARPAs stop tracking during these periods. The reason for this is that in most ARPAs, calculations are based on the relative motion velocity vector. In one particular ARPA, however, position and velocity of tracked targets are *stored* in true motion format, so that true motion vectors of targets do not need to be re-established after a change in relative motion.

A very good *check* on the smoothing process is the study of the *past history track*.

At very short range, due to rapid bearing changes, the tracker may lose the target.

(ii) *Computer Calculation Errors.* These are nearly always due to course and speed input errors.

  (a) *Effect on True Motion Velocity.* It has already been seen in section (1), that—with certain exceptions—the operator's true motion vectors do not possess the same reliability as relative motion vectors.

  (b) *Effect on Predicted Relative Motion Vectors.* As in (a) errors are introduced when course and speed input is not correct.

  (c) *Effect on PPC.* The effect of *speed input error* is illustrated in Fig. 15.23 (a) and (b). It is reassuring to see that, with the ships on collision courses, the PPC, although displaced, will remain on the heading marker.

On the assumption that the heading marker is correctly aligned, course input errors do not affect the PPC positions with respect to the heading marker. However, picture and heading marker will be disorientated inside the tube ; a correction has to be applied to obtain the true course to avoid a PAD.

(iii) *Vector Jumping*

  (a) This may occur when targets are close to each other and their two echoes are in the same tracking window. The two vectors may interchange and so will the digital information (*target* information *swop*) or sometimes they combine or, when in manual acquisition mode, one target may lose all its information while the other target may yield data for the first time, but they are the wrong data.

   Target swop should be overcome by "rate-aiding", the forecast of the target(s) predicted position ahead of the echo during the next scan (so that the proper vector can be drawn if the position is later confirmed), and by making the tracking window as small as possible after the initial acquisition.

  (b) It can also take place that while in automatic acquisition mode *false echoes* are received due to *side-lobe effect* or *indirect reflection* via superstructures on own ship. The remedy is to switch over to the manual acquisition mode or to put into action a minimum tracking and/or acquisition range.

K

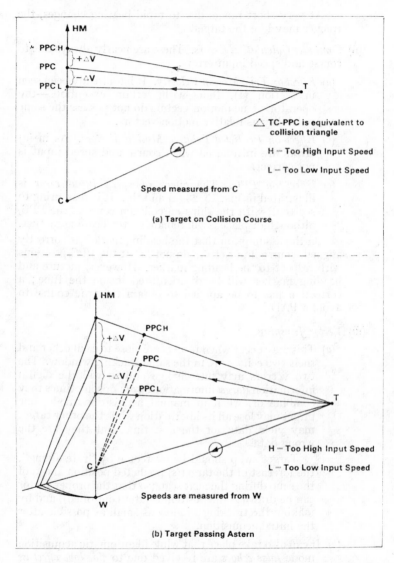

FIG. 15.23 (a) and (b)   EFFECT OF SPEED ERRORS ON PPC

(iv) *Spurious information owing to acquisitioning of rain and sea clutter echoes and to tracking information of land-based objects.*

This can happen while using the automatic acquisition mode. Not only does the observer get far too much unwanted information, it will also make the radar picture confusing to look at. Lastly it may saturate the tracking capacity of the computer and some of the targets may be dropped or ignored even though they are important to the observer.

In these cases one should go back to the manual acquisition mode or apply acquisition restriction for a minimum desired range and use the Area Rejection Boundaries or Zones (ARBs or ARZs).

Use of a 10 *cm. ARPA display* can be recommended to prevent computer saturation due to rain echoes (but keep on consulting a 3 cm. display if small targets can be expected nearby), although risk of target swop is increased as ships' echoes are "fatter".

(v) *Interrupted tracking of targets, loss of targets or even non-detection of targets.*

This will happen with low-level thresholds having been set too high. One may have to ask for technical advice, and in this connection it is wise to remember that with ARPA-navigation *consultation of a raw radar display should never be neglected.*

3. Errors in Interpretation

(i) *Misinterpretation of Display Presentation and Vector Mode.*

The combination of different display and vector (plus eventual history tracks) are so many that *mistakes are easily made in interpretation.* Sometimes spring-loaded switches are provided for certain vector modes and this can be helpful.

A brief guide is now provided :—

First the display presentation :—

(*a*) Relative Motion North-up Stabilized.
(*b*) Relative Motion Course-up Stabilized.
(*c*) Relative Motion Head-up Stabilized.
(*d*) True Motion North-up Stabilized.
(*e*) True Motion Head-up Double Stabilized.

Combined with the vector and "Past (History) Track" mode, the following Table is obtained.

ARPA	Display Presentation			
Data	Relative Motion (RM)		True Motion (TM)	
Vectors *or* Past Track	RM	TM	RM	TM
Vectors *and* Past Track	RM ╱╲ RM [TM RM] TM		TM ╱╲ RM [TM RM] TM	

[   ] Excluded.

TABLE 15.4

In the True Motion vector mode, using a Relative Motion display, a vector will be attached to the point representing own ship although the point remains stationary on the radar screen. Note also that in some cases the past track does *not* coincide with the afterglow (for example TM past track on a RM display).

(ii) *Misinterpretation of the Trial Manoeuvre (Simulation).*

Here, also, the type of display presentation has to be appreciated. With static simulation, showing the predicted situation immediately after the manoeuvre, it seems best to use a Relative Motion Display with Relative Motion vectors of moderate length. With dynamic simulation, showing the predicted developing situation up to thirty minutes *after* the manoeuvre has been carried out, it will be better to have a True Motion Display, for good understanding, plus Relative Motion vectors (if possible). Although "Simulation" will give guidance for a predicted safe manoeuvre, the observer should keep the "Rule of the Road" in mind, especially Rule 19, during poor visibility. The former prediction, which is based merely upon the other vessel keeping her course and speed, may clash with the latter requirement.

(iii) *Misinterpretation of the Input Speed (Velocity).*

In open sea the input speed to ARPA is generally manual sea speed or one-axis "water-locked" speed. In calm water— which is often the case during fog conditions—one can be reasonably certain from the true motion vector what the target's aspect will be. Near the coast or in estuaries, it is often advisable to use the *"Auto-Track"* or *"Echo-Reference"* facility (Chapter 4) if these are available. The true motion vectors will then show the ground velocity giving a good idea where the ships are *going* to (this arrangement, under restricted visibility conditions does *not* clash with Rule 19). This facility can be used with a True Motion *or* a *Relative* Motion Display.

Whatever the speed input, one must make certain what the type is—sea or ground speed, one-axis ; sea or ground speed dual axes (sea or ground velocity)—to appreciate the meaning of and to understand the interpretation of the true motion vectors. Also, during rough weather, one should realise that some vessels will have wind drift (leeway) superimposed on their directed motion and their real aspect may differ from the one shown on the display or read out digitally. Error in the speed or velocity input does not affect the accuracy of range, bearing and RM past track.

(iv) *Misinterpretation of Display Symbols.*

It is a pity that symbols (and the same is true for display controls) are not standardized, and that different manufacturers use different symbols (circles, triangles, squares, diamonds etc.) for the same message. Putting it in a different way : the same symbol on different ARPA often has different meanings. For example, depending on the ARPA make, a square symbol may indicate "Acquired", or "Stationary Target" or "Passing within the set CPA distance".

(v) *Misinterpretation of Data on Display which are using Points of Possible Collision (PPCs) and Predicted Areas of Danger (PADs).*

(*a*) The line connecting the echo (often emphasised by an "eyebrow") to the PPC, on its own, gives no indication of the speed of the target.

On the Sperry CAS display, heading marker and electronic bearing cursor consist of interrupted lines, and each short line and the length of the gap between the short lines represent the distance covered by own ship in six minutes (in addition the curzor has one-minute sub-divisions). With a change in speed the length of the lines will vary accordingly. The echo-PPC line has a fixed 6-minute vector superimposed on it. By comparing the length of this vector with one of the "chained" sub-divisions of heading marker or curzor, one is able to estimate the speed of the target vessel.

If the above arrangement is not provided (as in some other makes), one can measure the distance by means of cross-index from the spot representing own ship to the PPC and from the target's echo to the PPC. The ratio of these distances equals the speed ratio of own ship-to-target ship.

(b) The area sizes of the PADs are *determined by the operator*; it gives an indication of how far the operator wants to stay away from the target (minimum condition). A tangent line from the spot representing own ship to the PAD ellipse or hexagon does *not* give an indication of the present CPA.

This line indicates the heading to steer (ma; ntaining speed) to achieve the least *desired* CPA distance. CPA and TCPA should be read out on the window display in numerical form.

Even after altering course so that the heading marker coincides with the tangent line defined above, it is not possible from the display to estimate accurately the TCPA.

(c) The PPC is inside the PAD, but *not* usually in its centre. It is, however, close enough to accept the geometrical centre of the PAD as the PPC. But note that for faster targets there can be a PAD without a PPC.

(d) The PAD area does not change symmetrically with an operator's change in intended miss distance.

(e) The echoes of two target ships which are themselves mutually on collision courses, may have their PPCs widely separated on own ship's display. The reason for this is that eventual interception of each with own ship would take place at *different* times. In other words the distribution of the PPCs on own ship's display does not indicate either the existence or non-existence of collision dangers between other vessels, unless their PPCs actually overlap. In the latter case, it is an emphatic indication to Own ship to avoid the area where overlapping occurs.

Summary

In this Chapter a wide range of plotting devices has been discussed. On the one side of the spectrum is the simple plotting sheet—still used by many observers—and on the other side there is the ARPA, a sophisticated plotting aid, which, gradually, will be introduced on all ships of the medium and large tonnage class.

IMO has already adopted a Resolution on the *"Minimum Requirements for Training in the Use of Automatic Radar Plotting Aids (ARPA)"* which starts with the paragraph :—

Every master, chief mate and officer in charge of a navigational watch on a ship fitted with an automatic radar plotting aid shall have completed an approved course of training in the use of automatic radar plotting aids.

The contents of this course is published in IMO ARPA Publication, Appendix II, which one can find reproduced at the back of this book. Recently the Merchant Navy Training Board has issued a booklet, entitled "Training in the Operational Use of Automatic Radar Plotting Aids", which contains a course specification which is based on the IMO specification.

It is worthwhile reading through the specification ; two short sections are quoted below.

1. *The possible risks of exclusive reliance on ARPA.*

Appreciation that ARPA is only a navigational aid and that its limitations including those of its sensors, make exclusive reliance on ARPA dangerous, in particular for keeping a look-out ; the need to comply at all times with the basic principles and operational guidance for officers in charge of a navigation watch.

2. *Manual and automatic acquisition of targets and their respective limitations.*

Knowledge of the limits imposed on both types of acquisition in multi-target scenarios, effects on acquisition of target fading and target swop.

Reading through these it seems that raw radar displays, including a 10 cm. set, will remain as desirable and valuable aids.

CHAPTER 16

USE OF RADAR FOR ANTI-COLLISION

Timely and Intelligent Use of Radar and Radar Information

Radar for the Merchant Service is designed for what is known as " Surface Warning " and for anti-collision purposes its main use was during reduced visibility. With the faster ships, radar, however, is also now extensively used in clear weather as an extra aid for the look-out man. With the range-scale on 12 miles and the electronic centre off-set, strong echoes of ships can be detected up to 16 miles for an average bridge height and at a time when the ship has not yet been sighted visually. Another reason for earlier detection by radar is that the white echo pip against the dark background is often more conspicuous than the appearance of a ship against a grey sky and seas. By placing the curzor over the echo, a timely check can be kept on the bearing change.

When fog-banks are expected the radar set should be at least on " Stand-by " during daytime, making it ready for immediate use; but at night the set should be left on "Transmit ", as the vessel could well be steaming along near a fog-bank which is giving no visual indication of its presence. When approaching a fog-bank Rule 35 (Sound Signals in Restricted Visibility) must be adhered to, and radar must be used to see what is inside the fog-bank. Failure to employ the radar in such a case contravenes Rule 2 and blame accordingly has been attached to ships which did not comply with this rule near a fog-bank. Upon approach of the fog-bank, radar watch routine should be started, and inside the fog-bank, the observer should realize that some echoes on the screen might represent ships which are not in fog and may not exercise the same caution as his own ship. Unexplained manoeuvres by other vessels as observed from the radar screen might indicate the existence of a small vessel or vessels undetected by the ship's own radar, and a close watch should be kept on the suspected area.

Shipmasters have been blamed for not keeping a proper " look-out " because they were not using their radar on clear nights to detect the presence of the unlit oil-rigs with which their vessels collided.

It may, therefore, be said that it is always good practice, especially for fast ships, to keep the radar working. This also offers an opportunity to the officer of watch to maintain his plotting expertise, which is so important in cases when the visibility deteriorates and plotting becomes really essential. *At night, when in a region where fog-banks and/or unlit obstructions can be expected, the radar must be in continuous operation.* Previous Court Cases, by the way, have stressed that a shipmaster is considered to be at fault for not using radar provided for his ship and also for allowing the radar installation on his ship to remain in a defective condition for a prolonged period. At present the U.K. Government has made radar compulsory for all British ships of 1600 gross tons or more, following an IMO recommendation that at least one radar must be fitted in ships of 1600 g.r.t. or more, and at least two radars must be fitted to all ships of 10,000 g.r.t. or more, each capable of operating independently of the other.

Some important points to be kept in mind when using radar in reduced visibility, are the following:—

(*a*) The setting of the anti-clutter control on raw radar displays. Adjust, if possible, in such a way that echoes can be traced near the spot representing own ship. Be aware of over-suppression, as this will wipe off most of the echoes of ships nearby.

(*b*) The existence of blind and shadow sectors caused by objects on the ship itself. A slight " weaving " around the course is recommendable in such a case.

(*c*) The selection of range scale, taking account of:
   (i) The speed of own ship (the faster the ship, the greater the range scale).
   (ii) The accuracy of bearing and range observations (shorter range scales with the echo in the outer half of the screen yields an increased accuracy).
   (iii) The length of the " tadpole " tails (the shorter the range scale, the longer the tails). If a True Motion Display is available, this might entail off-centring the time-base and use of the Zero-Speed switch.
   (iv) The possibility of encountering small craft or ice growlers (easier discernable on the shorter range scales and, if possible, with long pulse selected).
   (v) The number of ships in the vicinity of own ship (a long-range scale can produce a confusing array of closely-packed echoes).
   (vi) The range at which most merchant ships are first detected (generally about 10–15 miles).

In addition to the above considerations, it should be remembered that a plot on a reflection plotter mounted on a True Motion display will become distorted if the range scale is altered during the plotting interval (see Fig. 14.5). This problem does not arise with Relative Motion radar.

Summarizing on this problem of selecting a range scale, it *is generally best to relate the range-scale used most of the time to the vessel's speed*—changing to shorter range scales now and again to obtain more accurate observations of bearings and ranges of any nearby objects and also to conduct a search for smaller objects.

(*d*) The obeyance of Rule 35 (Sound signals in restricted visibility) even if the screen is free from echoes on the longer range scales and one knows that the set is fully efficient.

(*e*) The obeyance of Rule 34 (*a*) (Manoeuvring signals) *only* when the other vessel is *in sight*.

Radar and the Collision Regulations

Under Part B (Steering and Sailing Rules) there are two sections which have a special bearing on this Chapter. These are Section I and Section III. The former deals with the conduct of vessels in *any* condition of visibility (Rules 4, 5, 6, 7, 8 and 10); the latter (Rule 19) discusses the conduct of vessels in *restricted* visibility.

Turning our attention to Section I first, it will be seen that **Rule 5** *specifically* deals with the importance of maintaining " *a proper look-out by sight and hearing as well as by all available means appropriate in the prevailing circumstances and conditions so as to make a full appraisal of the situation and of the risk of collision*". The word "specifically" is stressed because in previous Regulations this came under "the ordinary practice of seamen".

The inclusion of " as well as by all available means " refers obviously to a radar watch (the use of guard rings on the radar display will be helpful in this connection), but it also incorporates a V.H.F. R/T watch and the words "full appraisal" may be taken to include proper radar plotting procedures and active V.H.F. Radio-Telephony communication. Although this section deals with clear weather conditions *and* conditions of restricted visibility, the master is given some latitude in making use of radar and R/T information by the addition "appropriate in the prevailing circumstances and conditions".

Rule 6 introduces a new concept, namely *Safe Speed*. When, about 30 years ago radar was introduced on board ships, one of the greatest difficulties with which Mariners were confronted, was the term " Moderate Speed ". What, in fact, was a moderate speed using radar? A concise answer was not possible. It could be argued that a moderate speed, using radar, could in some cases mean " Full speed with engines on Stand-by ", but in other cases could mean a slower speed than a Mariner without radar might consider " moderate ". From the legal and philosophical point of view these arguments are quite correct but in the literary sense they are unsatisfactory.

A Safe Speed as defined in the 1972 Rules is not based only on the state of visibility (as in the 1960 Rules) but " Every vessel shall *at all times* proceed at a safe speed so that she can take proper and effective action to avoid collision and be stopped within a distance appropriate to the prevailing circumstances and conditions ".

Besides the state of visibility (i) the following factors should be taken into account in determining a safe speed:—

(ii) " the traffic density, including concentrations of fishing-vessels or any other vessels;"

(iii) " the manoeuvrability of the vessel with special reference to stopping distance and turning ability in the prevailing conditions;"

The manoeuvrability depends on the stern power of the vessel, the number and type of screws, the provision of a bow-thruster, the size of the ship and her loaded condition while the prevailing conditions are mainly governed by the wind and wave directions, wind force and wave height, and current and tidal conditions.

(iv) " at night the presence of background light such as from shore lights or from back scatter of her own lights;"

(v) " the state of wind, sea and current, and the proximity of navigational hazards;"

(vi) " the draught in relation to the available depth of water."

These factors, which determine a safe speed in general, are applicable to all ships. Vessels which use their radar need, in addition, take the following conditions into account:—

(i) " the characteristics, efficiency and limitations of the radar equipment;" The age and reliability of the equipment, the number of radars and displays, interswitching facilities, types of display presentations, plotting devices and facilities for automatic plotting etc., are all factors to consider.

(ii) " any constraints imposed by the radar range scale in use;" A constraint may be imposed on a particular range scale owing to strong radar or electrical interference, or for a very fast vessel the use of the 12-mile range scale (the 24-mile range scale is too small for effective plotting on a reflection plotter) for echo observation, might compel her to reduce speed.

(iii) " the effect on radar detection of the sea state, weather and other sources of interference;" Excessive " noise " due to wave, sea, rain-drops, snow crystals, other ships' radar pulses or electrical interference may swamp the signal and essential information may be lost.

(iv) " the possibility that small vessels, ice and other floating objects may not be detected by radar at an adequate range." This " possibility " can be a result of atmospheric conditions such as sub-refraction or it might be caused by a small reflection coefficient of the object.

(v) " the number, location and movement of vessels detected by radar;"

(vi) " the more exact assessment of the visibility that may be possible when radar is used to determine the range of vessels or other objects in the vicinity ".

In addition, we may say that the number of men for keeping radar watch and a plot, *and* their efficiency could influence the master's opinion about what is, or what is not a safe speed.

Rule 7 deals with the " Risk of Collision ". The Rule stresses again the use of " *all available means* appropriate to the prevailing circumstances and conditions to determine if the risk of collision exists ". This includes the listening to V.H.F. R/T messages of other ships and shore radar stations, but no guidance is given about actual active participation.

There is a very important last sentence in the first paragraph: " If there is *any doubt* such risk shall be deemed to exist." This might remove the possible element of indecision in a radar encounter.

Rule 7 (*b*) stresses the importance of making *proper* use of radar equipment, including early warning of collision risk on the longer range scales. It furthermore emphasizes the practice of radar plotting or " equivalent systematic observation of detected objects " (recording in writing and tabulation, automatic plotting aids).

Rule 7 (*c*) states : "Assumptions shall not be made on the basis of scanty information, especially scanty radar information." The omission of a plot, an incomplete plot or a plot based on an insufficient number of observations, in short, the determination of the position of another vessel without finding her movement, might be termed as "scanty". ARPA provides a solution here.

The last paragraph of Rule 7 states how risk of collision can be obtained from *compass* bearings and gives a warning that an appreciable change in bearing does not always indicate a safe passing (large vessel, or a tow, or a ship at close range; see also Fig. 14.4).

Bearings should be recorded as compass bearings and not as relative bearings as is so easily done on an unstabilized display. It is not possible to compare relative bearings when own ship is subject to yaw or makes alterations of course, and often the Master has been led to believe, that, by making a small alteration of course, the situation improved because the relative bearing changed and he did not realize that the change in the relative bearing was mainly due to own ship's alteration of course. If he had converted the relative bearings to compass bearings, he would have noticed that danger of collision after the alteration had become greater instead of less.

Rule 8 is headed "Action to avoid Collision".

Paragraph (*a*) states that, if the circumstances of the case admit, any action shall " be positive, made in ample time and with due regard to the observance of good seamanship ". The word " positive " in this connection means "effective" and bears no relationship to the conventional adoptations " positive and negative actions ", mentioned in certain papers about collision-avoidance (more about these later).

Paragraph (*b*) is an extension of paragraph (*a*), stating that " Any alteration of course and/or speed to avoid collision shall, if the circumstances of the case admit, be large enough to be readily apparent to another vessel observing visually or by radar; a succession of small alterations of course and/or speed should be avoided". The Rule requires substantial action in order *to make clear one's intention* to all vessels in the neighbourhood (" another vessel " is not necessarily the vessel for which avoiding action was taken) *both in clear weather as well as in fog*. This requirement should be kept in mind when an agreement is reached about collision-avoiding tactics between two vessels via V.H.F. R/T and also when using the 'Trial Manoeuvre' facility on ARPA.

The remainder of the Rule (paras. (*c*), (*d*) and (*e*)) emphasizes that an alteration of course, provided there is sufficient sea room, may be the most effective action to avoid a close quarters situation on condition that it is made in good time, is substantial and does not result in another close-quarters situation. It stresses the safe passing distance and warns that effectiveness of the action shall be carefully checked until the other vessel is finally past and clear. If necessary, or to allow more time to assess the situation, a vessel shall slacken her speed or take all way off by stopping or reversing her means of propulsion.

In short, what this Rule is saying is that if avoiding action for another vessel is going to be taken such action should be bold both in clear weather and in conditions of restricted visibility so that the intention of the vessel taking the action becomes readily apparent to other vessels in the vicinity. Seen in this light, an alteration of course is generally more effective than an alteration in speed.

There are some contributory factors for substantial action to be taken where radar navigation in fog is concerned. The *first*

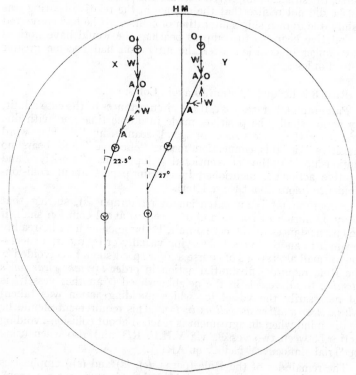

X altered course 45°. WA=WO . Change in OA
one-half.

Y altered course 90°. WA= ¹/₂ WO. Change in
OA about one-third.

FIG. 16.1 CHANGES IN THE DIRECTION OF THE
RELATIVE MOTION LINES ARE MUCH SMALLER THAN
THE COURSE ALTERATION CARRIED OUT BY THE TARGETS

factor is that for a collision encounter between two vessels meeting end-on or crossing—and each forward of the other's beam—an alteration of course or speed by one of the vessels shows up far less pronounced in the relative track (direction or rate) on the other vessel's display or plot when a *relative* presentation is used than if a true motion presentation were used. This is understandable when one remembers that the relative motion line is produced as the result of two vectors, of which only *one* is changed in this case.

Fig. 16.1 shows the effect of an alteration of course by another vessel on the relative motion display of own ship. The vessel on the port bow is on an opposite course and has the same speed as own ship. If she alters 45 degrees to starboard, then the OA line will change its direction by $22\frac{1}{2}$ degrees. The vessel on the starboard bow, also on a reciprocal course, but having a speed half of that of own ship, will cause a change of 27 degrees in the OA line if she makes an alteration of course of 90 degrees. It is quite obvious that the change in the direction of the OA line depends on the ratio of the ships' speeds. If the speeds are the same, then the change in the direction of the relative motion shown on own ship's plot is about one-half the alteration of course of the other ship.

The reverse is also true. If our own ship, for example, makes an alteration of course of 30 degrees, another crossing vessel, with approximately the same speed, forward of the beam, involving risk of collision, will observe a change in her relative motion line of about 15 degrees. To make, therefore, a course alteration—and this holds also for alterations in speed—readily apparent and on the assumption that other vessels in the vicinity use a relative motion display presentation, a substantial alteration is required by own ship.

The *second* reason for making substantial alterations is that errors in plotting and a wrong estimation of the direction of the relative motion line can easily take place (Chapter 14) especially when the display is unstabilized. The observer may, for example, conclude that the other vessel is on a collision course or will be passing on her port side while, in fact, the other ship, if she maintains her course and speed will be passing on her starboard side. If, in this case, own ship makes a *small* alteration to starboard, then, instead of improving the situation, the nearest approach between the two vessels will become even smaller. If later on, own ship makes a second alteration to starboard, and perhaps even a third one, then this may lead to collision. This type of action whereby one ship makes a succession of *small alterations of course* has become known as *The Cumulative Turn* and the majority of collisions in fog have been caused by this type of action. See Fig. 16.2.

In many of these collision cases, while one ship carried out the cumulative turn, the other vessel maintained her course and speed, simply because she had not detected the effect of the turn on her display. Although, for her, the bearing opened out, it did not open out *sufficiently* (Rule 7 (*d*) (ii) and Fig. 14.4) and in the final stages of the encounter it became steady. Hence the warning in Rule 8 (*b*) against a succession of small alterations of course and/or speed.

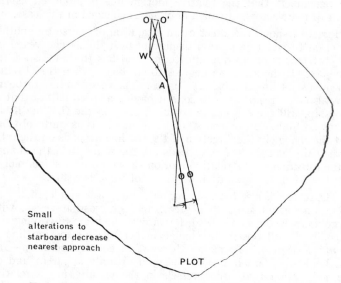

Small alterations to starboard decrease nearest approach

PLOT

FIG. 16.2   SMALL ALTERATIONS OF COURSE CAN BE VERY DANGEROUS

Rule 8 (*c*) states that if there is sufficient sea room, alterations of course *alone* may be the most effective action to avoid a close-quarters situation. The sea room, however, could be restricted by navigational dangers or a fair amount of traffic with ships crossing from different directions. In such cases substantial alterations of course may not be possible and may make the sitation even more dangerous when ships which, before the alteration, were not on collision courses, may after the alteration, involve a risk of collision. If it becomes necessary to avoid collision, a substantial reduction in speed must be made (Rule 8 (*e*)).

Paragraph (*d*) of Rule 8 emphasizes that the effectiveness of any action shall be carefully checked until the other vessel is finally past and clear. During restricted visibility and using one's radar, this means that after an alteration of course and/or speed, observations of the target should be taken at frequent intervals to see if the echo follows the predicted track at the predicted rate. If the echo deviates from the predicted track *away* from the centre of the plot, then the other vessel has taken action contributing to safety, but, if on the other hand the echo deviates from the prediction line *towards* the centre of the plot, then the other vessel has taken action which has cancelled out or partly cancelled out our avoiding action and in the majority of cases, the wisest thing to do is to reduce the speed of own ship to a minimum at which she can be kept on her course, or to alter course and to put the other ship right astern. On ARPA the history track should be watched for alterations of course or/and speed of targets.

Finally, Rule 8 makes a mention of " close-quarters situation " and " safe passing distance ". These concepts cannot be concisely formulated. It is obvious that their extent depends on many factors, such as weather conditions, state of visibility, type of vessel, manoeuvrability and if observations are carried out by visual means or by radar—which is far less discriminating than the naked eye in discerning changes in aspect. But even, when considering radar navigation in fog only, formulation does not come easily. For example, to pass another vessel at half a mile at three knots could be considered just as safe as passing a vessel four miles off at 15 knots. One must also take into account, when deciding what is a safe distance to pass, the direction in which vessels are shaping to pass. For instance, it could be quite reasonable and safe when overtaking a vessel to pass two miles off, whereas this could be unwise when passing a vessel on a reciprocal course with speed. In other words, it is really the relative speed and the direction of the target which should be considered when judging what is a close-quarters situation.

In thick fog, however, when there is plenty of sea room, it is practical to keep the *minimum* radius of the close-quarters situation at about three miles in order to allow for bearing errors, unsuspected manoeuvres of the target and to keep out the range of audibility of the other ship's sound signals so that delays owing to the application of Rule 19 (*e*) can be avoided.

Really, in order to assess the radius of the close-quarters situation, the Master must rely on intuition based on experience to give him the right answer (radar simulator courses are generally useful to accelerate this experience). CPA and TCPA data should be set on ARPA so that sufficient warning can be given to the O.O.W. when a close-quarters situation is approaching.

Rule 10 applies to Traffic Separation Schemes and is a highly-important addition to the 1972 Rules. It contains regulations adopted by IMO (see IMO-publication "Ships' Routeing Traffic Separation Schemes and Areas to be Avoided") but have become mandatory for all the published schemes. Additional paragraphs are included for small vessels and sailing ships, and exempted vessels.

The more important parts of the Rule state that ''a vessel using a traffic-separation scheme shall, so far as practicable, keep clear of a traffic-separation line or zone (*b* (ii)), normally join or leave a traffic-lane at the termination of a lane, but when joining or leaving from either side, shall do so at as *small an angle* to the general direction of traffic flow as practicable (*b* (iii)) and shall, so far as practicable, avoid crossing traffic-lanes, but if obliged to do so, shall cross as nearly as practicable at *right angles* to the general direction of traffic flow (*c*). It goes on: " (*d*) Inshore traffic zones shall not normally be used by through traffic which can safely use the appropriate traffic-lane within the adjacent traffic-separation scheme. However, vessels of less than 20 metres in length and sailing vessels may under all circumstances use inshore traffic zones."

In connection with the right-angled crossing, it is advised, if possible, to shape the new course well before the lane is reached, thus giving ships within the lane a *timely* indication.

The Rule dealing with Traffic-Separation Schemes (Rule 10) follows the Rule about Narrow Channels (**Rule 9**) and purposely so, as both can be grouped under Narrow Navigational Routes. There are, therefore, certain analogies between the two Rules, viz:

Rule 9 (*b*): A vessel of less than 20 m. in length or a sailing vessel shall not impede the passage of a vessel which can safely navigate only within a narrow channel or fairway.

Rule 10 (*j*): A vessel less than 20 m. in length or a sailing vessel shall not impede the safe passage of a power-driven vessel following a traffic-lane.

Rule 9 (*c*): A vessel engaged in fishing shall not impede the passage of any other vessel navigating within a narrow channel or fairway.

Rule 10 (*i*): A vessel engaged in fishing shall not impede the passage of any vessel following a traffic-lane.

Rule 9 (*g*): Any vessel shall, if the circumstances of the case admit, avoid anchoring in a narrow channel.

Rule 10 (*g*): A vessel shall so far as practicable avoid anchoring in a traffic-separation scheme or in its areas near its terminations.

One of the first traffic-separation schemes was instituted in the English Channel and was followed by a marked decrease in the number of collisions. Surveillance on the conduct of vessels is carried out by radar observation from Langdon Battery, Dover (Channel Navigation Information Service), light aircraft and fast launches.

Having discussed the relevant Rules of Part B (Steering and Sailing Rules), Section I of the Collision Regulation which applies to the conduct of vessels in *any* condition of visibility, we will skip Section II (conduct of vessels in sight of one another) and look at

Section III, which is applicable to vessels in *restricted* visibility. This section contains only one Rule, **Rule 19**, which has replaced the famous Rule 16 of previous Regulations, so well known to generations of seamen. As the Rule is so important, we will quote it in full (Italics are the Author's):

" (*a*) This Rule applies to vessels *not in sight* of one another when navigating in or near an area of restricted visibility.

(*b*) Every vessel shall proceed at a safe speed adapted to the prevailing circumstances and conditions of restricted visibility. A power-driven vessel shall have her engines ready for *immediate* manoeuvre.

(*c*) Every vessel shall have due regard to the prevailing circumstances and conditions of restricted visibility when complying with Rules of Section I of this Part.

(*d*) A vessel which detects by radar alone the presence of another vessel shall determine if a close-quarters situation is developing and/or risk of collision exists. If so, she shall take avoiding action in ample time, provided that when such action consists of an alteration of course, so far as possible the following shall be avoided:

(i) an alteration of course to port for a vessel forward of the beam, other than for a vessel being overtaken;

(ii) an alteration of course towards a vessel abeam or abaft the beam.

(*e*) Except where it has been determined that a risk of collision does not exist, every vessel which hears, apparently forward of her beam, the fog-signal of another vessel, or which cannot avoid a close-quarters situation with another vessel forward of her beam, *shall reduce her speed to the minimum at which she can be kept on her course*. She shall if necessary take all her way off and in any event navigate with extreme caution until danger of collision is over."

Paragraph (*b*) emphasizes the fact that the restricted visibility has to be taken into account when determining a safe speed. It also states that a power-driven vessel should have her engines on " Stand-by ".

Paragraph (*c*) refers to Section I of this part, again stressing the fact to allow for the additional circumstances and conditions of restricted visibility.

Paragraph (*d*) refers to ships which have their radar in working order and makes it *compulsory* for these ships to use their radar for determining if a close-quarters situation is developing, for assessing the risk of eventual collision and for taking avoiding action. The paragraph also places restrictions on certain manoeuvres: (i) Do not alter course to port for a vessel forward of the beam (it does not apply when one is overtaking a vessel); (ii) Do not alter course towards a vessel which is abeam or abaft the beam.

Note that the word "abeam" is not accurately defined within the Collision Rules and there is a choice of avoiding action (to starboard or to port) for a vessel abeam on starboard.

There are cases with a vessel forward of the beam and the bearing changing *slowly clockwise* in the initial stages that there is a reluctance to go to starboard and an inclination to go to port, especially when the rate of approach is fairly fast. One case is with a crossing vessel on the port bow. In such a case, with a close-quarters situation developing and the Master, feeling reluctant to alter course to starboard, should reduce the speed of own ship. In fact, as will be amplified later in this Chapter, this is quite a good manoeuvre and might be one of these cases where a reduction in speed is better than an alteration of course.

Another difficult case is when an echo of a vessel is detected fine on the starboard bow and shaping to pass apparently, as best as can be judged from radar, close to starboard. Again, there is often a reluctance—and this applies also to the other vessel if she is using her radar—to alter course to starboard to pass ahead of the oncoming vessel, especially if the rate of approach is fast. In this case stopping engines is *not* a satisfactory manoeuvre because the other vessel might not have detected own vessel—there is quite a possibility on a dark night that the other vessel has not yet entered the fog and is not aware either of own ship or the presence of fog—thus to stop engines and become immobile might place own ship in the path of the oncoming vessel and completely at her mercy. To stop in this case could only be justified in narrow waters or if one is hampered on both sides by other vessels. Thus if there is sea room the only answer here is to make a *very bold* alteration to starboard so as to put the echo of the other vessel abeam or even a little abaft the beam as quickly as possible. The advance of most ships when making this manoeuvre is usually much smaller than their corresponding head reaches when they make an emergency stop. Having made this manoeuvre the echo should be watched carefully. If the other ship keeps her course and speed, or if she alters course in a direction to support the alteration, or if she stops then all well and good. However, if the other vessel, contrary to Rule 19, alters in a direction which cancels out our own alteration it's not so good admittedly, but it's not so bad as the initial encounter—the relative motion will be much smaller than it was originally—and one can go easily " go on round " to put the echo right astern (complying with the second restriction of Rule 19 (*d*) and so make the relative motion even smaller.

For fast ships coming from abaft the beam which present a collision hazard to own ship if they maintain their course and speed, the best action is to alter course *away* from the vessel in such a way that the stern *keeps* pointing towards the danger. Rule 5 (Look-out) should be kept in mind (an extra look-out might be posted near the stern) and the frequency of sound signals should be increased.

Note that this Rule states " *shall* take avoiding action in ample time" (not "may" as in previous Regulations).

Paragraph (*e*) of Rule 19 is applicable to *every* vessel (not necessarily power-driven) and refers to the detection of other vessels *forward* of the beam of which the presence is detected either by hearing or by radar ("hears apparently . . .", or "which cannot avoid a close-quarters situation"). As this paragraph includes sailing vessels the expression "stop her engines" in previous Regulations is replaced by "reduce her speed to a minimum at which she can be kept on course".

The well-known proviso " a vessel, the position of which is not ascertained ", which has been used so many times in court cases by Masters to defend their action for not stopping engines has been changed to the more direct wording, " Except where it has been determined that a risk of collision does not exist." Such awareness could take place in the following cases:—

(*a*) A vessel forward of the beam, nearby, disappearing in fog, but own ship having determined by visual means up to a few moments before disappearance that there exists no risk of collision between the two vessels.

(*b*) A vessel forward of the beam, which together with own ship, is proceeding along a narrow channel or a traffic-lane.

(*c*) Where the intentions of both vessels have been firmly established by means of V.H.F. R/T contact.

Before leaving the Rules which are relevant to this book, mention should be made of **Rule 2**. It consists of two sections. Section (*a*) deals with the responsibility of owner, master and crew, and the requirement of good seamanship. Section (*b*) states that due regard shall be had to all dangers of collision and navigation and any special circumstances, including the limitations of the vessels involved, which may make a departure from the Rules necessary to avoid immediate danger. Its application is important in connection with Rule 19 : Common sense should go hand-in-hand with the obeyance of the Rule ; under special circumstances (for example, narrow channel with current, multi-target situation, large heavy vessel in traffic lane, etc.) a departure of the Rule may be necessary.

Summary

Rule 5 (Look-out by sight, hearing, radar, V.H.F. R/T) and Rule 6 (Safe speed) must be obeyed at all times, remembering that a safe speed has to be adapted to the prevailing circumstances and conditions when the visibility is restricted (Rule 19). Rule 35 (Sound signals) must be adhered to when the visibility is restricted, but Rule 34 (Manoeuvring and warning signals) are only applicable when vessels are in sight.

Guidance for a Master as to what action he should take during restricted visibility depends a great deal on radar information, V.H.F. R/T information and a radar plot. The use of radar—if a vessel is equipped with an operational set—is compulsory for assessment of the close-quarters situations and the taking of avoiding action (Rule 19(*d*)). The apparent motion line (*OA* line) on a radar plot can inform the Master whether danger of collision exists (Rule 7 (*b*)) and if a close-quarters situation is developing (Rule 19 (*d*)). Having taken avoiding action, the new *OA* line or predicted track will assist the Master in making sure that his action is having the desired effect (Rule 8 (*d*)).

As to the decision to reduce speed or alter course, the study of the direction of the *OA* line can be misleading in some cases, and the best policy for the Master is to study the proper motion and speed of the other vessel (direction and length of *WA* on the plot), *the position where the echo is at the moment when avoiding action is going to take place*, the action the Master in the other vessel might probably take, or will take (via V.H.F. R/T information) for own and for other vessels and to take account of the restrictions imposed on course alterations as laid down in Rule 19 (*d*). A few small wooden models representing ships can be of great assistance if they are placed on the plot at the final plotted positions (see Fig. 16.3).

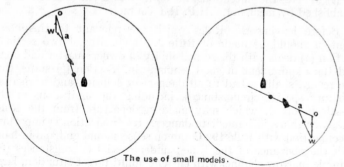

The use of small models.

Fig. 16.3 Planning Avoiding Action

If a close-quarters situation with risk of collision cannot be avoided, the Master should reduce the speed of his ship to the minimum at which she can be kept on her course and navigate with extreme caution until danger of collision is over.

It should also be advocated that a vessel altering course to starboard to avoid a crossing vessel forward of the beam on the starboard side, should generally, if it is practicable, put the other vessel on a *red* relative bearing.

Communication by V.H.F. R/T

V.H.F. R/T communication for collision-avoidance purposes presents an ideal way to make one's intentions clear to another vessel and to arrive at an unambiguous decision about avoiding tactics, so that supporting actions are guaranteed. Its use is extremely valuable in restricted visibility, but also in clear weather, and should be linked to radar information where available. It can also be used for broadcasting a general information report to ships in the vicinity when sailing in reduced visibility. Such a report could state the time, position, course, speed and manoeuvres to be carried out, not always necessarily used for collision-avoidance purposes.

Use of V.H.F. R/T is not explicitly advocated in the Rules, but implicitly (" by all available means ") its use is recommended. Widely employed in piloted waters, its use in open waters for collision avoidance is still restricted, but is worthwhile encouraging. One of the reasons is the fear of getting stuck with language difficulties. However, one can have a try and if language difficulties are arising, or, indeed any difference of opinion, the communication procedure *should be terminated at once* and action taken according to the Regulations using radar information where available.

English is the conventional language for use in sea and air navigation, but knowledge of the phonetic alphabet is essential both for British as well as for foreign vessels. INTERCO, at present, cannot really be used, as its vocabulary is not adequate enough to cope with certain concepts related to collision avoidance.

The procedure of transmitting a V.H.F. R/T message should be in a form which is simple, concise and have a logical sequence. The message itself should be brief but definite, as one wants to transmit in a minimum of time. The three basic steps in the procedure are:

(a) Identification.
(b) Ascertainment of movements and intentions.
(c) Disengagement.

Most of the responsibility for the first step is carried by the initiator, starting with the transmission of his ship's name or call-sign and type of vessel. This is followed by the position of his ship—

either as latitude and longitude, or as a range and bearing from a salient point—and the time in G.M.T. pertaining to the position. Course and speed are then announced, and range and true bearing of the target, obtained from the radar, are indicated. (*N.B. True* bearings should always be used.) Finally, the initiating ship asks for the target's name or call-sign. Care must be taken that the position of own vessel, time of position, and range and bearing of the target are correct.

Vessels in the vicinity can now set their range and bearing markers on their radar display according to the information received and should remember that a few minutes have elapsed since the observation was recorded. The vessel being called should then recognize the echo of the initiating vessel, and her answer should contain her name or call-sign, true heading and estimated speed in knots. Identification should now have been completed.

In the second phase, the initiating vessel repeats her course and speed, then clearly defines her intention and may ask the target to support her action. The latter should try to conform unless she has strong reasons to disagree or special circumstances and conditions (for example, difficulty in manoeuvring) make it difficult or impossible to follow the action proposed by the initiating vessel. Naturally the initiating vessel, before making the proposal, should have assessed the situation carefully with respect to dangers of navigation and other traffic, base his intentions on the Rules of the Road and plan his avoiding action with good seamanship.

The third step, or disengagement procedure, should indicate that both vessels have recognized that the action can be completed, stating the time when they resume their courses and/or speed and should include some form of courtesy signal.

To make the exchange briefer, the initiating vessel might already propose his action during the identification procedure, but this depends on the operator and the time available.

Here follows an example, starting with the initiating vessel:—

> Security, security, security.
> For collision avoidance, Channel 6.   } (Channel 16)
> This is S.S. Haminella, spelled Hotel, Alpha, Mike, India, November, Echo, Lima, Lima, Alpha. Medium tanker.
> Position 9·1 miles 236 degrees from St. Catherine Point at 1315 G.M.T.
> Course 094 degrees, speed 13 knots.
> Echo on radar zero, nine zero degrees, eight point two miles.
> What is your name or call-sign, course and speed?
> Time 1318 G.M.T.
> Over on Channel 6.

Haminella, Haminella, Haminella.
This is M.V. Hilversum. Hotel, India, Lima, Victor, Echo,
Romeo, Sierra, Uniform, Mike. General cargo, loaded.
I confirm your bearing zero nine zero degrees, range eight
point two miles.
My course is 265 degrees, speed 15 knots.
Time 1322 G.M.T.
Over on Channel 6.

Hilversum, Hilversum, Hilversum.
This is S.S. Haminella.
Course 094 degrees, speed 13 knots.
I am altering course to one five zero at 1327 G.M.T.
Maintain your course and speed.
Do you concur?
Time 1324 G.M.T.
Over on Channel 6.

Haminella, Haminella, Haminella.
This is M.V. Hilversum.
Concur with your proposal. Will maintain my course and
speed.
Time 1326 G.M.T.
Back to Channel 16.

Hilversum, Hilversum, Hilversum (on Channel 16).
Channel 6 please.
This M.V. Haminella.
Will resume my course 094 degrees at 1334 G.M.T.
Thank-you for your co-operation.
Time 1332 G.M.T.
Channel 16.

Haminella, Haminella, Haminella (on Channel 16).
Channel 6.
This is M.V. Hilversum.
You are resuming course at 1334 G.M.T. Agree.
Bon Voyage.
Time 1334 G.M.T.
Over to Channel 16 and Out.

All this needs some practice with careful observation of the echo
on the radar display. One can start in clear weather in open waters
with only a few echoes on the screen, so that confidence with the
procedure can be gained. If one gets muddled or cannot understand

the other vessel's message, cancel any proposal at once and terminate communication politely.

A "Standard Marine Navigational Vocabulary" has been published in two M. Notices (M 702 and M732) by the Department of Trade.

Positive and Negative Action

The words " Positive and Negative Action " originate from the writings of Calvert and Hollingdale (*The Journal of the Institute of Navigation*, Volumes 13 and 14), in which the mathematics of collisions in two dimensions was investigated. The meaning is based on the rotation of the sight-line in the horizontal plane with respect to the local meridian, where the sight-line is defined as the great-circle line connecting two vessels.

If there is no rotation of the sight-line, then obviously there is collision risk. If the rotation of the sight-line is in an anti-clockwise direction, it is said that the rotation and also the associated closest distance of approach are positive; if, on the other hand the sight-line rotates in a clockwise direction we speak of a negative rotation and a negative closest distance of approach. This is sometimes expressed as " an anti-clockwise rotation of the sight-line produces a positive miss distance ". Similarly, a clockwise rotation will yield a negative miss distance. Fig. 16.4 shows an illustration of how the rotation of the sight-line can be recognized on a relative motion radar display.

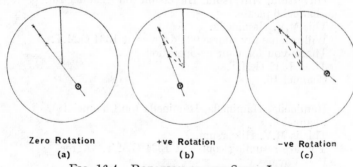

Zero Rotation   +ve Rotation   −ve Rotation

(a)   (b)   (c)

Fig. 16.4   Rotation of the Sight-Line

Suppose that own ship is taking avoiding action in cases 16.4 (*a*) and (*b*), and this results in starting a positive rotation in case (*a*) and an acceleration in the positive rotation in case (*b*), then it is said that own ship has taken *positive* action. Suppose that in case (*b*) the original nearest distance of approach was estimated to be one mile and that, as a result of the positive action by own ship, it was found that the nearest approach had increased to three miles *after* own ship had gone back to her original course and/or speed, then we may say that the positive contribution produced by the positive action of own ship is two miles, provided during that time *the target maintained her course and speed.* If the target meanwhile has also taken positive action, so much the better, and if this resulted in an increase of the miss distance from one to five miles (say) after both ships have resumed their original courses and/or speeds, then the *total* positive contribution is four miles.

In collision or near-collision cases the contribution by a vessel depends on:—

(*a*) The speed of own ship.

(*b*) The relative bearing of the threat.

(*c*) The amount of course and/or speed alteration.

(*b*) The time during which the action is held.

Also, the final miss distance is the sum of the original miss distance, plus the contributions (not to forget to attach the correct signs).

Danger arises when one vessel is taking positive action whilst the other vessel is taking negative action, so that the individual contributions partly or completely cancel out.   Or (see Fig. 16.4 (*c*)) that the original miss distance is negative and one of the vessels makes an insufficient positive contribution, while the other vessel maintains her course and speed.   In such a case the final miss distance may become zero, and this is exactly what happens in the notorious Cumulative Turns, where the sum of the total positive contributions by one vessel equals the original negative miss distance.

It can be easily checked by the reader that positive actions can be achieved by the following manoeuvres (Fig. 16.5):—

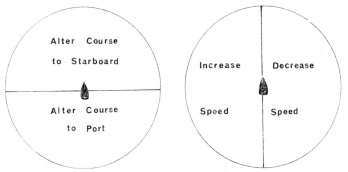

Fig. 16.5   Manoeuvres to Achieve Positive Action

Alter course to starboard when the other vessel is forward of the beam and to port when the vessel is abaft the beam.

Reduce speed when the other vessel is situated in the starboard semi-circle, and increase speed when the vessel is situated in the port semi-circle.

It can also be proved that *maximum* contribution is achieved by placing the other vessel on the port beam when altering course.

Looking at these diagrams it can be seen that positive action for a vessel near the port beam by an alteration of course alone will yield little contribution. Even for maximum contribution, the required alterations of course to bring the other vessel on the port beam are insignificant. One can understand now (see page 278) why a reduction in speed for a vessel near the port beam, especially when the bearing is changing slightly clockwise, is not a bad practice. It is quite true that a negative action is carried out, but if a substantial reduction in speed is ordered according to Rule 19 and if the other ship maintains her course and speed, the negative miss distance will increase rapidly and will soon reach a magnitude which can be considered safe—because the original miss distance was already negative—to resume the original speed. If the other vessel meanwhile takes positive action, it may be necessary to stop and also increase the frequency of sound signals.

If speed is reduced for a vessel on the port bow, it is really not advisable to alter course to starboard at the same time, because the former action is negative and the latter positive and a combination cannot contribute much to the miss distance.

Theoretically, if each ship in an encounter and on a collision course, made positive contributions according to the diagrams, risk for collision between these two (it is assumed that ships are point objects and that there are no other ships nearby) would be eliminated. However, although the taking of positive action forms one of the requirements of the Steering and Sailing Rules (hence the adopted sign in the convention), the actions prescribed in the diagrams do neither conform with the clear weather Rules 13, 15 and 17 (overtaking, give-way and stand-on vessels r.p.), nor do they comply with Rule 19. Furthermore, there are practical limitations to the positive actions :—

(*a*) For vessels approaching from the starboard beam, a substantial reduction in speed is required, and the de-celeration for many modern large ships is insufficient to produce quickly an effective contribution.

(*b*) For vessels on the port beam, a substantial increase in speed is required to produce a positive action, the more so because no alternative course alteration is available. Fast acceleration in speed is an exception in merchant vessels, and it may happen, quite often, that vessels, even in restricted visibility, have little reserve power available.

(c) Some of the course alterations have to be small to make a maximum positive contribution and would not give a clear indication on the radar or visually that action has been taken.

(d) An alteration of course for a vessel on the port quarter would not contribute very much to the miss distance, but it increases the rate of approach and is certainly not advisable.

The limitations could be overcome by taking negative action, advising the other vessel by V.H.F. R/T and requesting her to refrain from positive actions.

With these limitations, and others, in mind, the Royal Institute for Navigation formed a Working Party in 1970 in order to study Calvert's and Hollingdales' Rules and to see if they could be made adaptable to practical sea navigation. At the same time Captain K. D. Jones of Liverpool Polytechnic made an intensive study of the problem. The results were published in *The Journal of the Royal Institute of Navigation* (respectively Volume 25, page 105 and Volume 24, page 60). The Institute published a manoeuvring diagram for course alterations only, with a written guide-line for speed alterations. Both are intended primarily for use in avoiding a vessel detected by radar and out of sight. Captain Jones included also a diagram for speed alteration and his diagrams apply to clear weather condition *and* restricted visibility. Both publications allow negative action under certain conditions, but in these cases the rate of approach is decreased if the other vessel maintains her course and speed. There exist only minor differences between the diagrams from each source for the recommended course alterations; in fact the R.I.N. diagram for course alterations (published in Appendix VII of this book) contains information which does not contravene the new Regulations as no exact definitions are given in the Rules for the expressions "abeam" and "forward of the beam". The diagram and the accompanying Recommendations give a good guidance how Rule 19 can be interpreted.

Another very relevant publication to read is Captain Corbet's paper, " Collision Avoidance at Sea, with Special Reference to Radar and Radio–Telephony ", which was published in *The Journal of the Royal Institute of Navigation*, Volume 25, page 520.

Of course, all these diagrams need changes in the Rules, and may well be valuable material for future revisions in the Rules. But for the radar observer the concepts of positive and negative action are important, as they undoubtedly clarify actions under conditions of restricted visibility.

Choice of Presentation

The different types of display presentations have been discussed fully in Chapter 10 though the treatment was related mainly to navigational purposes. In this section we are concerned with anti-collision.

There is no doubt about it that the True Motion Display is much easier to understand than the Relative Motion Display even when this is stabilized. The appreciation of the latter display requires a lot of understanding and even an experienced observer can become confused when own ship alters her course or changes her speed so that kinks are introduced in the echo trails of the targets.

Relative Motion Display, Unstabilized, Head-Up

The bearing accuracy is not high and this affects the plot.

After a course alteration by own ship, it will take time before it is appreciated if the action has been effective. The echo of the target swings over the screen and its motion is broken up. Comparison between a bearing before and after the alteration is awkward. The two relative bearings are different and a correction has to be applied to the old bearing. This correction is the amount of course alteration plus eventual yaw. There is also blurring of the picture during the course alteration.

The interpretation is especially difficult when own ship is yawing and " tadpole " tails become distorted. Great care should be exercised while marking the echoes on the reflection plotter when the ship is yawing ; all the marks should be made for one particular heading.

It is worthwhile to mark and number the echoes on the Parallel Index with a chinagraph pencil before a course alteration is made. Turning the Parallel Index through the same number of degrees as the alteration of course will assist in re-identification of the targets. See Figs. 16.6 and 16.7.

Relative Motion Display, Stabilized Azimuth Ring, Course-Up

All bearings are true and unaffected by yaw. In all other respects, the behaviour of the picture itself is the same as that on the Unstabilized Display (Fig. 10.1).

Relative Motion Display, Stabilized, North-Up

The bearing accuracy is much better than that of the unstabilized display, especially when an electronic cursor is employed. Bearings are true and unaffected by yaw.

Changes in compass bearing can be readily observed, even after a course alteration so that one can quickly notice whether danger of collision is reduced or persists.

No blurring takes place during a course alteration or when yawing. See Figs. 16.6 and 16.7.

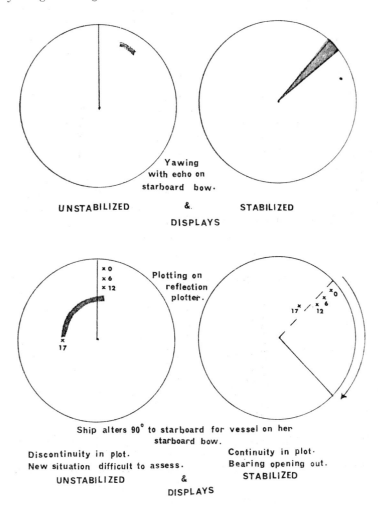

FIG. 16.6 THE EFFECT OF YAW AND ALTERATIONS OF COURSE ON UNSTABILIZED AND STABILIZED DISPLAY PRESENTATIONS

Relative Motion Display, Stabilized, Head-Up

This display can be used by Masters if difficulty is experienced in the interpretation when relating a North-up radar picture to the visual view. It is a Stabilized Head-up display with the heading marker moving when yaw is present or the ship alters her course. In the latter case the operator, after the vessel is steady on its new course, should bring the heading marker to the 'Up' position by manual means (Picture-Rotate or Picture-Shift Control).

Relative Motion Display, Stabilized, North-Up, Off-Centred

This display can be employed when there is True Motion Presentation with a Zero Speed Control.

An electronic bearing cursor must be used for bearings.

The warning range ahead is increased. This display can be very useful for observing the length and direction of the "tadpole" tail of a target. For example, one could use the six-mile range scale with a ten-mile warning range ahead. The movement of the "tadpoles" on this range scale is twice as quick and their length is greater than on the 12-mile range scale which otherwise (without off-centring) has to be used for the same warning range. This presentation also makes the Parallel Index technique easier.

True Motion Display

The electronic bearing cursor should be used and a close check should be kept on the bearing of target vessels.

This display provides a quicker detection of alterations of course and speed of targets when own ship is moving than does the Relative Motion Display (Stabilized or Unstabilized).

No tidal information should be fed in so that the motion of ships through the water is shown and their actual headings can be derived.

Great care should be taken to feed in the accurate log speed, as slight errors in speed input may introduce large errors in aspect of other vessels (see Fig. 14.2).

With ARPA which can be echo-referenced, the display presentation can be put in the Ground-Stabilized Mode, provided the operator knows what he is doing and understands the interpretation of vectors and tracks appearing on the screen (see also Chapter 15, page 77).

This display can be extremely useful in close-quarter situations and also in places where there is a lot of shipping.

No attempt should be made to switch over to the Relative Motion Display (for example, use of the Zero Speed Switch) when targets are close by. The time for resetting should be chosen with

forethought. If tubes have a centre-burn, one should be careful that
the heading marker does not run through the centre of the tube,
otherwise stationary objects dead ahead yielding echoes at the
centre of the PPI, would not be detected.

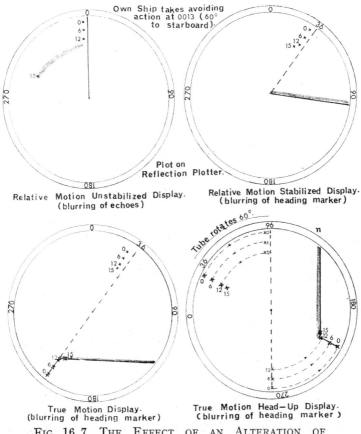

Relative Motion Unstabilized Display.
(blurring of echoes)

Relative Motion Stabilized Display.
(blurring of heading marker)

True Motion Display.
(blurring of heading marker)

True Motion Head—Up Display.
(blurring of heading marker)

FIG. 16.7 THE EFFECT OF AN ALTERATION OF
COURSE BY OWN SHIP ON THE MAIN TYPES OF DISPLAY

To determine the closest distance of approach quickly one can
make either use of the check (zero speed) switch or have the echo
pin-pointed down with range marker and electronic curzor, then,
after a few minutes run, have this point connected by a line on the
reflection plotter to the present position of the echo.   The latter

L

line yields the *OA* line and the distance between this line and Own Ship's position on the display can now be estimated (see Decca A.C. display).

True Motion Head-up Stabilized Display
Relative Motion Course-Up Stabilized Display

This is a true or relative motion presentation with the heading marker always in the " upward " position. Double stabilization (of tube and picture) is applied. There is no smear effect of the echoes when own ship alters course although the orientation of the picture changes. Bearings are true. See Fig. 16.7. On Relative Motion Displays the heading marker shows the course steered.

Figure 16.7 shows the effect of an alteration of course to starboard for a ship on an opposite course on the four main types of display presentation. The True Motion Displays are set to half the range scale of the Relative Motion Displays. Note that the True Motion displays need re-setting after the course alteration.

It is very useful if the ship is equipped with two displays, one with True Motion Presentation, one with Relative Motion, Stabilized Presentation. The displays should have independent time-bases, so that different range scales can be employed. See: Interswitching, Chapter 2. If the same transmitter is used and one display is switched to the short range scale and the other to a medium or long range scale, then some thought must be given to the duration of the transmitter pulse (long or short) one wishes to use. The latter factor is determined by the setting of the range scale on the master display (see remark at the end of Chapter 2: Display).

ARPA Displays

ARPA displays form a special category offering presentations both for picture mode as well as for plotting mode, which can not be reproduced on raw radar displays. Chapter 15 gives details about these presentations.

IMO Recommendation

Section 2, paragraphs 16–20 on the IMO Recommendation on Navigational Watchkeeping contains the following:—

The officer of the watch should use the radar when appropriate and whenever restricted visibility is encountered or expected, and at all times in congested waters, having due regard to its limitations.

Whenever radar is in use, the officer of the watch should select an appropriate range scale, observe the display carefully and plot effectively.

The officer of the watch should ensure that range scales employed are changed at sufficiently frequent intervals so that echoes are detected as early as possible and that small or poor echoes do not escape detection.

The officer of the watch should ensure that plotting or systematic analysis is commenced in ample time, remembering that sufficient time can be made available by reducing speed if necessary.

In clear weather, whenever possible, the officer of the watch should carry out radar practice.

Shore-based Surveillance Radar

Many shore-based surveillance radars have now been installed near narrow waterways, estuaries, rivers and other port approaches, and their operators can provide valuable advice, not only about navigational dangers, but also about collision risks. Furthermore, these stations are usually equipped with recording devices (sound, video and magnetic tape), so that dangerous situations and collisions can be analysed at a later date.

It is important for navigators and pilots to realise that properly-sited Shore-based Radar has a number of advantages over Ship-borne Radar:—

(i) It gives a perfectly ground-stabilized True Motion picture.

(ii) Bearing and range accuracy can be easily and continuously checked.

(iii) Scanner sites can be chosen to give an optimum view of important areas under surveillance. Shadow and blind areas, indirect and clutter echoes can be either avoided or at least limited to unimportant areas.

(iv) Larger scanners than those which can safely and economically be fitted on ships can be used on a land site. These scanners have much narrower beam-widths, and consequently, they give better bearing discrimination and accuracy, and stronger echo-returns can be achieved than with a shipborne radar scanner.

(v) Limits of navigational channels, etc., leading lines and track-lines can be produced on the radar screen. Bearing curzors with additional range control can be positioned at any point on the CRT.

(vi) Reliability of the equipment is likely to better than with Ship-borne Radar, which is subjected to difficult environmental conditions, particularly vibration. Duplication of components for standby back-up purposes can be more easily justified with a Shore-based Radar, which is serving many ships. In addition to this service, engineers are often on the site to maintain or repair the equipment, or can be called to deal with a fault in a very short time. Because the Shore-based Radar remains in a fixed position, deterioration in the equipment's detection performance is more likely to be recognised.

The main disadvantage compared with Ship-borne Radar is that it is more likely to fail to detect a small object, such as a small boat, near to a distant ship, than the ship's own radar.

It is the purpose of this section to make the ship's officer aware of the fact that valuable information in some areas can be obtained from Shore-based Radar stations by using V.H.F. R/T in case the ship's radar breaks down or develops a drop in its performance.

APPENDIX I

BRIEF HINTS ABOUT MAINTENANCE

The manufacturer must provide a Maintenance Manual with the equipment.

When the set is switched on, a check should be made on the ship's main voltage, a.c. input, meter indications, performance test, operation of the range rings, gyro-control (hunting) and correct operation on all range scales.

Dust and dirt should be removed from inside the set as this may lead to break down of high voltage supplies. Use bellows or a fine paintbrush to do this. See that the air flow inside the transmitter—produced by fan or natural convection—is sufficient to carry away the heat.

The external waveguide can be painted, but the reflector should never be painted inside and should occasionally be cleaned with fresh water to remove salt deposit. It is advisable now and again to wipe the perspex of the horn at the upper end of the waveguide with a piece of dry cloth, in order to remove salt, soot or dirt which must not penetrate into the waveguide. A glass-fibre dome round the aerial unit must be kept free from any deposits.

De-icing units sometimes are situated inside the aerial unit, but these must only be switched on in freezing weather. Drying heaters are often provided for the transmitter, displays and aerial pedestal so that condensation inside the equipment is prevented. These heaters are automatically switched on when the equipment is shut *down*.

If the vessel's main electrical power supply has been cut off for a prolonged period, for example during a dry-docking period, then one should not switch the radar on to test it immediately power is restored as there is a strong possibility that the equipment has become damp during the shut-down period. Ensure that the heaters are on (bulkhead switch supplying the power to the radar) and wait a few hours, to allow the set to dry out before switching on the set. Additional heaters in the radar compartment should also be switched on if they are available.

Chemical substances like silica gel crystals are placed in exposed transceivers. While these crystals collect moisture, they change in colour from blue to pink. Pink crystals have to be replaced or dried by baking.

Snow in the scanner should be removed before starting the set.

Sometimes the aerial driving gear is immersed in oil and its level can be checked by means of a nutwindow. Occasionally the oil has to be renewed, but care should be taken not to overfill.

Take the main fuses out and put them in your pocket when maintenance or repairs to the scanner unit are carried out.

Care should be taken with high tension components in trays. Short circuit condensers first by means of a screw-driver before touching them. The same applies when renewing the CRT; earth the H.T. lead by means of a screw-driver. Always check the safety-switches first.

A blown fuse means an overload. Overloaded resistors betray themselves by blistering.

Valve trouble is quite common. A *cold* valve can be an indication.

295

Study the block diagram in your Manual and find out where the main components are situated in the set. Get acquainted with the noise pitch of modulator and scanner motor. Logical reasoning will often tell you in which unit the fault lies. For example, failure in the power unit, transmitter and time-base will remove the complete picture; failure in the receiver only will *not* remove the range rings. Read carefully the pages in the Manual about fault finding. Often a chart is provided. Label components when you take them out.

Periodical attention should be paid to the wear of brushes on motor alternator, extractor fan, aerial driving motor and to the cleaning (with a cleaning spirit) of magslips, deflector coils, commutators, sliprings (de-icing and performance indicator) and heading marker contact (see Appendix III). Examine periodically chassis for overheating or leaky components, loose waveguide connections or omissions of shims. Check the grease pots of the alternator every six months.

Echo-box performance is lowered by tarnishing of inside surfaces (funnel smoke).

Modern radar sets almost exclusively use printed circuit boards (PCBs) throughout. They are easily replaced by sliding defective boards out and putting new ones in until they click into place. Enough spare PCBs of the most important circuits have to be carried on board ; the defective ones can be returned to the manufacturer for repair.

Maintenance on PCBs on board by the electronic officer can be done by means of an "extender" board which fits into the vacant rack of the PCB to be tested. The latter is then coupled to the extender and adjustments can be carried out with the PCB brought clear of its adjacent boards.

APPENDIX II

ELEMENTARY GENERAL PRINCIPLES

To understand the broad outline of the action of a radar set one must know something about the most elementary principles concerning electricity, electronics and the electromagnetic wave theory. These principles are explained very simply in this Appendix.

Electricity and Electrons

Electronic Structure

As most of us know, the smallest portion of matter which still displays the properties of the substance is known as the molecule. The molecule can be broken up again in units all of the one kind (elements) or various kinds (compounds). These minute particles are known as atoms and their structure is fascinating. They consist of a relatively heavy nucleus with a very tiny body or bodies whirling around them. Inside the nucleus, particles were found (*protons*) which carried an electrical charge and the nature of this charge was identified by scientists as *positive* (the existence of positive and negative electricity had been known 150 years before this investigation). Besides the protons, particles with a mass approximately equal to that of the proton, but carrying no electrical charge were discovered and these were called *neutrons*. A *negative* electrical charge was discovered to exist on the very small particles circulating around the nucleus and these particles were named *electrons*.

The simplest atom is that of hydrogen which consists of one proton around which one electron moves in a fixed orbit. The structure of most atoms is more complex, with a central positive nucleus surrounded by planetary electrons revolving at high speed in definite orbits which are elliptical or eccentric in character. These orbits exist in groups or *shells*, and each member of any one group moves in the type of orbit characteristic to that group. See Fig. AII. 1.

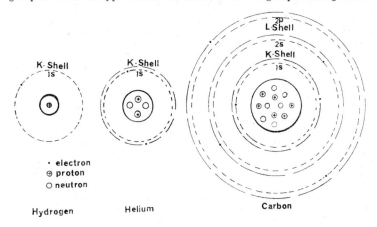

FIG. AII.1 ELECTRONS, PROTONS AND NEUTRONS

297

Except for the simple hydrogen atom, the nucleus is made up of protons and neutrons in approximately equal numbers. The nucleus therefore contains practically the whole of the atomic mass, and carries a positive charge equal in magnitude to the sum of the charges carried by the extra-nuclear electrons.

The elements may be arranged in a Periodic Table proceeding step by step from a simpler element to the next higher. At each step, the nuclear charge of the atom is increased by one unit and at the same time an electron is inserted in the outer part of the electronic envelope. There are about 100 different types of atoms in existence including some which are produced artificially. The number of protons inside the nucleus determines the type of atom and its place in the Periodic Table and this number varies between one and about 100. The removal, however, of one or more electrons does not change the property of the atom nor of the substance, but it will alter the electrical distribution in an atom and inside the substance, though the latter will not display electrical properties if the distribution is merely disorganized. A positively charged atom has a deficient number of electrons corresponding to the number of protons; some of the electrons have been removed. Such an atom is known as a *positive ion*. An atom which carries too many electrons in accordance with its nuclear charge is called a *negative ion*.

There is a maximum to the number of electrons which can be in each shell. The first or inner shell contains a maximum number of two electrons, the second eight, the third 18 and the fourth 32, etc.

The space occupied by an atom is, except for the small nucleus, practically empty. The planetary electrons are separated from the nucleus by much greater distances, relatively, than are the planets of the solar systems.

Just as the electron is considered as a particle which carries a negative charge, so in the class of solid materials known as "Semiconductors" modern theory postulates the existence of a "particle" which carries a positive charge and is termed a *hole*. It is in fact a mobile vacancy which should normally be occupied by an electron.

Conduction

Atoms are not always stable systems and in some substances electrons become easily detached from the parent nucleus due to collisions from other electrons or because they come under the influence of the nucleus of the neighbouring atom. The latter nucleus being positively charged will exert a pull on the negatively charged electrons and may "steal" an electron from its neighbour. The outcome of this is that electrons continually change and exchange orbits around nuclei but as these motions are completely spontaneous and not governed by any actions from outside, no electrical change in the substance is detected.

Discipline in the motions can, however, be established by connecting the substance to the terminals of a cell or battery. The positive terminal of the cell will exert a pull on the *free electrons* and these will drift towards that terminal. The much heavier nuclei remain behind but, in turn, assist in dragging the electrons along and draw fresh electrons from the negative terminal of the battery. The result is a drift of electrons all around the circuit from the negative terminal though the substance, then towards the positive terminal and through the battery back to the negative terminal, so completing the circuit. This motion of electrons constitutes an electrical current and will continue until the energy of the battery is exhausted. See Fig. AII. 2.

Conductors are substances which possess free electrons and therefore can carry electricity from one part to another part of the substance. Most metals are conductors, silver being the best conductor known.

In *insulators*, on the other hand, very few free electrons are available and little exchange takes place between the electrons in their orbits. Examples are rubber, mica, paper, silk, etc.

In the *semi-conductor* the conduction is excellent in one direction but very poor in the other, opposite, direction. The flow of current by "holes" in these substances is naturally in the opposite direction to that of the electrons and therefore "holes" go from a positive to a negative terminal. Examples are copper oxide, selenium and germanium. They are extensively employed in radar technique for rectifiers and transistors and more will be said about them at a later stage.

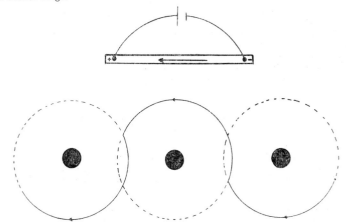

FIG. AII.2 EXCHANGE OF FREE ELECTRONS IN A CONDUCTOR

Up to this point we silently assumed that the substances we referred to were of the solid state. There are, however, liquids and gases which also conduct and where the conduction of electricity is of a different nature. These types of substances have in their molecules atoms of *different* kind (known as *compounds* in contrast with *elements* where the molecules have atoms of the *same* kind) and for some definite reason—the explanation cannot be given in this kind of book—the atoms of one kind become positive ions whereas the atoms of the other kind become negative ions *when the substances are in solution.* Salts and acids in fluid form are examples of this type of conductor.

If the terminals of a battery are now connected to two plates which are immersed in the solution then *two* streams of carriers will move in opposite directions inside the liquid. The negative ions will move towards the plate connected to the positive terminal and the positive ions will move towards the plate which is connected to the negative terminal of the battery. In this case, it is impossible to state, without a further definition, in what direction the current is flowing. Is it the direction of the positive or the negative ions? In the practical interpretation of electricity the direction of the current is taken to indicate the direction in which the positive ions are moving in spite of the fact that it is known that in solid conductors only the negatively charged electrons are in motion. The direction so defined is commonly called the *conventional* flow. If later, for special reasons, we want to refer to the direction of flow of electrons then this should be mentioned.

Units

In connection with the different units the water analogy, though not always perfect, is often employed and it is certainly helpful for the beginner. It means that, to some extent, an electrical flow can be compared with a flow of water through a closed system. A pump is responsible for the continuous flow by exerting a force on the water particles. The force must be kept on continuously, otherwise the flow will come to a stop due to resistance inside the pipes; the force should also be of constant character to ensure a steady flow. Between any two points of the system there must exist a pressure difference, otherwise the water would simply not flow between the pipes. The amount of difference in pressure is dependent on the length of piping between the points, the cross-section of the pipes and the material of which the pipes are composed; these three items determine the total frictional resistance between the points. Note also that the rate of flow, for example, expressed in litres per minute, must be the same at any point in a single pipe system, otherwise a burst would take place. The speed of the water, of course, will depend upon the cross-section of the pipe, but in the type of electricity, discussed in this Appendix, we will not be concerned with speed, only with the *rate of flow*. The power of the pump is a function of the pump's force and the rate of flow against frictional resistance.

Coulomb.

A coulomb carries a quantity of charge equal to the charge of $6 \cdot 28 \times 10^{18}$ electrons. It can, for example, be compared with a litre of water.

Ampere.

The flow of one coulomb per second past a given point in a circuit is called an ampere or amp. It is a unit measuring the rate of flow, equivalent, say, to litres per minute. In electricity the rate of flow is called the *current* (I).

Volt.

The electrical *pressure difference* or *Potential Difference* (*p.d.*) between two points in a circuit is measured in volts (V). The pressure difference created *inside* the battery (pump in our analogy) which is responsible for the flow of electricity throughout the circuit is known as *Electromotive Force* (*e.m.f.*) and it is also expressed in volts. Note that a battery or any generator does not produce electricity but produces an e.m.f. which is responsible for the potential difference between its terminals.

Exact definitions of the volt can be found in an electrical text book.

Ohm.

The resistance of a conductor—which depends on its length, cross-section and material—is measured in units called ohms (Ω or symbol omega). If the potential difference applied between two ends of a conductor is *one volt* and this is exactly right to produce a current of *one amp*, then it is said that the resistance of the conductor is *one ohm*. Resistance in conductors is caused by the obstruction to the flow of electrons by other electrons and fixed atoms. For many conductors (excluding semi-conductors) the resistance (R) increases markedly with temperature and the resistance indicated on an instrument, for example, is, unless otherwise stated, for a temperature of 15° C.

Watt.

The amount of energy or work to raise one pound weight vertically through one foot is called a foot-pound. For a pump to raise one pound of water through a height of six feet would require work of six foot-pounds. This means that the water has been pumped through a distance which corresponds to a pressure difference equivalent to six feet of water.

The unit of work in electricity is called the *Joule* and it is equivalent to the amount of work required to raise one coulomb up through a potential (pressure) difference of one volt.

Power is the rate of doing work, per minute or per second. For a pump it is given in horse-power (550 foot-pounds per second) and it tells us at what rate energy in one form is converted to another form.

In electricity the *watt* (*W*) is the unit of power. If energy is absorbed from the circuit at the rate of one joule per second, then that particular part of the circuit is rated as one watt. A heater rated as 1000 watt, for example, takes 1000 joules of electrical energy per second and converts it into heat.

Sometimes measurements in units are large and it saves some noughts to insert the prefix k for kilo and M for mega; in other cases measurements expressed in units are small or very small and we may insert the prefix m for milli and μ for micro. For example, one kV is 1000V, one MΩ is 1,000,000 ohms, one mA equals 0·001 Amps and one μV equals 0·000001 V.

Relationships between the Units

Ohm's Law.

Returning to the water analogy, if we make the length of the pipe longer between two points, i.e. increase the resistance and want to keep the rate of flow the same, then we have to increase the difference in pressure between the points. The same applies if we want to increase the rate of flow but do not alter the resistance. This means that the pressure difference is directly proportional to the resistance *and* the rate of flow between the points.

In electricity we may say that for most metal conductors:—The potential difference (p.d.)=constant × resistance (R) × current (I).

As a result of the definition of the ohm, the constant becomes unity because 1 volt=1 ohm × 1 amp.

Ohm's Law states that:—

p.d. between two points=current × resistance between the points.
$$p.d.=I \times R \ Volts$$

If, for example, a current of 4 amps goes through a heater having a resistance of 60 ohms, then the potential difference across the heater is 240 V.

The same law can be applied to calculate the electromotive force.

Electromotive force=current × total resistance in the circuit.
$$e.m.f.=I(R+r) \ Volts$$

where R is the total external resistance in the conductors and r represents the internal resistance of the battery or dynamo.

The above two formulae show that the p.d. across the terminals of a battery while under load is always less than the e.m.f. by an amount of Ir Volts. This internal voltage drop has no useful purpose and it is known as the *lost voltage*.

Watts, Volts and Current.

It is not difficult to understand that the power in a circuit or the rate of conversion of energy depends on the potential difference across the circuit and the rate of flow. The relationship is:—

Power=p.d. × current (in watts)
i.e. $W=V \times I$

In the example above, the heater with a potential difference of 240 V across the terminals and a current flow of 4 amps can be rated as 960 watts, i.e. it consumes 960 joules of energy every second.

Decibels.

The decibel is a logarithmic unit used for expressing the power gain or loss in a circuit or in an electromagnetic field.

Loss (Gain)=10 $\log_{10} P_1/P_2$, where P_1 and P_2 are the powers concerned. Thus—3 db means that there is a power loss of one-half.

Electronics

Until now we discussed the existence of mobile or non-mobile electrons but always in connection with the nucleus. It is, however, possible in liquids, gases or a vacuum that there is a motion of completely free electrons separated from any kind of atom. The branch of science which makes a study of this special type of electrical transfer is named "electronics".

Valves

Thermionic Emission

When platinum or tungsten is raised to a white heat or certain oxides to a dull red heat electrons break through the surface of the hot substance. When the escaping electrons leave the substance it becomes positively charged and so electrons are pulled back again. An equilibrium will be reached and a cloud of electrons, known as *space-charge*, will then hover around the hot substance.

The emitting substance is called the *cathode* and, in some cases, is coated on a wire (filament or heater), which is made to glow when connected to a low tension source. In the majority of cases, however, cathode and heater are separated (indirect heaters).

Diode Valve.

Instead of allowing the electrons to collect around the filament, they can be made to be attracted by a positively charged metal plate, called the *anode*. A high tension source maintains the anode at a positive potential, so assisting the electrons to bridge the gap between cathode and anode, while at the same time providing the cathode with the necessary electrons.

A diode valve is formed by putting a glass envelope over the cathode and anode and evacuating the space inside to decrease resistance to motion and to avoid burning up the filament. It is called a valve because it is a one-way conductor. Electrons can only go from cathode to anode and never in the reverse direction. Note that in diagram AII. 3 the direction of the *electron flow* is shown, not the conventional positive to negative direction of flow.

The diode valve is used as a rectifier, i.e. converting an alternating current into a pulsating direct current source (this ripple can be eliminated by the use of capacitors and inductors).

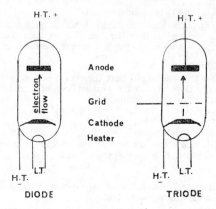

FIG. AII. 3   DIODE AND TRIODE VALVES

Triode Valve.

The triode valve has a *grid* inserted between anode and cathode but *nearer to the cathode than to the anode*. Because the grid is so close to the cathode, an electric charge on the grid easily affects the flow of electrons through the valve. This means that the triode valve acts as an adjustable valve. When the grid is made sufficiently negative, the valve is shut (cut-off); when the grid is sufficiently positive, the valve is completely open and there is maximum emission (saturation current). See Fig. AII. 3.

Generally the grid is kept at a constant slightly negative potential with respect to the cathode. This is called *grid-bias*. Oscillations applied to the grid are superimposed on the grid-bias (Fig. AII. 4).

Triode Valve as an Amplifier.

One of the uses of a triode is as an amplifier.

Between the values of complete cut-off and saturation current, small fluctuations in potential applied between the grid and cathode cause deep imprints in the anode current which, when applied across a high resistance, produces a voltage difference at the anode-end many times greater than that applied to the grid (note that the output signal is 180° out of phase with the input signal; when the input signal at the grid is going more positive, the anode-end of the load resistor becomes more negative). An incoming signal can be amplified in such a way. The output, similarly, can be amplified in stages by repeating the process.

The circ n t is shown in Fig. AII4.

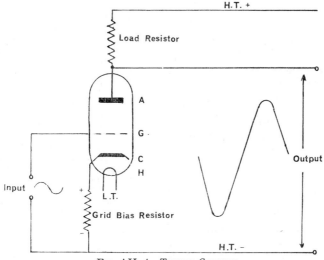

FIG. AII. 4   TRIODE CIRCUIT

Transistors

For many purposes, instead of valves, transistors can be used. They are much smaller in size than a valve, they are more robust than valves, no heater is needed and no high tension is required. They are made from Germanium or Silicon.

Let us consider the germanium transistor. The germanium atom has 32 electrons which are arranged in shells around the nucleus. The outer shell (N shell, 4p) is not completely filled. In fact, there is room for four more electrons. In pure germanium these vacant spaces are filled by sharing two electrons between two neighbouring atoms so that each atom is linked to four adjacent atoms. An illustration is shown in Fig. AII. 5; an electron from atom *A* goes to a vacant space of atom *B* and vice versa. This results in the formation of a crystal lattice. Owing to this lattice construction hardly any free electrons are available and the conductivity is very low. The only free electrons which are present come from broken lattice links due to thermal agitation.

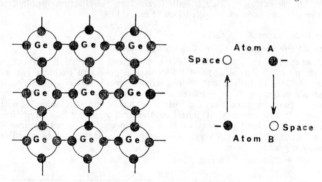

FIG. AII.5   CRYSTAL LATTICE OF GERMANIUM

It is, however, possible to obtain a higher conductivity by the introduction of an impurity i.e. another atom. If the other atom is an atom with only three vacant spaces in the outer electron shell, then it will only fit into the lattice if one electron is pushed out (Fig. AII. 6). This means that a free electron is available. Such an impurity atom is known as a *donor* because it donates one electron to current flow. The arsenic (atomic number 33) or antimony atoms are examples of donors. Germanium adulterated by donors is called *n-type germanium* (n for negative). The impurity atom itself becomes positive because it has lost an electron.

Alternatively, an atom (gallium, atomic number 31, aluminium or indium) may be introduced in the germanium. Such an atom has five vacant spaces in the outer shell. If this is made to fit in the lattice, one vacant space will remain. This vacant space is termed the *hole* and exerts an attracting force which may capture an electron and thereby creates another hole in the neighbouring atom whilst filling its own. Such a chain process may go on throughout the germanium. The direction in which the holes are created is opposite to the direction in which the free electrons are moving. Hence, as is already said before, the process of successive hole creation is equivalent to a flow of positively charged carriers.

The impurity atom itself having lost a hole and gained an electron will become negative. This type of atom is known as an *acceptor* atom because it accepts an electron (Fig. AII. 6). Germanium containing acceptor atoms is called *p-type germanium* (p for positive). Note that the overall charge in p-type and in n-type germanium is neutral.

The p-n Junction.

Consider a piece of n-type germanium placed against a piece of p-type germanium. See Fig. AII. 7(a).

One would expect that the electrons would flow from the n-type to the p-type and holes from the p-type to the n-type. However, as soon as this starts then the n-type becomes positive and the p-type negative and this will at once stop any further flow and the position can be compared with a battery placed across the junction (this imaginary battery is shown in dotted lines in the diagram). The direction of the electric field across the boundary (from + to −) is indicated by an arrow and this field forms a barrier across the boundary which prevents any movements of the holes against the field.

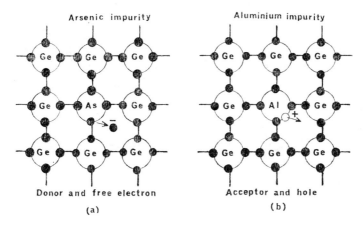

FIG. AII.6 DONOR AND ACCEPTOR

Now suppose that an actual battery is connected across the junction, the positive terminal to the n-type, the negative terminal to the p-type. This will only increase the field and the potential barrier and practically no current will flow through the junction.

If, however, the positive terminal is connected to the p-type and the negative terminal to the n-type then the newly established field will act in opposition to the internal field and the potential barrier will drop so that an interchange of holes and electrons can take place. Both cases are illustrated in Fig. AII. 7(a) and (b). In the first case it is said that the battery is connected in the *reverse direction*, in the second case in the *forward direction*.

This consideration shows that the p-n junction can be used as a *rectifier*. Appreciable current flow (conventional) can only take place across the junction from the p-type to the n-type.

The Junction Transistor.

The junction transistor consists of three slices of germanium placed in the arrangement n-p-n or p-n-p. (Fig. AII. 8).

One side is called the *emitter* where the current is generated, the other side is known as the *collector*. The middle is known as the *base*. The external connections are soldered to the germanium in the neighbourhood of the impurities. The emitter is biased in the forward direction with respect to the base, the collector is biased in the reverse direction with respect to the base.

The symbols are shown in the diagram below. It will be seen that the emitter, base and collector can be compared with the cathode, grid and the anode of a triode valve; a difference in the p-n-p type is that the "cathode" (emitter) is connected to the positive terminal and the "anode" (collector) to the negative terminal of the battery.

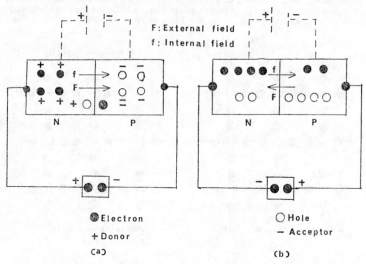

F: External field
f: Internal field

● Electron   ○ Hole
+ Donor   — Acceptor
(a)   (b)

FIG. AII.7   THE P-N JUNCTION

Fig. AII. 9 shows the action of the p-n-p type. The positive potential applied to the emitter will repel the holes against the internal field and drive them from the p-type across the junction into the n-type material (base).

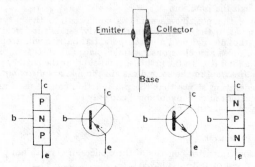

Emitter   Collector

Base

FIG. AII.8   THE JUNCTION TRANSISTOR

There is little field strength in the base but the holes will continue to drift slowly towards the collector region. A few holes in the base region will combine with electrons and it is therefore important to make the width of the base small. There is no barrier across the junction of the base and the collector as the external and the internal field act in the same direction. In fact, there is a steep drop in potential across this junction. On reaching the collector the holes become neutralized by electrons from the battery.

Voltage fluctuations applied between emitter and base will change the emitter current and cause proportional changes in the collector current.

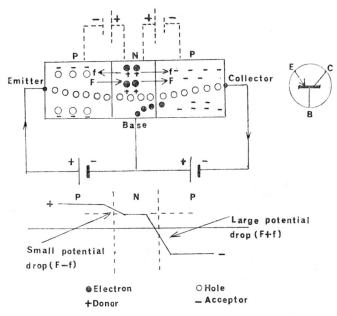

FIG. AII.9   ACTION OF THE P-N-P TYPE TRANSISTOR

The junction transistor is used as an *amplifying device*. It is not a current amplifier like the valve amplifier, but a volts and power amplifier. The current gain *a* is less than unity and is usually of the order of ·98 as the current through the emitter section is nearly the same as the current through the collector section apart from a very small leakage inside the base. The volts drop between emitter and base is, however, small (external and internal field are in opposition) compared with the volts drop between base and collector where the internal and external field are in collaboration. This means that the wattage (volts × amps) produced across collector-base is much greater than the wattage across emitter-base. Typical values of emitter-to-base and collector-to-base resistance are 500 Ohms and one Megohm respectively. This represents a resistance gain of 2000, a voltage gain of about 1960 and a power gain of the order of 1900.

The battery to drive the electrons and holes needs only to be a few volts and this is another advantage of the transistor over the valve where a high tension is always required.

Fig. AII. 10 shows the common-emitter arrangement for a transistor used as an amplifier. The emitter is common to the input and the output circuits. The input signal is applied to the base, and the collector is the output electrode.

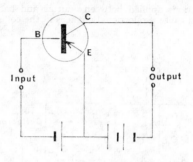

Common Emitter

Fig. AII.10 Simple Transistor Circuit

Integrated Circuits

Miniaturization techniques have been developed by extending the planar technique to transistors and resistors. The latter technique enables thousands of transistors to be made together on a thin slice of silicon (base or substrate) about 3 cm. in diameter and 0·3 mm. thick. Such a polished slice of n-type silicon is then placed in a furnace for several hours and steam is passed over it, causing a thin layer of oxide on the exposed surface. This serves to protect the surface against the ingress of dirt or moisture. Windows are then cut in the layer of oxide by means of special photo-lithographic processes. This is shown in Fig. A II. 11 where only three windows are shown, but in reality there may be as many as 3000 windows on one slice.

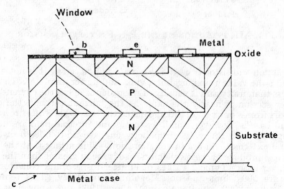

Fig. A II. 11 Silicon Planar Type Transistor (N–P–N)

The slice is re-heated and impurities are injected through the windows. These impurities diffuse with the substrate. One type of impurity results in the formation of n–type material and the other type causes a layer of p–type material. The various wavers or dice in a slice are bonded to a metal case which also acts as the collector connection and the base and emitter wires are welded on.

By extending the planar technique many transistors and other components can be diffused into a single silicon slice with dimensions typically $1 \cdot 5 \times 1 \cdot 5 \times 0 \cdot 3$ mm. Such a circuit is known as an Integrated Circuit and they have a further refinement, namely the addition of an extra plane. The n–type slice of sillicon, now called the " epitaxial " layer is made very thin in order to increase the resistivity of the collector—and preventing break-down of the crystal structure from high voltage application—and the complete assembly is fused to the substrate.

Integrated circuits nowadays are used extensively in radar circuitry and in the microprocessor, the central processing unit (cpu) of the microcomputer.

Transmission of Signals

Synchros and Servomechanism

Synchro and servomechanisms serve different purposes, but they are often used in combination.

Basically a synchro consists of a stator and a rotor and is used for electrical transmission of angular position information, for example, scanner rotation.

A servomechanism receives and obeys an order, for example, from a synchro:

(a) It senses the error between the actual signal and the signal as it *ought to be*;

(b) It corrects the error and changes the actual signal to the signal as it ought to be.

There are two types of transmission:

(i) Torque transmission. No servomechanism is employed.
(ii) Control transmission. Here the output may be required to drive a mechanism which may have an inertia too large for torque transmission and so a servomechanism is used at the receiver end.

Torque Transmission

Both the rotor in the transmitter and in the receiver are energized by a.c They both produce an alternating flux causing induced voltages across the

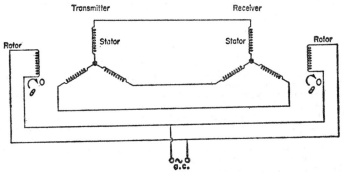

Fig. AII.12   Torque Transmission

three stator windings in transmitter and receiver. If the relative positions of the rotors with respect to the stators in both transmitter and receiver are the same then no current will flow in the leads connecting the two stators. Should, however, the relative position of the rotor in the receiver differ from the relative position of the rotor in the transmitter then different voltages will be induced in the receiver and in the transmitter stator windings. A current will start flowing between the two stators which produces a magnetic field in the stator windings of the receiver. According to the motor principle this field will produce a torque on the rotor in the receiver to turn it until the voltages across the windings of both stators are equal. When this is the case, the rotor in the receiver is re-aligned with the rotor in the transmitter. See FIG. AII.12.

Control Transmission.

In order to make the system suitable for heavier loads the servomechanism is introduced which is linked to the stator in the receiver by means of what is called a *control transformer* (Fig. AII.13). The output of the control transformer is electrical (not mechanical as in the synchro receiver with torque transmission) in the form of a voltage induced in the rotor winding.

Only the rotor in the transmitter is fed by a.c.

The voltages induced by the alternating magnetic flux across the stator windings in the transmitter are also supplied, through interconnection, to the stator windings in the receiver (control transformer). The voltages cause identical alternating currents to run in the two stators and the direction of the alternating magnetic flux in the stator of the receiver is a reproduction of the direction of the magnetic field in the stator of the transmitter. The direction of the latter magnetic field depends on the position of the rotor in the transmitter.

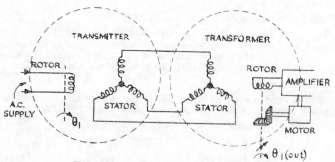

FIG. AII.13　CONTROL TRANSMISSION

The alternating magnetic flux in the receiver induces a voltage in the rotor of the receiver by *transformer action*. At this instant a servomechanism takes over. The induced voltage is called the *error signal*, its magnitude giving an indication of the degree of misalignment of the rotor. This voltage is amplified and then drives a motor (*controller*) which is geared to the load *and* to the rotor (*feedback*). When the rotor windings in the control transformer (receiver) are at right angles to the direction of magnetic flux the output of these windings becomes zero, the motor stops and the rotor comes to rest. Any subsequent change in the position of the rotor in the transmitter will change the direction of the magnetic field in the receiver and the rotor in the receiver will turn to a new null position.

As the amplification factor of the amplifier can be made quite large the rotor in the receiver can be made to operate substantial mechanical drives remote from the transmitter.

Note that in torque transmission the rotor in the receiver aligns itself in line with the magnetic flux; in control transmission, the rotor is aligned at right angles to the direction of magnetic flux.

Fig. AII. 14 shows the type of control transmission where the sum or the difference of two angles can be transmitted. A *differential transmitter* is inserted between the synchro transmitter and the control transformer. The differential transmitter has a three-phase star connected rotor and stator. Its stator is connected to the stator of the transmitter and its rotor to the stator of the control transformer.

If the rotor of the transmitter is turned through an angle θ_1 and the rotor of the differential transmitter turned through an angle θ_2, the angle $\theta_1 + \theta_2$ or $\theta_1 - \theta_2$, depending on whether they are moved in the same or opposite directions, will appear at the receiver end (control transformer).

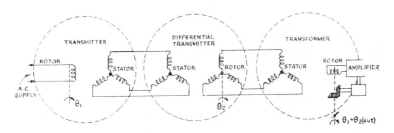

FIG. AII.14   DIFFERENTIAL CONTROL TRANSMISSION

Electromagnetic Wave Theory

Having discussed the principles of electronics, let us now turn to the concept of electromagnetic radiation and ways of propagation.

The needle of a small pocket compass will suffer a deflection from the magnetic meridian when brought in the vicinity of a magnet. It aligns itself according to the pull of an imaginary string attached to its North-seeking pole. By moving the compass gradually, in the direction of the exerted force, it is possible to plot the direction of the pull. The resulting locus is indicated by the name *Magnetic line of force*. The direction of a magnetic line of force is represented by the direction of the North-seeking pole of a small pocket compass. The magnetic lines of force can be clearly demonstrated by sprinkling iron filings onto a piece of cardboard covering a strong magnet. It will be seen that the filings arrange themselves according to special patterns, and each filing seems to be pulled into line by an invisible string which has its origin in one of the poles. It seems as though a materialisation of the magnetic lines of force (flux) were taking place. Their density is an indication of the strength of the field.

Suppose now that a pocket compass is placed near to an electromagnet which can be made into a magnet at will by closing a switch in the electrical circuit (Fig. AII. 15). When the switch is open, no field due to the magnet exists, and the needle of the compass points towards magnetic North. If, however, the switch is closed, the needle will swing to one direction which is the direction of the line of force at that place. A question arising from the result of this experiment is whether there is a time lapse between the moment of closing the switch and the moment when the needle starts swinging, or whether these actions happen instantaneously. It can be proved that a time interval occurs between the two actions, though that interval is extremely short. In other words, it will take some time before the lines of force which are radiated from the magnet have reached the needle of the compass. A propagation of the lines of force does exist.

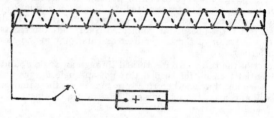

FIG. AII.15   ELECTRO-MAGNET

We could have carried out the experiment in a vacuum with the same result which shows that magnetic lines of force are propelled through space, even with no medium present, though it can be found that their velocity in this case is slightly different from what it is in air.

There exists a close affinity between magnetism and electricity, and a property of magnetic lines of force is that they are always accompanied by *electric lines of force*, which are at right angles to them. An electric line of force is defined as the direction in which a small positively charged particle would move. This relationship between magnetic and electric lines of force gave rise to the term "Electromagnetic field".

The wave is said to be horizontally or vertically polarized depending on whether the electric field is in the horizontal or vertical plane. For purposes of reflection 3 cm. radar sets have their radar waves horizontally polarized. The 10 cm. radar set, mainly for technical reasons (construction of the slotted waveguide) frequently uses waves which are vertically polarized.

Let us turn back to the experiment described above and replace the switch and battery by a small a.c. generator. Instead of a constant deflection, the needle of the compass will acquire a swinging motion from one direction to exactly the opposite direction depending upon the polarity of the magnet (it is assumed here that the strength of the earth's magnetic field is very small compared to that of the magnet and that the frequency of the a.c. generator is very low). The magnetic lines of force will pull the needle in one direction, then slowly thin out when the current diminishes in the circuit and completely disappear when the current stops flowing. The needle then returns to its original position, i.e. pointing to the magnetic North. When the current reverses the needle follows the direction of the lines of force, which are now leading in an opposite direction, and then comes to a momentary standstill when maximum current flows. A repetition of the process then follows.

What actually happens is that a train of compressions and rarefactions of lines of force are passing over the needle of the compass. During the compressions, the density of the lines of force is large, and so is the pull at the compass needle. In the middle of a rarefaction no lines of force due to the magnet exist, and there is no pull at the needle, apart from the force caused by the earth's magnetic field. A reversal of the lines of force takes place after the passing of a rarefaction. Remembering that the density of the lines of force is an indication of the disturbing force on the compass needle, a graph has been plotted in Fig. AII. 16, showing the variation of the magnitude and direction of the force against time. A wave form is obtained.

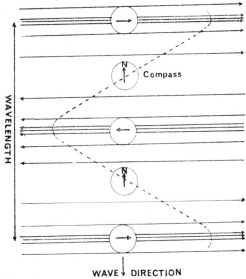

WAVE ↓ DIRECTION

FIG. AII.16   ELECTRO-MAGNETIC WAVE PRINCIPLE

The distance between two peaks in the same direction is called the *wavelength*. It is the length of one cycle of the wave.

Marine radar sets use a wavelength of 3 or 10 centimetres. Most British manufacturers produce sets which operate predominantly in the 3 cm. waveband or X-band but lately the sale of 10 cm. equipment is on the increase.

The *frequency* of a wave is the number of cycles which pass a stationary observer in one second.

If f cycles pass the observer in one second, then the total length of the wave train is $f \times \lambda$ where λ denotes the wavelength, or the length of one cycle, and f is the frequency. But the distance covered per second is the velocity of the wave and this for an electromagnetic wave is 3×10^{10} cm. per second. Hence we obtain the formula:—

$$Velocity = Frequency \times Wavelength$$

For the 3 cm. wavelength, for example, the frequency is $\dfrac{3 \times 10^{10}}{3}$ i.e. 10^{10} cycles per second or 10,000 megahertz (MHz) where a megahertz is equivalent to one million cycles per second.

Generally the transmission in this wave band is done between 9300 and 9500 MHz (X-Band). This corresponds to a wavelength of between 3·2 and 3·1 cm.

In the same way it can be calculated that for 10 cm. transmission a frequency of 3000 MHz is required.

Generally the transmission in this wave band is done between 3000 and 3246 MHz (S-Band). This corresponds to a wavelength of between 10·0 and 9·2 cm.

Logic and Computation

Logic can be defined as the science of reasoning. In Control Systems one is often concerned with high-speed switching circuits acting either electronically or pneumatically. The type of logic used here is called "Switching Logic". One or more discrete signals are processed and a single output is obtained. This involves a kind of computation and generally only two types of signals occur : Logic ZERO and logic ONE. To understand the digital computation, a brief introduction to Number Systems is given.

Number Systems

The most common number system is the Decimal System, which is based on counting techniques on 10 fingers and its origin can be traced far back into man's history.

In all number systems the various marks or symbols carry different weights and in most systems the *position* of the symbol within the number is important, for example in number 2060 the zeroes carry different weights.

The number of discrete characters (coefficients or marks) used in writing down a number is known as the *Radix* or *Base* of the system. The radix number of a decimal system is 10, a binary system 2 (coefficients 0 and 1), a hexa-decimal system 16 (coefficients 0, 1, 2 .... 9, a, b, c, d, e, f).

By convention, marks which have the least effect on the total are placed on the right, while marks which have a greater effect on the total value of the number are placed on the left. If we therefore consider two marks, the mark on the left is called the *most significant* digit and the mark on the right the *least significant* digit.

If one can think of numbers divided into columns which are assigned a weight, then in order to obtain the weight of the next more significant column on the left, multiply the value of the column by the radix. For example in Table AII.1, the three is positioned in the column carrying a weight of 1,- the four in the column carrying a weight of 10 and the six in the column carrying a weight of 100.

1000	100	10	1
	6	4	3
		1	2
1	7	2	3

TABLE AII.1

It follows that numbers can be expressed as a function of the powers of their radix. For example:

Decimal value   *Radio*

Decimal value		Radio
62711	$6\times10^4+2\times10^3+7\times10^2+1\times10^1+1\times10^0=62711$	10
23	$1\times\ 2^4+0\times\ 2^3+1\times\ 2^2+1\times\ 2^1+1\times\ 2^0=10111$	2
102	$1\times\ 3^4+0\times\ 3^3+2\times\ 3^2+1\times\ 3^1+0\times\ 3^0=10210$	3
7231	$1\times\ 8^4+6\times\ 8^3+0\times\ 8^2+7\times\ 8^1+7\times\ 8^0=16077$	8
442847	$6\times16^4+c\times16^3+1\times16^2+d\times16^1+f\times16^0=6\mathrm{cldf}$	16

In the same way fractional numbers can be represented :

| 627·11 | $6\times10^2+2\times10^1+7\times10^0+1\times10^{-1}+1\times10^{-2}=627\cdot11$ | 10 |
| 5·75 | $1\times\ 2^2+0\times\ 2^1+1\times\ 2^0+1\times\ 2^{-1}+1\times\ 2^{-2}=101\cdot11$ | 2 |

APPENDIX II 315

The Binary System

This system is very common in electrical, electronic and pneumatic systems. For example, a simple electric switch has two states, open (0) or closed (1). Similarly, a valve or transistor can be in the conducting state (1) or non-conducting state (o), and there are numerous examples such as these. The words "True" (1) or "False" (0) are also often used.

A *b*inary dig*it* is usually known as a *bit*. For example, the number 101100 has 6 bits. The number of bits in a binary system is greater than the number of digits of the equivalent number in the decimal system. If there are n digits to represent a decimal number and there are m bits to represent the same number in the binary system, then $10^n = 2^m$ or $m = n/\log 2$.

For example, a 5-digital number in the decimal system can be represented by $5/0{\cdot}301$ i.e. $16{\cdot}6$ or 17 bits.

To translate decimal numbers into binary numbers, divide successively by 2 and record the remainder after each step. The first remainder is the least significant bit and the last remainder the most significant bit. For example :

$$
\begin{aligned}
&457\\
457/2 &= 228 + 1\\
228/2 &= 114 + 0\\
114/2 &= 57 + 0\\
57/2 &= 28 + 1\\
28/2 &= 14 + 0\\
14/2 &= 7 + 0\\
7/2 &= 3 + 1\\
3/2 &= 1 + 1\\
1/2 &= 0 + 1
\end{aligned}
$$

Number is 111001001

To convert this number back into decimals, place the powers of 2 below the bit positions and add.

```
1   1   1  0  0  1  0  0  1
8   7   6  5  4  3  2  1  0   (powers of 2)
256+128+64    +8       +1  =  457.
```

Addition, Multiplication etc.

Addition and subtraction follow the same rules as in the decimal system, but it has to be realised that $1+1=0$ with 1 to carry to the column on the left. If 1 has to be subtracted from 0, i.e. 0—1, borrow one digit (bit) from the column on the left and together with the 0 it now makes 10 (2). The column on the left, of course, has lost a 1. Here are some examples :—

Example 1, addition.

```
      1  0  1  1  0  1      45
  +   1  0  1  1  1  1      47
                           ---
      0  0  0  0  1  0
carry 1      1  1     1
      ---------------------
      1  0  1  1  0  0  0
carry             1
      ---------------------
      1  0  1  1  1  0  0      92
```

Example 2, subtraction.
borrow and remainder

```
                                 1
                            10  10  10
                  1   1   1   0   0   1   57
              —                 1   1   1   1   15
                 ─────────────────────
                  1   0   1   0   1   0   42
```

In computers, subtraction is often carried out by the following manipulation :—

216 — 45 = 216 + (99 —45) — (100 — 1), thus

```
        *           indicates column to the left of the most
                    significant digit of the complement.
      216
   +   54           complement of 99
      270           borrow 1 from the column indicated by the asterisk
                    and add to the least significant digit.
   +    1
      171
```

Here are some examples for the decimal system :—

```
45672—4234=41438        45672—14234=31438        95672—4234=91438
    *                       *                         *
   45672                    45672                     95672
+   5765 (compl. 9999)  + 85765 (compl. 99999)   +   5765 (compl. 9999)
  ─────                   ──────                     ──────
   51437                   131437                    101437

+  → 1                  +  →  1                   +  → 1
  ─────                   ──────                     ──────
   41438                    31438                     91438
```

In the binary system an equivalent operation is carried out by using the 1's complement which means that the number to be subtracted is taken away from a series of 1's. To do this, simply substitute 0 for 1 and vice versa.

Start by placing the asterisk, showing the column to the left of the most significant bit of the number to be subtracted (or the complement and include the zero on the left). Add the complement, borrow 1 from the column indicated by the asterisk and add to the least significant bit.

Examples :—

```
1011011 — 10101 = 1000110        1011011 — 1000110 = 10101
  91        21       70            91        70        21

     *                                *
   1011011                          1011011
+    01010  complement            + 0111001  complement
  ───────                          ────────
   1100101                          10010100

+  ┕━━➤1                          + ┕━━━➤ 1
  ───────                          ────────
   1000110                            10101
```

Boolean Algebra

It is convenient to represent logical problems in control systems by equivalent algebraic expressions. Logic devices and circuits are merely the hardware used to solve the problems. Logical algebra was invented by the Reverend George Boole (1815–1864), from whom is derived the name "Boolean Algebra".

There are two basic types of logic systems, Combinational Logic and Sequential Logic. Combinational logic systems are those not requiring a memory, the basic functions being known as AND, OR and NOT. Sequential systems such as counters., shift registers, sequence controllers, all require some form of memory device.

With logic, only two signal levels are used, Logic 1 and Logic 0. For an output f, a signal is either present (logic 1) or not present (logic 0), as for instance in a light switch. However, a logic 0 signal does not necessarily mean that it has an actual numerical value of 0 but it generally refers to the lower of two signal levels. The higher level is then known as logic 1 and the system is said to work in *positive logic* (in general use to-day). Sometimes the lower of the two levels is taken as logic 1 ; this is known as *negative logic*.

Basic Function 1—The OR Function

The OR function is defined as a device having two or more inputs and one output. An output is present only when one or more inputs are on.

This can best be illustrated by two water pipes A and B, joining into a single pipe f. Each pipe has a stop valve (Fig. AII.17). From the figure it can be seen that valve A, OR valve B, OR both, must be open to give the required value f.

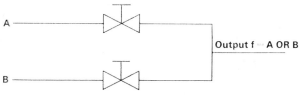

A

Output f = A OR B

B

FIG. AII.17   BASIC PRINCIPLE OF *OR* FUNCTION

The sign for OR is either $+$, or $\mathbf{v}$, so we can write $A+B=f$, or $A \mathbf{v} B=f$. We shall use the $\mathbf{v}$ sign so as not to confuse the OR function with numerical addition. If the value 1 is used when the valve is open and the value 0 when the valve is closed, then applying all possible states to the valves, the following combinations are obtained :—

For A closed, B closed : $f = 0 \mathbf{v} 0 = 0$
For A closed, B open : $f = 0 \mathbf{v} 1 = 1$
For A open, B closed : $f = 1 \mathbf{v} 0 = 1$
For A open, B open : $f = 1 \mathbf{v} 1 = 1$

The above can be expressed in tabular form known as a TRUTH TABLE, thus :

Inputs A v B		Outputs f
0	0	0
0	1	1
1	0	1
1	1	1

TABLE AII.2

As there are only two possible operating states (on and off), for n input variables the possible combinations of settings must be 2^n. In the above example there are two input variables and 2^2 or 4 possible combinations. With three inputs, for example, $f = A \vee B \vee C$, the possible number of settings would be 2^3 or 8.

The Truth Table for $A \vee B \vee C$ would be :—

Inputs A ∨ B ∨ C			Output f
0	0	0	0
0	0	1	1
0	1	0	1
1	0	0	1
1	1	0	1
1	0	1	1
0	1	1	1
1	1	1	1

TABLE AII.3

The symbols used for logic elements are laid down by the British Standards Institution. The OR symbol and the symbol for $A \vee B \vee C$ are shown in Fig. AII.18.

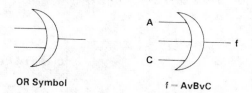

OR Symbol $f = A \vee B \vee C$

FIG. AII.18   ELECTRONIC SYMBOLS

Basic Function 2—The AND Function

The AND function is defined as a device having two or more inputs, all of which must be 'ON' to give an output.

Again the water pipe analogy can be used, except that this time the stop valves are arranged in series.

FIG. AII.19   BASIC PRINCIPLES OF *AND* FUNCTION

It can be seen from Fig. AII.19 that both valves, A AND B, must be open to give an output f. Therefore A AND B = f, or more conveniently, $A.B = f$. The "dot" is the logic AND connective. The Truth Table for $f = A.B$ would again have $2^n = 2^2$ or 4 possible combinations of inputs.

| Inputs | | Output |
A	B	f
0	0	0
0	1	0
1	0	0
1	1	1

TABLE AII.4

And or ff = A.B.C :—

| Inputs | | | Output |
A	B	C	f
0	0	0	0
1	0	0	0
0	1	0	0
0	0	1	0
1	0	1	0
1	1	0	0
0	1	1	0
1	1	1	1

TABLE AII.5

The symbol for the AND function is shown in Fig. AII.20.

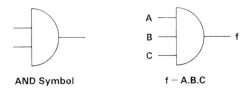

AND Symbol   **f = A.B.C**

FIG. AII.20   ELECTRONIC SYMBOLS

Basic Function 3—NOT

The NOT function is the process of logical inversion in which the output signal is NOT equal to the input.

As there are only two signal states, namely 0 and 1, an input of 1 gives an output of 0 and conversely, an input of 0 gives an output of 1.

This operation is also known either as logical complementing, or logical negation, and is represented by placing a bar over the variable, i.e.:—

$$f = \text{NOT } A = \overline{A}$$

The Truth Table for $f = \text{NOT } A = \overline{A}$ is shown in Table AII.6.

Input A	Output $f = \overline{A}$
0	1
1	0

TABLE AII.6

The symbol is illustrated in Fig. AII.21.

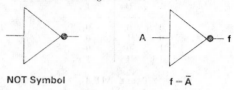

NOT Symbol $f = \bar{A}$

Fig. AII.21   ELECTRONIC SYMBOLS

Electronic Applications

Logic functions can be performed electronically by relays, valves (diodes and triodes) and transistors.

OR Gates

These are illustrated in Fig. AII.22.

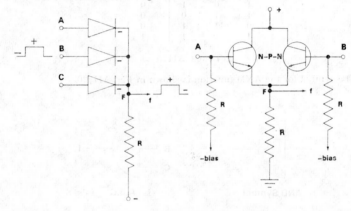

Inputs		Output
A	B	$f = A \vee B$
0	0	0
0	1	1
1	0	1
1	1	1

Fig. AII.22   *OR* GATES

In the diode OR gate the cathodes—which are all interconnected are biased through the resistor R in such a way that the common negative voltage is more negative than the lowest level of any of the input voltages. With no input signals the diodes are conducting but the current is relatively small and the output, being connected directly to the input signals, yields the same i.e. zero signal level. When one Or two, OR all three input signals become highly positive with respect to the cathode (1), there will be a strong increase of current through R and the potential of point F—where the output signal is taken off—will rise simultaneously and display an output of 1.

The transistor OR gate shows two N-P-N transistors in parallel. The output is taken from the common load resistor across the emitter and earth. This type of arrangement is called "emitter-following" and ensures that the output signal is in phase with the input signal. That means, that if the input at A, OR the input at B, OR both, become positive with respect to the emitter, current flows and one, or both, transistors are conducting, and the output at F becomes positive with respect to earth as electrons always flow from negative to positive. In other words, the output is 1. If *both* input signals are at their lowest level (0), both transistors are cut off and no current is flowing so that the output signal at F remains at ground (earth) level and records a zero.

Note, by the way, that if negative logic were used and the bases were positively biased with respect to the emitter so that in the normal condition both transistors were conducting, only a negative voltage applied to *both* the bases would have made the output signal go negative. This leads to the conclusion that a gate can be used either as a positive OR gate, or as a negative AND gate. The reverse is also true. If a gate is operating as a positive AND gate, it can also act as a negative OR gate. This can also be easily seen from the Truth Table, by replacing the zeros by one, or vice versa.

AND Gates

These are illustrated in Fig. AII.23. In the diode AND gate, the diodes are reversed as compared with the arrangement in the diode OR gate in Fig. AII.22. The diodes are normally conducting (zero level) unless a positive voltage, equal to the positive anode voltage, is applied to all the inputs of A AND B AND C.

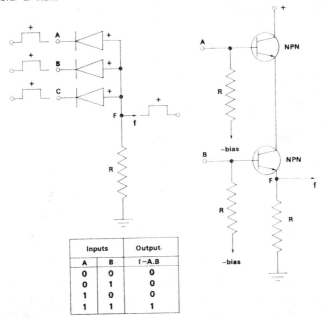

Inputs		Output.
A	B	f = A.B
0	0	0
0	1	0
1	0	0
1	1	1

FIG. AII.23 *AND* GATES

The transistors are connected in series, instead of in parallel as shown in the transistor OR gate in Fig. AII.22.

Multivibrators

A multivibrator consists of two amplifier-inverters (valves or transistors) coupled back-to-back, i.e. the output of one (taken from the load resistor of the anode or collector end) drives the input (at grid or base) of the other and vice versa. The output of the first stage—which differs 180° in phase with the input signal of the first stage—is connected to the input of the second stage. Hence the output signal of the second stage differs 2 × 180° with the input signal of the first stage, i.e. the two signals are in phase and there is 100% positive feed-back. Such a condition will generally lead to oscillations of some kind. Whether these oscillations are sustained or have to be maintained by the application of an external trigger pulse, depends on the type of coupling between the two stages.

The multivibrator is included in this section on account of its usefulness in carrying out logic operations and binary calculations such as the generation of binary numbers, counting, and the timing and delaying of arithmetical calculations.

There are three types of multivibrators :—

Type 1: The bi-stable multivibrator or flip-flop. As the name implies, it has two stable states : Either stage 1 is conducting (ON) and stage 2 is non-conducting (OFF) or the other way round. To change the state of the flip-flop, the circuit has to be reset by a "reset" pulse which turns either the conducting stage to OFF, or the non-conducting stage to ON.

The coupling between the two stages is *by resistors only.*

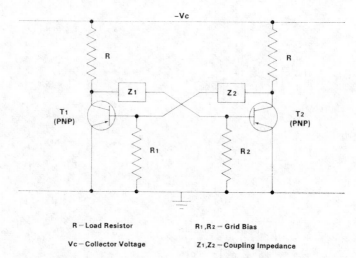

R — Load Resistor R_1, R_2 — Grid Bias

V_c — Collector Voltage Z_1, Z_2 — Coupling Impedance

Fig. AII.24   Basic Multivibrator Circuit (P-N-P Transistor)

Type 2: The astable multivibrator or free running multivibrator. The circuit oscillates from one unstable state to the other and produces a series of (nearly) rectangular pulses at the load resistor of the collector.

The coupling between the two stages is *by capacitors only*. If the components of the two stages are identical, a square-wave generator is obtained.

Type 3 : *The mono-stable or one-shot multivibrator*. The circuit does not need resetting as it has only one stable state and returns to this state after being temporarily disturbed. The input trigger pulse at the grid or base of the stable stage is short, but the single output pulse is square or rectangular in shape and has the same polarity as the input pulse. This means that the trailing edge of the output pulse is delayed with respect to the trailing edge of the input pulse.

The coupling between the two stages is *mixed*, one by means of capacitor, the other by means of a resistor.

Fig. AII.24 shows the basic configuration of a P-N-P transistor.

The Bi-stable Multivibrator or Flip-Flop

Fig. AII.25 shows the bi-stable multivibrator using P-N-P transistors. Although the circuits are symmetrical, it is extremely unlikely that its operation would settle down to a state where both transistors are conducting and drawing equal collector currents, as slight changes in the transistor characteristics and resistors due, for example, to temperature fluctuations, would upset this (unstable) equilibrium. Suppose that a slight increase of collector current takes place at stage 1 transistor. This will increase the voltage drop across the load resistor and point A will rise in potential, i.e. it becomes less negative.

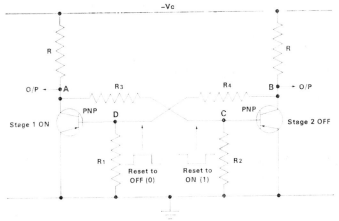

FIG. AII.25   BASIC FLIP-FLOP (BI-STABLE)

Point C, at the base of stage 2 transistor becomes more positive, owing to the coupling R3, and opposes the forward emitter bias (base negative w.r.t. emitter). The current through this transistor and the collector load resistor is now reduced so that the potential at point B becomes more negative, which, via coupling R4, makes point D at the base of stage 1 transistor more negative. This will result in increasing the forward bias of this transistor resulting in an increase in the current through the transistor and its load resistor, and a positive rise at point A, and so on. This action will continue at a very rapid speed (measured in micro-seconds) until stage 1 transistor is fully conducting, or ON, and stage 2 transistor is cut off, or OFF.

M

The flip-flop is now in a stable state with point A at practically earth potential (resistance of transistor is negligible compared with the load resistor) and point B at a voltage equal to the negative supply. To reset the flip-flop to its former stable state, a positive reset triggering pulse is required at the base of stage 1 transistor in order to switch it to OFF, or a negative reset pulse at the base of stage 2 to switch it to ON.

The logic symbol for the flip-flop is shown in Fig. AII.26, where the left-hand stage is known as the ZERO side, and the right-hand stage as the ONE side.

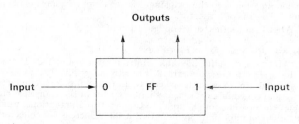

FIG. AII.26   LOGIC SYMBOL FOR FLIP-FLOP

The stable states of a flip-flop can be defined as follows :—

ZERO STATE : Zero side base input = 0 and collector output = 1.
              One side base input = 1 and collector output = 0.
ONE STATE :   Zero side base input = 1 and collector output = 0.
              One side base input = 0 and collector output = 1.

Similar analyses can be carried out for the N-P-N transistor flip-flop, and for the triode valve flip-flop.

Thr Memory Function or SET-RESET Flip-Flop

Each transistor in Fig. AII.25 acts as an inverter, or NOT gate, and if an OR gate could be inserted between the base of stage 1 (zero side) and resistor R4, and also between the base of stage 2 (one side) and the resistor R3, a memory element, or Set-Reset Flip-Flop, is obtained. Such an element only responds to logic "one" signals and cannot be reset via the same input line which caused the stable state. This is made clear in Fig. AII.27.

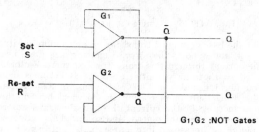

FIG. AII.27   MEMORY FUNCTION

If (Fig. AII.27) a logic "one" is applied to the S-line, output Q is set to the "one" level, and any successive signals, whether "zeroes" or "ones", applied to the S-line cannot alter the state or the output. Only the application of a "one" signal to the R-line can cause output Q to be reset to zero.

Hence in order to alter the state of the memory device, successive "one" signals have to be applied *alternately* to the set and reset lines.

The Astable Multivibrator or Free Running Multivibrator

This is shown in Fig. AII.28, using P-N-P transistors and two coupling capacitors. The capacity of the capacitors and also the resistances of the load resistors at the collector ends are relatively small, while the resistance(s) of R_1 and R_2 are very high. This results in a very quick charging of the capacitors —from the negative supply voltage via one load resistor and the opposite

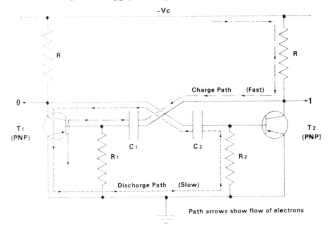

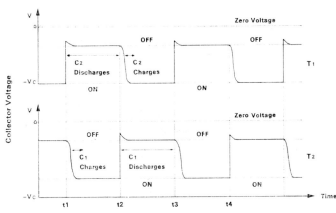

FIG. AII.28   FREE RUNNING MULTIVIBRATOR

transistor (which has a negligible resistance)—and a very slow discharge—to earth via a base resistor and the opposite transistor.

Hence, when the zero-side starts conducting and the "lower" end of the collector has a rise of potential in the positive direction, this rise is immediately transmitted via C2 to the base of the one-side ("immediately" because there is no resistor involved and the electrons on the right-hand side of the capacitor travel directly to earth via the emitter of T2). This causes a decrease in collector current of T2, and the lower end of its load resistor becomes more negative. Capacitor C1 must charge now to the higher negative collector voltage and it does so via the load resistor. As the value of the load resistor is relatively small so that the time-constant RC1 is brief, the collector voltage of T2 rises very fast with a slight 'exponential' rounding off (Fig. AII.28). The charging of the capacitor C1 makes the base of T1 negative so that T1 becomes fully conducting with its collector voltage being close to earth's potential. Simultaneously transistor T2 is driven to "cut-off" (time t_1.)

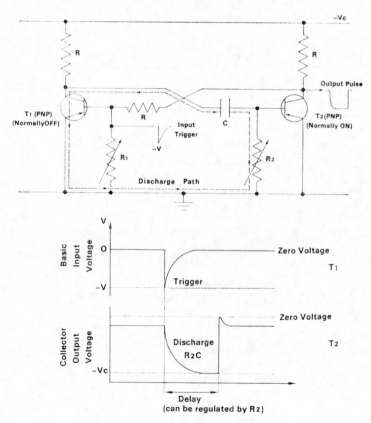

Fig. AII.29   One-Shot Multivibrator

APPENDIX II

There follows now an *inactive* or *quiescent* period (t_1 to t_2) with C2 slowly discharging and the electron flow going from its left-hand plate, via the collector, base, emitter of T1 and resistor R2, back to the right-hand plate (the electrons cannot travel in the direction of the arrow of transistor T2). The discharge is slow owing to the high resistance of R2.

At time t_2 the base of transistor T2 has lost sufficient positive charge (becoming more negative) for the transistor to start conducting again and the process is now reversed with respect to the previous condition. The result of the switching condition at t_2 is to turn transistor T2 ON and transistor T1 OFF.

From t_2 to t_3, while capacitor C1 is discharging, the circuit again is in a quiescent state and this will last until the T1 base has reached a voltage that will turn T1 ON again. Thus the cycle is repeated with the transistors being switched ON and OFF, alternately, with relative long rest periods in between, each with a duration which is dependent on the time-constants R1C1 and R2C2.

The Mono-stable or One-Shot Multivibrator

The basic circuit is shown in Fig. AII.29 with one resistance coupling and one capacitance coupling between the collector of one transistor and the base of the other. Without further analysis it should now be understood that transistor T1 (with resistance coupling to the base) is normally OFF, and transistor T2 (with capacitance coupling to the base) is normally ON. When a negative trigger is applied to the base of T1 it will be turned ON and T2 will be turned OFF. Between t_1 and t_2 the circuit will be in an unstable state, the duration depending on the time-constant R2 C, after which the circuit will flop back again to its initial stable state. There is therefore a time delay (t_2-t_1) after the input pulse before the output pulse is produced. This delay can be changed by varying the circuit constants.

Applications of the Flip-Flop

An application of the flip-flop is shown in Fig. AII.30, where two numbers A (the *addend*) and B (the *augend*) are added by means of the half-adder. The word "half-adder" implies that it only adds the addend to the augend, but it does not carry the carry digit generated by the *previous* (lower order) addition.

The binary addition table of two numbers is illustrated in Table AII.7.

Addend A	Augend B	Sum	Carry
0	0	0	0
0	1	1	0
1	0	1	0
1	1	0	1

TABLE AII.7

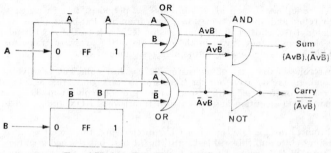

FIG. AII.30   FLIP-FLOPS + HALF-ADDER

By substituting the numbers for A and B given in Table AII.7, in the 'Sum' and 'Carry' expressions (Fig. AII.30), verify for yourself that the output answer is correct.

APPENDIX III

FUNCTION OF COMPONENTS OF A RADAR SET

Conventional Units

The main units of a marine radar set are:—

Transmitter;
Pedestal Unit, including the scanner;
Mixer;
Receiver;
Display

A block diagram showing the different components, is illustrated in Fig. AIII. 1

Two auxiliary units are included in the diagram:—

Time-base Unit;

Range Unit.

Transmitter

The Transmitter is the producer of short square-shaped powerful pulses. It comprises the Modulator, Trigger Circuit and Magnetron.

The function of the *Modulator* is the collection—via a network of inductors and capacitors—of power from the mains which it then releases at regular intervals to the Magnetron where the cm. wave pulse is created.

There are several types of Modulators. Some are designed as an ordinary (hard) valve, others are based on magnetic principles and consist of a coil filled with an iron core which is easily magnetically saturated. The most common design is a valve, filled with hydrogen (thyratron) or an inert gas. The latter type is known as a "trigatron" and is basically a diode with an umbrella-shaped anode. Between anode and cathode is a third electrode which is connected to the *Trigger Circuit*, a circuit containing valves (multivibrators) which in turn operate to produce a sharp rise and fall in potential. When there is no transmission, electrical energy is stored up at the anode of the trigatron. At regular intervals the trigger circuit applies a direct voltage pulse to the third electrode in the trigatron. When this takes place, the gas in the modulator becomes ionized, a spark jumps over and the valve is "fired". The energy released—graphically represented as a square-shaped pulse—is fed via a transformer into the cathode of the magnetron. The number of times the modulator fires per second, is known as the pulse-repetition frequency or p.r.f.

The *Magnetron* (Fig. AIII. 2) is a diode valve with a series of cylindrical cavities set in the wall of a cylindrical block of copper, which acts as the anode. It operates between the poles of a powerful magnet. When the square-shaped pulse of electrical energy reaches the cathode of the magnetron, then, under the influence of the sudden electric field, and drawing energy from the

329

magnetic field, the electrons inside the magnetron burst into violently swirling oscillations inside the cylindrical cavities. The result is the generation of a Radio Frequency Pulse (R.F. Pulse) of very short wavelength. The output is taken off by means of a pick-up loop (or probe) situated in one of the resonant cavities.

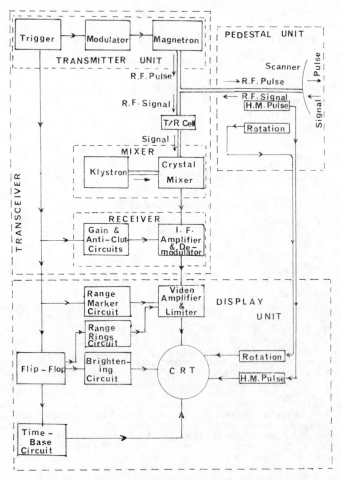

FIG. AIII.1   RADAR BLOCK DIAGRAM

Usually in marine radar the magnetron transmits in the 3 or 10 cm. wave band. The frequency of transmission depends on the cavity size. The higher the frequency required, the smaller the diameter of the cavities. During the transmission heat is developed which must be conducted away. Owing to the relatively small diameter of the magnetron for the very high frequencies, additional cooling flanges are provided. Excessive heat is sometimes carried away by an air flow produced by an extractor fan but in other cases natural convection is introduced.

The duration of transmission for a cm. set varies between 0·05 and 1 microsecond. In radar theory we take the velocity of the radar pulse to be 300 metres per micro-second. Hence the length of the pulse leaving the scanner varies between ·05 × 300, i.e. 15 metres and 300 metres respectively. The choice of pulse-length, as pointed out in Chapter 1 is related to the range and range discrimination. The relationship with range also applies to the p.r.f. which can be between 500 and 4000 pulses per second.

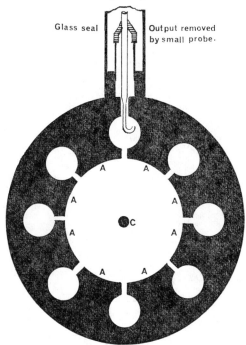

Glass seal   Output removed by small probe.

FIG. AIII.2 MAGNETRON

The *peak* power of transmission is stated in the Manual. The *mean* power is far less. If, for example, the peak power is rated as 75 kilowatt (kW), the length of transmission is ·5 micro-second and the p.r.f. is 1000 pulses per second which means that the time between successive transmissions is 1000 micro, seconds, then the mean power is $\frac{·5}{1000}$ × 75,000 i.e. 37·5 watts.

Let us now have a closer look at the radar block diagram (Fig. AIII. 1). The transmitter is synchronized with the *Time-base Circuit* via the *Flip-Flop*. The Flip-Flop is a system of transistors which produce one long pulse each time it is triggered by a short one. This means that as soon as transmission starts, the time-base circuit generates a current waveform which will deflect the spot over the PPI (trace). Hence the time-base circuit is connected with the CRT or, in the cases where the PPI deflection coils are stationary, this waveform signal is conducted to the scanner unit where it is superimposed on the signal representing the scanner rotation and the combined signal is then fed to the CRT.

At the same time, the Flip-Flop performs two further functions:—

(a) It initiates the *Range Rings Circuit*, which then starts generating short electric pulses at predetermined equal intervals so that their recording on the screen corresponds with equal range intervals. These pulses are amplified in the receiver.

(b) It initiates the *Brightening Circuit* which produces a brightening pulse equal in length to the time-base sweep and it is this pulse which makes the trace on the PPI visible to the operator.

An illustration is shown in Fig. AIII. 3, where the length of pulse transmission is 0·5 micro-second and the length of time-base sweep and brightening pulse corresponds to 300 micro-seconds (24 M range scale). The time between transmissions is 1000 micro-seconds which corresponds to a p.r.f. of 1000. p.p.s. Hence we see that between successive transmissions the spot on the PPI is doing work for a period of 300 micro-seconds and remains at rest for a period of 700 micro-seconds during which no echoes are recorded.

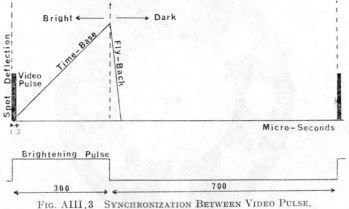

FIG. AIII.3 SYNCHRONIZATION BETWEEN VIDEO PULSE,
BRIGHTENING PULSE AND TIME-BASE

In modern radar installations a new technique, known as *Interscan*, has been evolved by making the spot on the PPI paint a line or lines of any desired length on the screen *when it is not used for recording echoes* (in the example quoted above this would be during the 700 micro-second period). This painting can be done many times per second and *the direction of its trace does not depend on the position of the scanner*. It means that the human eye will see a line *continuously* displayed and the brightness of this line will remain uniform during a scanner revolution. Such a line can be employed as an electronic bearing cursor and by varying its length can also be used for the measurements of ranges.

The interscan period can also be used for re-generation of the time-base plus re-cycling of the echoes (if a storage device for the signals is available) so that a brighter display is obtained.

The Interscan is generated independently of the rotating mainscan by means of a second rotatable deflection coil, situated inside the off-centring coils.

The radar block diagram also shows that the transmitter is connected to the Pedestal Unit via a *waveguide*. The waveguide is a hollow tube, made of copper for good conduction and its cross-section may be either rectangular or circular. The dimensions of the cross-section depend on the wavelength used. Through this waveguide the R.F. pulse enters the scanner unit.

This idea of transmitting electromagnetic energy along the inside of a hollow tube is by no means new. If a current flows along a conductor, the magnetic energy is transmitted through the surrounding medium, assumed to be air. It is a property of an electromagnetic wave that, as the wavelength decreases, the current tends to seek the surface of the conductor (skin effect). For waves of small wavelengths, such as the cm. waves, the core of the conductor is completely left out. The current flows along the outside, and the magnetic lines of force are propagated between the walls of the tube.

Waveguides should not have sharp corners, and their run must not be too long (not longer than 15 metres), otherwise the pulse and signal attenuate too much. They can be kept short by having the transmitter and the mixer (R.F. Head) externally fitted below the scanner, as is the case with some sets. Such a set will be cheap to run, as the magnetron power called for can be less, while the set will still retain the same performance as any other set with a longer waveguide. They are, however, a headache when repairs in the unit have to be carried out.

Pedestal Unit

The Pedestal Unit comprises the *Scanner Unit* and sometimes the *Echo-box*. The *Magnetron* and *Mixing Crystal* can be part of the Pedestal Unit when the transceiver is mounted partly externally.

The Scanner Unit consists of three parts:—

    (*a*) Aerial or Antenna;

    (*b*) Aerial Motor;

    (*c*) Synchro Transmitter.

(*a*) *Aerial.*

Cm. and mm. radar sets employ an aerial which consists of a *parabolic metal plate* or a *slotted waveguide*. The dimensions of an aerial should be large in relation to the wavelength used. The width of a marine radar aerial is large compared with its height. This ensures a narrow horizontal beam-width and a much wider vertical beam-width. The aerial width varies between $1 \cdot 2$ and $3 \cdot 7$ metres.

Parabolic aerials have a horn or flare placed at their focal point. This horn is the upper end of the waveguide and rotates round with the reflector. By means of a small piece of waveguide it is connected to a rotating joint which forms the link between the fixed and the moving guide. The horn radiates the energy which is re-radiated by the plate.

It is impossible to make a directional aerial without producing side lobes, as these are due to subsidiary maxima of field strength in directions other than the main beam. But in practice, the horn is not a point-source, as it should be in theory, and it also gives rise to a certain amount of scattering effect (re-radiation of secondary unwanted sources). This will *increase* the side-lobe effect as compared with the ideal parabolic reflector or with the slotted waveguide aerial.

A well-known type of aerial was the so-called *Cheese* type, illustrated in Fig. AIII. 4. The plate is in the form of a parabolic cylinder. The protective plates also help to shape the beam. The horn forms an obstruction and may increase the side-lobe effect. To overcome this defect, the *Tilted Parabolic Cylinder* type was designed where the reflector is tilted with respect to the

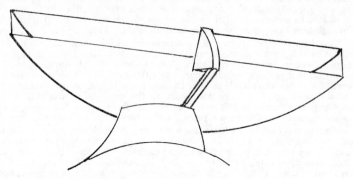

FIG. AIII.4   CHEESE TYPE SCANNER

horn and the re-radiated wave has a reasonably clear path in front of it (see Fig. AIII. 5). Sometimes balancing fins are fitted at the extremeties of this reflector so that continuous and uniform rotation is maintained in high relative wind speeds.

The reflector plate can be made of meshed material, provided the size of meshing bears a close relationship with the wavelength used.

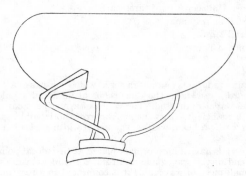

FIG. AIII.5   TILTED PARABOLIC CYLINDER

In the majority of radar sets there is a common horn and waveguide both for the outgoing pulse as well as for the returning signal. Receiver *and* transmitter are connected to this common waveguide. In other words, there exists a duplexing arrangement for transmission and reception. Some radar sets, however, have a cheese-type scanner which is sliced by a horizontal plane into two parts. Transmission and reception in such a case are performed separately by the upper and lower parts of the parabolic mirror, which forms the backplate of the "cheese". Part of the transceiver is incorporated in the pedestal unit and two short waveguides lead to transmitter and receiver (mixing crystal) separately.

Another type of aerial is the *slotted waveguide*. In this type an end section of the guide about 1·8 m. (or 3·6 m.) long, with slots cut in the wall of the guide, is turned in a horizontal plane. The slots are sometimes cut obliquely, but vertically polarized aerials are in use with horizontal slots cut in the broad face of the waveguide at both 3 and 10 cm. Also polarizing is sometimes done in the window of slotted waveguide aerials or sometimes by blocks alongside the slots. The side of the waveguide with the slots is placed inside a horn which runs along the length of the waveguide. The slots in the guide interrupt the current and a strong magnetic field is created which radiates outwards. Overall radiation takes place which is not the case with other aerials where the centre of the plate receives stronger radiation than the extremities. The horizontal length of the scanner and the vertical cross-section of the horn determine, respectively, the extent of the horizontal and vertical beam-width. Side-lobe effect is little. There exists however, a limited horizontal width of scanner (about 6 m.) beyond which the radiation becomes less efficient. This explains why large scanners used for Harbour Radars, employ the parabolic plate shape.

Slotted waveguide aerials are boxed in a sealed case of toughened perspex or fibre-glass in order to keep the slots free of obstructional matter and to make the waveguide watertight.

An example of the slotted waveguide is shown in Fig. AIII. 6.

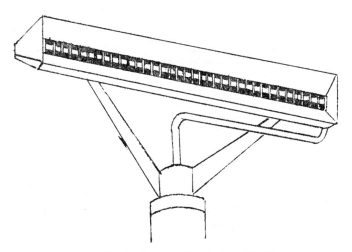

FIG. AIII.6   SLOTTED WAVEGUIDE AERIEL

(b) *Aerial Motor*.

The aerial motor drives the aerial round through a chain of gears at a uniform speed.

Sometimes the aerial unit is enclosed in a fibre-glass weathertight dome. The size and power of the driving motor can then be reduced and the cost of components is cheaper.

(c) Synchro Transmitter.

The function of the synchro transmitter (see Appendix II: *Transmission of Signals*) is to transmit the rotation of the scanner to the sweep on the display so that the trace rotates in unison with the scanner and its true or relative indication on the screen corresponds to the direction of the aerial.

In the pedestal unit a rotor having a single winding is mechanically coupled to the aerial driving motor. Surrounding the rotor are three windings (stator) spaced 120° apart in the shape of a star or a delta. When the rotor turns, alternating voltages are induced in the stationary windings and their resulting magnetic field is related to the direction of the rotor (and the scanner). The stator coils in the pedestal unit are connected to three similar coils in the display unit (synchro receiver).

From this point, either one of two arrangements can take place:—

(i) The voltages induced in the stator coils of the display unit produce a field which will turn a rotor either via torque transmission or control transmission (True Motion systems). See Appendix II: *Transmission of Signals*. This rotor in the display unit rotates at the same rate as the rotor in the scanner unit. Coupled to this rotor are the deflector coils which now rotate in synchronism with the scanner round the neck of the PPI. *The time-base circuit in this arrangement*, which causes the spot on the screen to move outwards, *is directly connected to the deflector coils of the PPI* (see Fig. AIII. 1).

(ii) There is no rotor in the display unit but the stator coils themselves form the *stationary* deflection coils. Their *magnetic field rotates* and this affects the direction of the electron stream in the CRT. With this arrangement *the time-base waveform is fed as a sawtooth current in the rotor of the synchro transmitter of the pedestal unit*. The time-base waveform is superimposed on the signal which determines the angular position of the aerial and the combined signal is fed to the deflection coils of the CRT.

The result in both cases ((i) and (ii)) is the same. The trace is painted across the face of the PPI indicating a direction corresponding to the direction in which the scanner is pointing at the moments of transmission and reception.

There are radar sets which do not use synchro transmission. Here the aerial driving motor operates an alternator. This alternator drives a similar motor coupled to the deflector coils which then should turn in synchronism with the aerial driving motor.

Turning back to the block diagram (Fig. AIII. 1), we see that there is another connection between the pedestal unit and the display unit. This connection is marked *Heading Marker Pulse*. The action of the pulse is to brighten the trace when the aerial radiates energy in the direction of the ship's head. The final result is a painted line on the screen. This line, called the Heading Marker represents the heading of the vessel.

There are many methods to initiate the pulse. Sometimes a cam is mounted on the driving shaft of the aerial motor which, during each revolution touches a small roller in the scanner housing when the aerial points dead ahead. A contact is then established, a circuit completed and a pulse, meanwhile amplified, is fed to the display unit. If the heading marker shows up only as a faint line, then the fault may lie in dirty contacts. In some sets an automatic cleaning apparatus has been provided.

Fig. AIII. 7 shows a very simplified diagram. By loosening the stud **screws** and turning the split ring the heading marker can be aligned on initial **installation** or re-adjusted when it appears that a heading marker error is in **existence.**

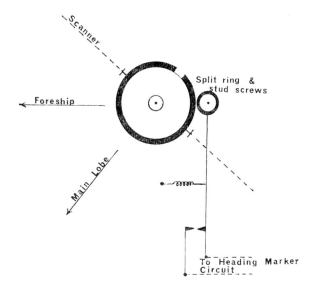

FIG. AIII.7   HEADING MARKER CONTACT

The *Echo-box Unit*, which is used as a Performance Indicator, is sometimes, as already stated, part of the Pedestal Unit. Its purpose is to check the *overall performance* of a radar set, especially important when no echoes are present. It consists of a small box, named the echo-box. In some cases it is attached directly to the scanner plate and rotates *with* the scanner. This is illustrated in Fig. AIII. 8. When the box is opened, a small part of the electromagnetic energy is fed straightway to the box via a short piece of waveguide through a small aperture giving access to the box. The result is an artificial echo on the display having the shape of a sun. In other cases, the echo-box is attached to the *stationary* part of the pedestal unit and then displays a "feather" on the screen. Often the box contains drying material (dessicator) so that its action cannot deteriorate.

The box can be opened by means of a switch—marked *Performance Indicator* or *Performance Monitor*—on the display unit. The switch operates a relay which lifts a resistor out of the guide, or in other cases energizes a polarized ferrite switch and in doing so aligns the plane of polarization of the ferrite with the plane of polarization of the radiated field which then is transmitted through.

As the aperture in the box is small, the wave is trapped inside for some time, reverberating between the walls of the box. The term "ringing" is used, denoting the relatively slowly weakening oscillations inside the box. The outleaking electromagnetic energy is collected by the horn and fed to the receiver, and thence to the display. A paddle rotates at a high speed inside the box and its function is to prevent the distortion of the frequency (the resonance frequency of the box is sensitive to temperature changes), so that the re-radiated wave retains its frequency. Failure of the paddle breaks up the pattern on the display.

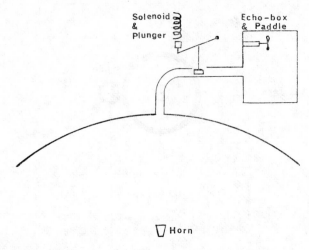

FIG. AIII.8   ECHO-BOX INSIDE SCANNER UNIT

This "ringing" of the echo-box extends the returning echo considerably. Even when the echo-box is placed in the scanner unit and pulse transmission has ceased, its echo keeps going on. Its performance therefore does not depend on the minimum distance at which targets are detected. This is in contrast with superstructures close-by like masts and funnel, whose echoes are not recorded on the screen.

The echo-box is sometimes placed well astern of the pedestal unit, for example on the funnel or the main mast. A small aerial (di-pole) picks up the radiation which is reflected via a parabolic reflector into the box. The unit is not always connected to the display and the echo plume is then a permanent feature of the screen. A vibrating reed inside the box prevents the lowering of frequency.

Sometimes the echo-box is replaced by a neon tube. When the scanner sweeps over the tube, the gas inside is ionized and causes the plume to appear on the display. The extent of the plume gives a check on the transmitted power. See Fig. 3.7. Instead of the plume a separate meter can be used to measure the current (a function of the output peak power) through the neon tube. Note from Fig. 3.7 that the output from the neon tube, although converted into a waveform does not use the radar receiver. The receiver, in this case, can be tested by means of a separate tuned cavity inside the transmitter unit which collects the energy so giving rise to a response on the PPI in the form of a sun.

The latest type of Performance Monitor employs a transponder located adjacent to the scanner. The transponder re-radiates a calibrated signal back which can be made visible on the display as a bright sector (assuming the performance is up to standard).

Mixer

If we follow the block diagram (Fig. AIII. 1) we see that the returning echo or R.F. signal, via the same waveguide as that used by the pulse, branches off to the *mixer*. This R.F. signal is very weak and needs amplification and one would expect that valves could be employed for this purpose. However, the transit of an electron inside a valve is slow compared with the frequency of the cm. wave. The electron flow inside the valve cannot follow the frequency. Hence control over the input signal is lost and ordinary valves are unsuitable for amplification of signals of very high frequency. A solution is found by *lowering* the frequency while maintaining the characteristics of the waveform variation and *then* to apply amplification. This lowering of frequency can be achieved by *mixing* the incoming wave with another wave which has *nearly* the same frequency (or wavelength). The latter wave is produced by a local oscillator, called the *Klystron*.

An example may make this clear. Suppose we have two tuning forks which have pitches i.e. frequencies of 450 and 460 vibrations per second respectively. It will be noticed that when both forks are sounded together (mixing), the resultant intensity waxes and wanes in cycles, which can be proved to have a frequency of 460 − 450 i.e. 10 vibrations per second. This lower frequency of 10 vibrations per second is known as the *beat frequency*. If we now change the frequency of one fork from 460 to 455 vibrations per second, by loading it, for example with a very small quantity of wax, then the beat frequency will decrease from 10 to 5 (455 − 450) vibrations per second. In other words, a much lower frequency is produced, but having variations similar to those of the tuning fork which has changed in frequency. This process is known as 'heterodyning'.

The klystron can be compared with one of the tuning forks. It is a valve consisting of a cathode opposite which is a highly negative reflector plate. Between the cathode and the reflector are two cavities, kept at a positive potential and separated by a gap. The electrons drift from the cathode, group in bunches through the gap, are then repelled by the reflector, retrace their path and fall back in step with the outgoing stream. Amplified oscillations are set up between the gap and imparted to the cavities. The output is taken off from the cavities and passed to a crystal where the actual mixing with the R.F. signal takes place. The crystal called *Crystal Mixer* has also rectifying (smoothing out) qualities.

All modern radars have two crystal mixers in parallel (*balanced mixer*). See Fig. 3.6. Such an arrangement practically eliminates the amount of noise introduced in the klystron.

After the mixing the resultant frequency is known as the *Intermediate Frequency* or I.F. It has retained its pulse character and can now be amplified.

If, for example, the signal has a frequency of 9400 MHz and the frequency of the klystron output is 9370 MHz, then the I.F. is 30 MHz.

It follows that the I.F. can be altered by changing the frequency of the klystron. This can be done by screwing a plug or slug into the walls of the cavity of the klystron, or by changing the reflector voltage. This procedure is called *Tuning*.

In many sets the tuning can be done manually. In other sets an *Automatic Frequency Control (A.F.C.)* unit keeps the tuning adjusted. This unit is a feed-back unit tapping off part of the I.F. output from the crystal mixer and feeding it into two channels, one tuned slightly above, one tuned slightly below I.F. The difference in the output signals from the two channels is rectified and applied as a potential to the reflector of the klystron. This potential will change when the I.F. starts drifting and will re-tune the klystron.

Note that after the mixing the signal is no longer transmitted via a waveguide.

See Fig. AIII. 1. The I.F. signal is then conducted by means of a co-axial cable.

Sometimes the klystron is replaced by a *Gunn Device Transmitter* (named after J. B. Gunn, who, in 1963, discovered the phenomenon now called the Gunn Effect). Basically this consists of a very thin slice of n–type gallium arsenide contained between a cathode and an anode. Gallium arsenide displays a remarkable property when an electrical field is applied to it. When the electrical field is increased in strength, the electrons inside the gallium arsenide initially increase in velocity (as is normal), but then, at a certain critical value of the field, the electrons suddenly increase their mass and their velocity drops. The field applied to the Gunn device is above this threshold value, and as a result electrons coming into the crystal at the cathode are excited to low-mobility heavy electrons, forming a concentrated bunch or so-called " domain ". These domains drift towards the anode, thereby screening the anode, so that the field throughout the remainder of the crystal falls in intensity below the threshold value. During the transit time, therefore, production of heavy electrons stops. As the domain reaches the anode and is absorbed by the anode, the field at the cathode is being built up again until it exceeds the threshold value and a new domain is established.

This action repeats itself continuously and current pulses are produced which are used to induce oscillations in a cavity or waveguide resonator. The transit time of the domains through the crystal determines the frequency of the pulses. In fact the frequency is inversely proportional to the layer thickness. By making this layer very thin, radar pulses of suitable frequency are generated.

A final word must be said about the crystal mixer. The rectifying action of such a crystal can be damaged when too much power passes through it. We say, in such a case, that the crystal "burns out" though in actual fact no visible change occurs. To protect the crystal from the powerful R.F. pulse produced by the magnetron, a safety device is inserted in the waveguide leading to the mixer. This device is the *T/R (Transmit/Receive) Cell* which acts like a shutter which is automatically closed when transmission takes place, but otherwise is open. See Fig. AIII. 9. Passive T/R cells without a primer have lately been introduced.

Therefore the receiver is out of action during transmission.

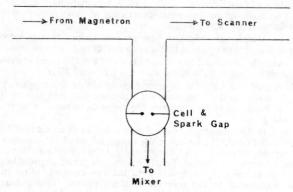

FIG. AIII.9   T/R CELL

The T/R cell consists of a cavity filled with water vapour. When the magnetron is pulsed, a spark is produced across the electrodes. The spark ionizes the gas and a short is effected across the opening, so preventing energy reaching the receiver.

The blocking of the receiver during transmission is one of the factors governing minimum range. If a radar set, for example, transmits for 0·25 micro-second, its scanner radiates a pulse of a length equal to $0·25 \times 300$, i.e. 75 metres. When the tail end of the pulse has left the scanner and the front part of the pulse has travelled 75 metres, the blocking to reception is removed. Hence echoes from objects at a range of 37·5 metres return to the scanner 0·25 micro-second after transmission. Just at that moment transmission stops and reception commences. Such echoes have the first opportunity to be fully recorded. The minimum range for such a set transmitting during 0·25 micro-second would be 37·5 metres. Owing to the slight delay which elapses before the T/R cell is ready to receive after transmission has ceased, the minimum range stated above will be slightly more.

In some sets, as has been discussed already, magnetron and mixer are mounted externally in which case the exit for the outgoing pulse and the entrance for the incoming signals are separated—an example of this arrangement is the "double-cheese" type of scanner. A T/R cell is really not required here and the above consideration for minimum distance does not apply. In fact, it will be much shorter. Even so, crystals in these sets "burn out" sometimes, especially when another radar transmitter is close-by. Hence a protective gas cell might also be of use here.

Receiver

In the receiver, the I.F. signal is fed to a series of amplifiers (gain) and further to a detector or *demodulator* which smoothes out the signal. After this process, the signal receives a new name, namely the *Video Signal*. Pulses from the Range Unit are also fed in for amplification.

The output of the receiver can—depending on the setting of a switch on the display unit—be fed into a *Differentiator Unit* (not shown in block diagram Fig. AIII. 1) or Peaker, which incorporates a circuit of capacitors and resistors. Its function is to sharpen the rectangular echo pulse (signal) to a peaked one. The leading edge of the waveform is accentuated against a trailing tail. Its effect is to "chop-up" masses of echoes caused by rain, snow or hail.

Another way to overcome the difficulty caused by rain clutter is *Circular Polarization*. Use is made of the fact that rain drops are spherical and will therefore reflect equally, radiated energy without regard to its direction of polarization, whereas a ship, being of a much more irregular shape, will usually accentuate radiation of one particular polarization.

Normally radio waves can be regarded as vibrations occurring in either a vertical or a horizontal direction as the wave moves forward in space. Such waves are described as vertically or horizontally polarized, respectively.

A circularly polarized wave is one in which the direction of polarization rotates during this forward motion, so that if at a given interval the plane of polarization is vertical, then, a quarter of a cycle later, or what is the same thing, a quarter of a wavelength further on, the polarization will take place in the horizontal plane.

A target illuminated by such radiation will reflect energy in the same way as it would if the radiation were of successively varying polarization.

If the target structure is identical in all planes, each component of the circularly polarized wave will be reflected equally and the wave reaching the receiver will also be circularly polarized. A rainstorm, consisting as it does of spherical drops is such a target, but a ship which is without these regular features, will nearly always reflect more energy of one particular polarization.

The radar equipment is required to discriminate between echoes from targets such as ships, and those of rain and to remove the latter from the display. The technique involves the generation of plane polarized waves, the conversion of these to circular polarization, the conversion of the reflected wave back to plane polarization and the detection of this wave in the receiver.

The converter used for each of the above transformations is known as the *quarter-wave plate* and one such plate normally functions for both transmitted and received waves.

A quarter-wave plate consists of a number of parallel metal strips, accurately dimensioned and spaced, with their short dimensions in the direction of propagation of the wave and their long dimension making an angle of 45° with the direction of polarization of the incoming wave. The dimensions of the plate are so chosen that the component of the wave at right angles to the strips emerges from the plate a quarter of a wavelength *after* the component along the strips. These two waves emerging from the plate together meet the requirement for circular polarization.

When this circularly polarized wave is reflected from a symmetrical object such as a group of rain drops, it will appear to be rotating in the same direction as the transmitted wave, when viewed from the scanner, or, if viewed along its direction of propagation, it will seem to rotate in the *reverse* direction.

This wave now passes through the quarter-wave for the second time and, as before, the component normal to the strips is changed in phase relative to that parallel to the strips, with the result that the emerging wave is again plane polarized at 45° to the direction of the plates, but now *at right* angles to that of the original transmission. This direction of polarization cannot be accepted by the receiver waveguide and thus no echo from the rain storm is seen.

With an echo of a ship, however, the two components will not be equal to amplitude and the wave emerging from the plate, going to the receiver, will not be at 45° to the plates. This means that it will not be at right angles to the acceptance direction for the wave guide and therefore an echo will be received.

Marine radars use two scanners, back-to-back. One is used for the radiation of plane polarized waves, while the other one uses a series of slots in its waveguide which act as the quarter-wave plate and produces the circular polarization. A waveguide switch, operated by remote control from the display, decides which slotted waveguide and which scanner should be used.

Application of *anti-clutter* initiates a negative pulse from trigger of modulator—synchronized with transmission—which applied to the grid of a valve suppresses the gain. Restoration of gain is achieved when the pulse leaks away via a resistor.

Some radar sets have a switch on their display unit marked *Lin. Log.* which selects the linear receiver circuit or the logarithmic receiver circuit. See Fig. AIII. 10.

Strong I.F. signals might have their amplitude restricted owing to the saturation level in amplifier stage no 4 when the linear path of the receiver is employed; by switching over to logarithmic action, a parallel path is formed for the signal via demodulator 2. At the output end the video signal is the sum of the saturated signal from A4 *and* the unsaturated signal from A3. For a still stronger signal, amplifier stage no. 3 might become saturated and if the linear path were used A3 and A4 contribute little to the amplification of the signal. When the LOG. setting is selected, the video signal is the sum of the output of D1, D2 and D3.

A graph is shown in Fig. AIII. 10 showing the output signal plotted against

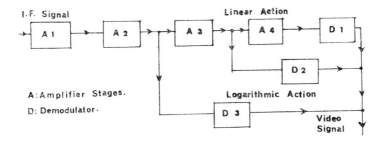

A: Amplifier Stages.

D: Demodulator.

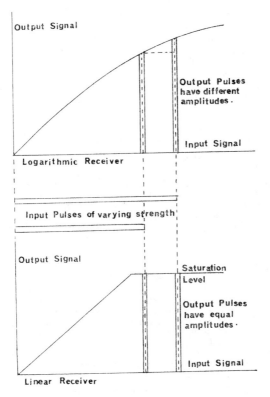

FIG. AIII.10   LOGARITHMIC RECEIVER

the input signal for both the logarithmic and the linear receiver. The graph shows that the logarithmic receiver can, in certain cases, display the difference in echo strength between two strong signals which otherwise would yield the same echo intensity on the screen. Examples are echoes of ships in strong sea clutter and two echoes, one from a small object near the ship and a second one from a strongly reflecting object some distance away from the ship.

There is one point in connection with a certain property of the receiver which needs mentioning. Operators not familiar with electronics are often puzzled when they look through the instruction manual and are confronted by expressions such as "I.F. Band". It is hoped that the following explanation will make this clearer.

We know already that the signal consists of a pulse and its waveform is rectangular in shape. It is a law of nature that it is impossible to transmit a wave pulse in just that form. Water waves, for example, cannot have the shape of blocks. It is possible, by mixing waves of different frequencies (or wavelengths) to obtain a wave shape which approximately resembles the rectangular form. Hence to produce and cope with such a shape, the receiver must be able to pick up different frequencies; in other words it must not be tuned to one single frequency, but to a *band* of frequencies.

The shorter the pulse, the more frequencies should be available to build it up. In fact, it can be proved that the width of the frequency band is inversely proportional to the duration of the pulse. If, for example, the receiver bandwidth is 10 MHz on the short range scales for the short pulse and if the long pulse has a duration of five times the short pulse, then on the longer range scales the receiver bandwidth must be changed to 2 MHz.

Display

The echoes are displayed on the Cathode-Ray Tube or CRT. The CRT is an electronic valve with several anodes. It contains:—

(*a*) The *gun*, comprising cathode, grid and one or more pierced anodes which fire electrons at

(*b*) the *screen*, which is transparent and acts as the final anode. The inside of the screen is coated with a special chemical so that the electrons paint a *spot* when they hit the screen. Materials used to obtain a short afterglow (or persistence) contain zinc, for example Willemite and they cause a green display; screen coating materials possessing a long afterglow are made of Magnesium Fluoride and these give the screen an orange phosphorescence.

In colour radar several phosphor layers (from red- to blue-yielding phosphorescence) are applied to the screen and their excitation depends on the depth of penetration of the electron stream (a function of the anode voltage of the tube).

The coating is sprayed with an extremely thin film of aluminium to which an extra high tension is applied, so that it forms a good conducting path for the electrons, which can then be carried away along the walls of the tube by means of another conducting layer known as "Aquadag" coating (colloidal graphite coating). Another great advantage of the aluminium film is that some lightrays (which form the picture) although originally scattered backwards towards the inside of the tube, are reflected and return in the direction of the observer so that the brightness is increased.

The extra high tension applied to the rear of the screen also has two purposes:—

(i) It makes the screen act as the final accelerating anode;

(ii) It avoids *secondary emission* by which is meant a dislodging of the secondary electrons in the screen by the bombardment by the stream of primary electrons.

The intensity of the electron stream (and of the spot) is called the *brilliance* and it is governed by the *brightening pulse* which is applied as a positive pulse to the grid or as a negative pulse to the cathode. This pulse extracts the electrons from the cathode.

The *leading* edge of the modulator pulse in the transmitter triggers off the square wave generator (or flip-flop). See radar block diagram Fig. AIII. 1. The square wave generator responds with three simultaneous actions:—

(a) Operates the time-base circuit which sets the spot into motion so that it moves uniformly over the screen, thereby describing a *trace, sweep* or *time-base*.

(b) Produces the brightening pulse so that the trace becomes visible.

(c) Initiates the oscillation of the fixed range marks generator, so that zero range coincides with pulse transmission. The result is the production of pips on the screen of the A-Scan or rings on the PPI.

When the spot has reached the edge of the screen, the time-base voltage generator quickly returns to zero, the brightening pulse is suppressed to a low or zero intensity and the oscillation of the range unit dies away. On the screen the spot returns, at an extremely high speed in semi- or complete darkness, to its starting point (this movement is known as "fly-back").

The number of sweeps per second drawn across the face of the tube equals the number of transmissions per second which again equals the p.r.f.

The CRT is a dull-emitter valve i.e. the filament of the cathode is treated with barium, strontium or calcium oxide, as a result of which electrons are emitted in large numbers at relatively low temperatures.

Because the tube is exhausted it has to be shaped to prevent it imploding. This is the reason why, at present, the face of the tube has to be curved. The dimension of the tube is limited because of physiological considerations. It is thought at present that a 16-inch diameter tube is the largest which the human eye can accommodate when the trace rotates at 20 r.p.m.

Two types of CRT can be utilized:—

A: A-Scan or Short Persistence Tube.

The trace is horizontal and diametric from left to right when facing the screen. The electron stream is amplitude-modulated i.e. the strength of an echo can be derived from its amplitude. The pips die away quickly.

It is used in marine radar sets for testing and monitoring the different components and also for tuning up the Automatic Frequency Control (AFC) Unit.

B: Plan Position Indicator or PPI, also known as a Long Persistence Tube.

The trace is rotated round in unison with the rotation of the scanner and echoes previously recorded are retained during a period of at least one scanner revolution. In many displays the trace is from the centre to the edge. The electron stream is intensity-modulated, i.e. the strength of an echo can be derived from its brightness.

Another division can be made according to the construction:—

I: Electrostatic Tube (Fig. AIII. 11).

Voltage control is used.

The brightening pulse is applied to the *grid* which has the shape of a hollow cylinder and the grid therefore controls the *brilliance* of the tube. Three *anodes* with different potentials *focus* the beam by means of electrostatic lines of force and the small apertures in the annular discs. The anodes also *accelerate* the electrons. The horizontal *Y-plates* move the beam *vertically*. The signal (echo-pulse) and the range marks are applied here. The vertical *X-plates* move the beam *horizontally* and are connected to the time-base generator.

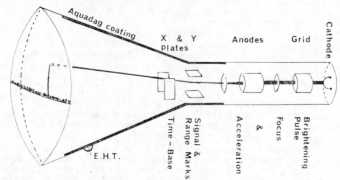

FIG. AIII.11 ELECTROSTATIC TUBE (A-SCAN)

The X and Y plates are known as *deflection plates* and note that they are named according to the direction of the deflection they cause and *not* according to the position they occupy inside the tube.

II: Electromagnetic Tube (Fig. AIII. 12).

Current control is used.

Signal, range marks and *brightening pulse* are applied to the *grid* as positive pulses. The brightening pulse is often, instead, applied to the *cathode* as a negative pulse.

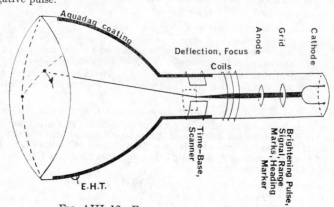

FIG. AIII.12 ELECTROMAGNETIC TUBE (PPI)

The *anode accelerates* the electrons.
The magnetic field of the *external focus coil focuses* the beam.

The *deflection coils*, spaced 120° apart around the neck of the tube, have a two-fold function:—

(a) They cause the spot to move *from the electronic centre towards the edge* and are therefore connected to the time-base generator.

(b) They cause the trace to move in *step with the scanner* and are therefore linked up with the pedestal unit. Either the coils move mechanically or their magnetic fields rotate (see Pedestal Unit section).

The gun (not shown in diagram) is often bent, so deflecting slightly the lighter electrons while trapping the heavier ions. This prevents, for a limited period, a burn developing in the centre of the screen.

The heading of the ship is shown on the screen by the *heading marker* flash. The information is supplied by the heading marker pulse from the pedestal unit which increases the brightening pulse when the scanner radiates energy ahead in the direction of the fore-and-aft line.

The deflector coils can be turned mechanically, by means of gears, through a certain pre-determined angle. This is done by a control on the display unit (*picture-rotate* or *picture alignment* control) and it turns the picture and heading marker bodily round.

From the description given above, it should already be clear that generally the A-Scan is an electrostatic tube and the PPI an electromagnetic tube, though there exist minor modifications and combinations, for example, electrostatic focusing is often combined with electromagnetic deflection.

Characteristics of electrostatic tubes compared with those of electromagnetic tubes are their great sensitivity, their long necks and comparatively small screens, low power but high voltage requirements and the great number of *internal* electrodes which makes construction more difficult and expensive.

Time-base Unit

In the electrostatic tube a voltage is applied between the X-plates which increases gradually, then suddenly drops back to the original value. The broad principle is that a square pulse is applied to the grid (base) of a valve (transistor) which then becomes conductive and *slowly* charges a capacitor which is connected to the X-plates. At the end of the pulse the valve (transistor) cuts-off and the capacitor discharges *quickly* through a resistor.

In the electromagnetic tube the current in the deflection coils rises linearly, then suddenly falls to zero (sawtooth). This is done by applying—as in the electrostatic tube—a sawtooth voltage pulse but *on top* of a rectangular pulse across the terminals of the deflection coil. The rectangular pulse is introduced to off-set the back-e.m.f. caused in the deflection coil by the linearly rising current (electromagnetic induction).

Range Unit

The fixed range marks generator can be a tuned-anode oscillator. The oscillations are then converted to square pulses followed by a peaking or differentiating process. The result is a series of short electric pulses at pre-determined equal intervals, so that their recordings on the screen correspond to equal range intervals.

If, for example, the distance between marks (rings) should represent two miles, then the frequency of the number of oscillations generated per second $= \dfrac{1,000,000}{4 \times 1852/300}$ i.e. 40·5 KHz (velocity wave is 300 metres per micro-second, one mile is 1852 metres, one KHz is 1000 cycles per second and the difference in distance covered by two pulses causing echoes of objects two miles apart in the same bearing from the scanner, is four miles).

The variable range marker generator produces only one pip during the cycle from one transmission to the next. The time at which this pip is generated depends on the potential of the grid of a valve (base of a transistor) and can be regulated by means of a variable resistance or potentiometer.

Bearing Marker Unit

This unit consists of an encoder assembly which drives the bearing read-out directly, and a circuit which generates the bearing marker line. In the encoder assembly is fitted a gear-driven six-track digital encoder disc (see Fig. 15.11), whose angular position can be turned through 360° by the bearing marker control. Brushes are fitted in the assembly to convey the angular information on each track to the digital read-outs.

Gyro-Stabilization

The simplest way to provide gyro-stabilization is the case when the deflector coils round the tube are *stationary in the unstabilized condition*. A compass repeater motor is then coupled via simple gearing to the deflection coils. The coils will then rotate round the tube when course is altered.

When the deflection coils *turn round* the tube in the *unstabilized condition*, then the problem is more complicated. One can either fit a separate synchro transmission system between the gyro transmitter and the neck of the tube, or—and this is the usual method—place a Differential Transmitter (see Appendix II, *Transmission of Signals*) between the Pedestal Unit and the Display (see also Fig. AIII. 13). As discussed before, the basic transmitter consists of a rotor containing a single-phase winding positioned between a stator possessing a three-phase winding. The differential transmitter has a *three-phase winding on rotor and stator*. This means that *two* types of information can be transmitted, one determined by the *position* of the rotor of the differential transmitter, the other by the *electrical field* from the stator of the differential transmitter. The stator can be linked-up with the scanner unit, the rotor with the transmitter of the gyro-compass (see Fig. AIII. 13).

True Motion or Track Indication

A simplified diagram is shown in Fig. AIII. 13.

The True Motion mechanism consists of a resolver which is a complex gearing arrangement with two input and two output shafts. One input shaft is driven by a step-by-step motor. This motor—the "Distance Run" motor, when the presentation switch is set to TRUE MOTION LOG is operated by the output of an electrical circuit whose input is provided by the ship's log. When the presentation switch is set to TRUE MOTION MANUAL, the log output is simulated by a low frequency oscillator, which is controlled by the SPEED control on the true motion panel.

The second input of the resolver is coupled via a differential gear to the same compass repeater as that used for azimuth stabilization purposes. This differential gear enables the difference between the ship's head and the course made good to be corrected by the COURSE MADE GOOD Control, to which the second shaft of the differential is coupled.

The two output shafts of the resolver are coupled via clutches (for resetting purposes) to two shift potentiometers which are connected to the separate N-S and E-W Shift coils mounted outside the deflection coils around the neck of the PPI.

The resolver translates the "course" and "distance run" from the information received from the compass motor and the "distance run" motor, into two mechanical components, representing the N-S and E-W component of the

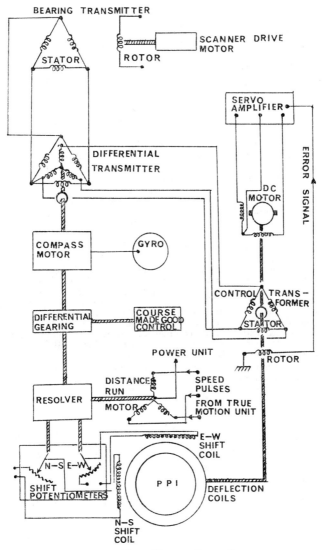

FIG. AIII.13   TRUE MOTION BLOCK DIAGRAM

"distance run". The rotation of the two output shafts rotate the potentio-meter wipers and so produce a shift in the PPI picture in a direction and speed corresponding to that of the ship. When the potentiometer wipers reach the limit of their traverse, switches are operated which arrest the true motion presentation and light the reset warning lamp on the true motion control panel.

Radar sets without true motion use synchro systems with torque trans-mission in order to synchronize the rotation of the PPI deflection coils with that of the scanner (see Appendix II).

Radar sets with true motion facilities use synchro systems with control transmission for this purpose and a servomechanism is introduced in the display unit (see Figs. AII. 13 and 14 and AIII. 13).

The electronic bearing marker is produced by a switch in the deflection coil system. Once per revolution of the deflection coil this switch is closed for a very short moment at a bearing determined by the setting of the calibrated bearing marker control. As the switch closes it applies a pulse to the video circuit and thus brightens the trace on the CRT at the selected bearing.

The block diagram may differ for the various types of radar sets. Some-times the design is all-electronic. The course resolver is a sine-cosine potenti-ometer, where the ship's speed is resolved into Northings and Eastings components, generating voltages which are integrated, amplified, and fed to the appropriate deflection coil.

Conclusion

This Appendix dealt with the components of the *Conventional Marine Radar Set* with *Gyro Stabilization* and *True Motion*. The discussion concerning the functions and lay-out of units was very general and extremely simplified. Only block diagrams were dealt with and circuitry was completely omitted. Readers who are interested in greater detail and actual circuits should refer to standard textbooks about radar technique. The purpose of this Appendix is to make the operator familiar with the *main* components. It will help him in identification for maintenance purposes and to give sound preliminary advice to an expert if repairs have to be carried out.

Sailors are not usually electronic experts and the best radar set for the observer and operator is that one which seldom breaks down.

Extracts from the Marine Radar Performance Specification, 1968

A Performance Specification for a General Purpose Shipborne Navigational Radar Set

1. Object of the Equipment

To provide an indication in relation to the ship of the position of other surface craft and obstructions and of buoys, shorelines and navigational marks in a manner which will assist in avoiding collisions and in navigation.

2. Range Performance

(a) Under normal propagation conditions, when the radar aerial is mounted at a height of 50 ft. above sea level, the equipment shall give a clear indication of:

  (i) Coastlines

   At 20 nautical miles when the ground rises to 200 ft.

   At 7 nautical miles when the ground rises to 20 ft.

  (ii) Surface objects

   At 7 nautical miles on a ship of 5000 grt.

   At 3 nautical miles on a fishing vessel of length 30 ft.

   At 2 nautical miles on a General Purpose Conical buoy. (The General Purpose Conical buoy used in type-testing is a 10 ft. diameter heavy duty general purpose buoy fitted with a conical cage and lights, this having an echoing area of approximately 10 sq. metres.)

(b) The equipment will be considered as capable of fulfilling the above conditions if under normal propagation conditions:

  (i) For equipment operating in the 9300–9500 megahertz band the echo of a GPC buoy, whose range from the radar scanner is 2 nautical miles, is visible on the radar for 50 per cent of the scans when the aerial is mounted at a height of 50 ft. above sea level.

  (ii) For equipment operating in the 2900–3100 megahertz band the performance is at least 5 db above that specified in (i).

3. Minimum Range

When the radar aerial is mounted at a height of 50 ft. above sea level the echo of a GPC buoy shall be visible on the radar for 50 per cent of the scans down to a range of 50 yards on the most open scale provided, when it shall be clearly separated from the transmission mark. Without subsequent adjustment of the controls, other than the range selector control, the echo of the buoy shall remain visible at a range of one mile.

4. Range Accuracy

The means provided for range measurement shall be fixed electronic range rings or fixed electronic range rings and variable range marker. Fixed range rings shall enable the range of an object, whose echo lies on a range ring, to be measured with an error not exceeding 1½ per cent of the maximum range of the scale in use, or 75 yards, whichever is the greater. The accuracy of range measurement using a variable range marker, if provided shall be such that the error shall not exceed 2½ per cent of the maximum range of the displayed scale in use or 125 yards, whichever is the greater. Direct reading of the range shall be provided. No additional means of measuring range shall be provided whose accuracy is less than that required of the variable range marker.

5. Range Discrimination

On the most open scale appropriate, the echoes of two GPC buoys shall be displayed separate and distinct when the buoys are on the same bearing 50 yards apart and at a range of 1 mile.

6. Bearing Accuracy

Means shall be provided for obtaining quickly the bearing of any object whose echo appears on the display. The accuracy afforded by this means of measurement shall be such that the angle subtended by any two objects whose echoes are separate and distinct and appear on the edge of the display may be measured with a maximum error of 1°.

7. Bearing Discrimination

To achieve adequate bearing discrimination and minimise the occurrence of spurious echoes due to radiation outside the main beam, the aerial radiation pattern in the horizontal plane shall conform to the following specification as a minimum requirement, the figures relating to one-way propagation only:

		Position relative to maximum of main beam	Power relative to maximum of main beam
Main beam – – – – –		± 1°	− 3 db
		± 2·5°	−20 db
Side lobes – – –	within	±10°	−23 db
	outside	±10°	−30 db

8. Roll

The performance of the set shall be such that targets remain visible within the range limits laid down in paragraphs 2 and 3 when the ship is rolling ±10°.

9. Radio Frequency and Polarisation

The radiation which shall consist of either horizontally or vertically polarised waves must be substantially confined within the bands internationally agreed for marine radar below 9500 MHz and must not cause harmful interference to services outside these bands. A facility for introducing circular polarisation may be added if desired; it must be possible to switch it to linear from the display position.

Note: At present, racons in the United Kingdom only operate on 9300–9500 megahertz horizontally polarised waves.

10. Scan

The scan shall be continuous and automatic through 360° of azimuth at a rate of not less than 20 rpm in relative wind speeds of up to 100 knots. The aerial must start and run satisfactorily in this wind.

11. Display

The equipment shall provide a satisfactory main plan display of not less than $7\frac{1}{2}$ inches effective diameter, and it must be possible, if necessary by simple switching, to operate this as a satisfactory relative motion plan display. The display shall be capable of being viewed in indirect light by two persons simultaneously without undue restriction of the angle of view. Optical magnification may be employed provided the display so magnified remains within the accuracy limitations of other clauses of the Specification. When optical magnification is employed to attain the minimum diameter of display permitted in this clause, the magnification device shall be a fixed component of the display.

12. Scale of Display

The equipment shall provide at least six scales of display whose ranges of view shall be either (a) $\frac{1}{2}$ or $\frac{3}{4}$, $1\frac{1}{2}$, 3, 6, 12 and 24 or more nautical miles, or (b) $1\frac{1}{2}$, 3, 6, 12, 24 and more than 24 nautical miles. Range scales in addition to the six specified may be provided but where provided must comply with the clauses specifying range and bearing accuracy. On a $\frac{1}{2}$ or $\frac{3}{4}$ nautical mile range scale range rings shall be displayed at intervals of $\frac{1}{4}$ nautical mile. Six fixed range rings shall be provided on each of the other scales of display; but where a variable range marker is provided on any scale of display the fixed range rings available on that scale may be at intervals of $\frac{1}{4}$ or 1 or 4 nautical miles as appropriate.

Where the facility of a continuously variable range of view is provided, the maximum range of view so obtainable shall be 6 nautical miles. On this variable range scale fixed range rings at intervals of $\frac{1}{2}$ nautical mile and a variable range marker shall be provided.

Positive indication of the range of view displayed shall be given except that, where a continuously variable range scale is provided, it need only be given at the mandatory ranges of view covered by that range scale. Positive indication of the interval between range rings shall be given.

The natural scale of display shall not be less than 1 : 933,888.

13. Heading Indicator

Means shall be provided whereby the heading of the ship is indicated on the display with a maximum error not greater than $\frac{1}{2}$°. The thickness of the displayed heading line shall not be greater than $\frac{1}{2}$°. Provision shall be made for switching off the heading line and shall take the form of a spring-loaded switch or equivalent device which cannot be left in the " heading marker off " position.

14. Azimuth Stabilisation

Means shall be provided to enable the display to be stabilised to give a constant north-upwards presentation when controlled from a transmitting compass. This facility may be either an integral part of the equipment or an additional component which shall be capable of being readily fitted to the equipment. When a transmitting compass is used to control the orientation of the picture or of any part or parts of the display unit, the accuracy of alignment with the compass transmission shall be within $\frac{1}{2}$° with a compass rotation rate of 2 rpm.

The equipment shall operate satisfactorily for relative bearings when the compass control is inoperative or not fitted.

15. Sea or Ground Stabilisation

Where this facility is incorporated in an equipment the following additional conditions must be met:

(i) *Transference of speed*

By timing the movement of the trace origin over a distance of not less than 50 per cent of the radius of the scale in use or half an hour, whichever gives less time, the error in speed when compared with the speed calculated from the input signals or the setting of the manual speed control shall not exceed 5 per cent or $\frac{1}{4}$ knot, whichever is the greater. This accuracy shall apply in using any of the range scales provided for true motion.

(ii) *Transference of course*

By measuring the movement of the trace origin over a distance of not less than 50 per cent of the radius of the scale in use or half an hour, whichever gives less time, the error in course when compared with the compass input or the setting of the manual course control shall not exceed 3°.

(iii) *Circular re-setting*

The motion of the trace origin must stop when 25–40 per cent of the radius of the display remains between it and the edge of the displayed picture. Automatic resetting is acceptable.

(iv) *Re-set warning*

When re-setting is by manual control, a visual warning shall be provided on the display unit. An audible warning may also be provided which can be switched off when not required.

16. Operation

(i) The set shall be capable of being switched on and operated from the main display position.

(ii) The set should be suitable in all respects for operation by the officer of the watch. The number of operational controls shall be kept to a minimum and all of them shall be accessible and easy to identify and use.

(iii) The design shall be such that it is impossible for an operator to cause damage to the equipment by misuse of the controls.

(iv) The time required for the set to become fully operational after switching on from cold shall be no longer than four minutes. A standby position shall be provided so that the set may be brought to the fully operational condition from the standby position within one minute.

17. Power Supply

(i) The set shall be capable of normal working under the following variations of ship's power supplies:

AC Variation from nominal voltage $\pm$ 10 per cent

Variation from nominal frequency $\pm$ 6 per cent

DC Variation from nominal voltage:

110/220v DC$+$10 per cent, $-$20 per cent

24/32v DC$+$25 per cent, $-$10 per cent

(ii) Provision shall be made by means of fuses, overload relays, or other means, for protecting the equipment from excessive current or voltage.

18. Electrical and Magnetic Interference

(i) *Electrical interference*

All reasonable and practicable steps shall be taken to ensure that the equipment shall cause a minimum of interference to radio equipments such as are customarily installed in merchant vessels. Spurious radiation in the frequency band 50 KHz to 300 MHz should be such that the maximum excursion of the voltage waveform at the output of a test receiver (as measured on a cathode ray oscilloscope) does not exceed twice the maximum excursions caused by the continuous receiver noise of the test receiver at ambient temperature.

(The test receiver characteristics and the conditions of test are detailed in Appendix A to the Specification, which is not included in this extract).

(ii) *Magnetic interference*

Each unit of the radar equipment shall be marked with the minimum distances at which it should be mounted from a standard and a steering magnetic compass. These distances shall be understood as being measured from the centre of the magnet system in the compass bowl to the nearest point of the unit in question, and are defined as the distances at which each unit, whether working or not, will not impair the functional use of the compasses. These distances shall be determined by the Admiralty Compass Observatory in accordance with the standards specified in CD pamphlet No. 11D. The term " unit " is defined as any part of the equipment supplied or provided by the manufacturer.

19. Performance Check

Means shall be provided for the operator to determine readily a drop in performance of 10 db or more, while the set is being used operationally, relative to the calibration level established at the time of installation. The necessary calibration information shall be immediately available to the operator in the form of a calibration label fixed on or near to the display unit.

20. Climatic and Durability Testing

The equipment shall comply with the provisions of Appendix B—a performance specification for the climatic and durability testing of marine radar equipment. Those parts of the equipment that are intended to be installed in a working space in the ship, e.g. radio office, chart room, enclosed bridge or wheelhouse, shall be regarded as " Class B " equipment. Those parts of the equipment that are to be installed in exposed positions shall be regarded as " Class X " equipment. Each unit of equipment shall be marked " Class B " or " Class X " as appropriate.

21. Mechanical Construction

In all respects the mechanical construction and finish of the equipment shall conform to good standards of engineering practice, and the design shall be such that all parts of the equipment are readily accessible. Mechanical noise from all units shall be at a level acceptable to the testing authority. Units which will be mounted in or near operational areas of the ship must not add appreciably to the noise level in these areas.

22. Protective Arrangements

(a) *High voltage circuits*

The radar set shall incorporate all such isolating switches, door switches, means for discharging capacitors and/or other approved devices as are

N

necessary to ensure that access to any high voltage is not possible. Alternatively, the equipment shall be so designed that access to high voltage may be gained only by means of a tool such as a spanner or screwdriver; and a warning label shall be prominently displayed within the equipment.

The term "high voltage" shall be taken as applying to all circuits in which the direct and alternating voltages (other than radio frequency voltages) combine to give instantaneous voltages greater than 250 volts.

(b) Radio frequency radiation

The power level at which radio frequency radiation becomes a hazard to personnel is specified in *Safety Precautions relating to Intense Radio Frequency Radiation*, issued by the GPO, and published by HMSO. The radiation level from the radar must be measured and if necessary a minimum safe distance from the aerial must be stated in the handbook, which shall include instructions regarding precautions against radio frequency radiation hazard, as detailed in Appendix C (not included here).

(c) X-radiation

External X-radiation from the equipment in its normal working condition must not exceed the limits laid down in British Standard 415: 1957.

When X-radiation can be generated inside the equipment above the level of BS 415, a prominent warning notice must be fixed inside the equipment drawing attention to an appropriate paragraph in the handbook, along the lines of Appendix D (not included here).

23. Manufacturers' Arrangements

Where limitations are known to exist, the manufacturer shall propose for consideration by the testing authority the maximum and minimum distances by which units of the equipment must be separated in order to comply with the requirements of this Specification. The manufacturers shall take such measures as are necessary to ensure that any such limitations as the testing authority may approve as reasonable, and any other limitations which that authority may impose, are observed in installing the equipment on merchant ships.

24. Maintenance and Operating Manuals

The manufacturer shall cause to be provided with each equipment, such information as is necessary to enable competent members of a ship's staff to operate and to maintain the equipment efficiently. Maintenance manuals shall, where the type-testing authority deems it necessary, incorporate in a prominent position the instructions regarding special precautions against hazard from radio-frequency and X-radiation reproduced in Appendices C and D (not included here).

25. Anti-Clutter Devices

Satisfactory means shall be provided to minimise the display of unwanted responses from both precipitation and the sea.

(Extract from Board of Trade *Marine Radar Performance Specification*, 1968, and reproduced by permission of the Controller of H.M. Stationery Office.)

Explanatory Note:—

The ratio between two powers can be established from the following formula:

Number of decibels $= 10 \log_{10} \dfrac{P_1}{P_2}$ where P_1 and P_2 are the powers concerned.

Thus:		
	$-$ 3db	Power ratio $= \frac{1}{2}$
	-10db	Power ratio $= 1/10$
	-20db	Power ratio $= 1/100$
	-23db	Power ratio $= 1/200$
	-30db	Power ratio $= 1/1000$

IMO SPECIFICATIONS FOR ARPA

ARPA Appendix I(a)

PROPOSED CARRIAGE REQUIREMENTS FOR AUTOMATIC RADAR PLOTTING AIDS

1. An automatic radar plotting aid of a type approved by IMO and conforming to performance standards not inferior to those adopted by the Organisation shall be fitted on :

(1) each ship of 10,000 g.r.t. or more, the keel of which is laid or is at a similar stage of construction on or after 1 January 1984 ;
(2) existing tankers of 40,000 g.r.t. or more from 1 September 1984 ;
(3) existing tankers of 10,000 g.r.t. or more from 1 January 1985 ;
(4) other existing ships of 40,000 g.r.t. or more from 1 January 1986 ;
(5) other existing ships of 20,000 g.r.t. or more from 1 January 1987 ;
(6) other existing ships of 15,000 g.r.t. or more from 1 January 1988 ;
(7) other existing ships of 10,000 g.r.t. or more from 1 January 1989 ;

except that automatic radar plotting aids fitted prior to 1 January 1984 which do not fully conform to the performance standards recommended by the Organisation may, at the discretion of the Administration, be retained until 1 January 1991.

2. The Administration may exempt ships from this requirement, in areas where it considers it unreasonable or unnecessary for such equipment to be carried, or when the ship will be taken permanently out of service within 2 years of the appropriate implementation date.

Except that automatic radar plotting aids fitted prior to (1 January 1983) which do not conform to the performance standards recommended by the Organisation may be retained until (1 January 1990), (at the discretion of the Administration).

The Administration may exempt from this requirement, in areas where it considers it unreasonable or unnecessary for such equipment to be carried, or when the ship will be permanently taken out of service within 2 years after the appropriate implementation data, provided that it is acceptable to the Governments of the States to be visited by the ship.

ARPA Appendix I(b)

PERFORMANCE STANDARDS FOR AUTOMATIC RADAR PLOTTING AIDS (ARPA)

1.   **Introduction**

1.1   The Automatic Radar Plotting Aids (ARPA) should, in order to improve the standard of collision avoidance at sea :

1. reduce the work-load of observers by enabling them to automatically obtain information so that they can perform as well with multiple targets as they can by manually plotting a single target ;

2. provide continuous, accurate and rapid situation evaluation.

1.2   In addition to the General Requirements for Electronic Navigational Aids, the ARPA should comply with the following minimum performance standards.

2. Definitions

2.1   Definitions of terms used in these performance standards are given in Appendix 1.

3. Performance Standards

3.1   *Detection*

3.1.1   Where a separate facility is provided for detection of targets, other than by the radar observer, it should have a performance not inferior to that which could be obtained by the use of the radar display.

3.2   *Acquisition*

3.2.1   Target acquisition may be manual or automatic. However, there should always be a facility to provide for manual acquisition and cancellation. ARPA's with automatic acquisition should have a facility to suppress acquisition in certain areas. On any range scale where acquisition is suppressed over a certain area, the area of acquisition should be indicated on the display.

3.2.2   Automatic or manual acquisition should have a performance not inferior to that which could be obtained by the user of the radar display.

3.3   *Tracking*

3.3.1   The ARPA should be able to automatically track, process, simultaneously display and continuously update the information on at least :

1. 20 targets, if automatic acquisition is provided, whether automatically or manually acquired ;
2. 10 targets, if only manual acquisition is provided.

3.3.2   If automatic acquisition is provided, description of the criteria of selection of targets for tracking should be provided to the user. If the ARPA does not track all targets visible on the display, targets which are being tracked should be clearly indicated on the display. The reliability of tracking should not be less than that obtainable using manual recordings of successive target position obtained from the radar display.

3.3.3   Provided the target is not subject to target swop, the ARPA should continue to track an acquired target which is clearly distinguishable on the display for 5 out of 10 consecutive scans.

3.3.4   The possibility of tracking errors, including target swop, should be minimised by ARPA design. A qualitative description of the effects of error sources on the automatic tracking and corresponding errors should be provided to the user, including the effects of low signal to noise and low signal to clutter ratios caused by sea returns, rain, snow, low clouds and non-synchronous emissions.

3.3.5   The ARPA should be able to display on request at least four equally time-spaced past positions of any targets being tracked over a period of at least eight minutes.

3.4 *Display*

3.4.1 The display may be a separate or integral part of the ship's radar. However, the ARPA display should include all the data required to be provided by a radar display in accordance with the performance standards for navigational radar equipment.

3.4.2 The design should be such that any malfunction of ARPA parts producing data additional to information to be produced by the radar, should not affect the integrity of the basic radar presentation.

3.4.3 The size of the display on which ARPA information is presented should have an effective display diameter of at least 340 mm.

3.4.4 The ARPA facilities should be available on at least the following range scales :

      1. 12 or 16 miles ;

      2. 3 or 4 miles.

3.4.5 There should be a positive indication of the range scale in use.

3.4.6 The ARPA should be capable of operating with a relative motion display with "north-up" and either "head-up" or "course-up" azimuth stabilisation. In addition, the ARPA may also provide for a true motion display. If true motion is provided, the operator should be able to select for his display either true or relative motion. There should be a positive indication of the display mode and orientation in use.

3.4.7 The course and speed information generated by the ARPA for acquired targets should be displayed in a vector or graphic form which clearly indicates the target's predicted motion. In this regard :

      1. ARPA presenting predicted information in vector form only should have the option of both true and relative vectors ;

      2. an ARPA which is capable of presenting target course and speed information in graphic form, should also, on request, provide the target's true and/or relative vector ;

      3. vectors displayed should be either time adjustable or have a fixed time-scale ;

      4. a positive indication of the time-scale of the vector in use should be given.

3.4.8 The ARPA information should not obscure radar information in such a manner as to degrade the process of detecting targets. The display of ARPA data should be under the control of the radar observer. It should be possible to cancel the display of unwanted ARPA data.

3.4.9 Means should be provided to adjust independently the brilliance of the ARPA data and radar data, including complete elimination of the ARPA data.

3.4.10 The method of presentation should ensure that the ARPA data is clearly visible in general to more than one observer in the conditions of light normally experienced on the bridge of a ship by day and by night. Screening may be provided to shade the display from sunlight but not to the extent that it will impair the observers' ability to maintain a proper lookout. Facilities to adjust the brightness should be provided.

3.4.11 Provisions should be made to obtain quickly the range and bearing of any object which appears on the ARPA display.

3.4.12 When a target appears on the radar display and, in the case of automatic acquisition, enters within the acquisition area chosen by the observer or, in the case of manual acquisition, has been acquired by the observer, the ARPA should present in a period of not more than one minute an indication of the target's motion trend and display within three minutes the target's predicted motion in accordance with paragraphs 3.4.7, 3.6, 3.8.2 and 3.8.3.

3.4.13 After changing range scales on which the ARPA facilities are available or re-setting the display, full plotting information should be displayed within a period of time not exceeding four scans.

3.5 *Operational warnings*

3.5.1 The ARPA should have the capability to warn the observer with a visual and/or audible signal of any distinguishable target which closes to a range or transits a zone chosen by the observer. The target causing the warning should be clearly indicated on the display.

3.5.2 The ARPA should have the capability to warn the observer with a visual and/or audible signal of any tracked target which is predicted to close to within a minimum range and time chosen by the observer. The target causing the warning should be clearly indicated on the display.

3.5.3 The ARPA should clearly indicate if a tracked target is lost, other than out of range, and the target's last tracked position should be clearly indicated on the display.

3.5.4 It should be possible to activate or de-activate the operational warnings.

3.6 *Date requirements*

3.6.1 At the request of the observer the following information should be immediately available from the ARPA in *alphanumeric* form in regard to any tracked target:
  1. present range to the target;
  2. present bearing of the target;
  3. predicted target range at the closest point of approach (CPA);
  4. predicted time of CPA (TCPA);
  5. calculated true course of target;
  6. calculated true speed of target.

3.7 *Trial Manoeuvre*

3.7.1 The ARPA should be capable of simulating the effect on all tracked targets of an own ship manoeuvre without interrupting the updating of target information. The simulation should be initiated by the depression either of a spring-loaded switch, or of a function key, with a positive identification on the display.

3.8 *Accuracy*

3.8.1 The ARPA should provide accuracies not less than those given in paragraphs 3.8.2 and 3.8.3. for the four scenarios defined in Appendix 2. With the sensor errors specified in Appendix 3, the values given relate to the best possible manual plotting performance under environmental conditions of plus and minus ten degrees of roll.

3.8.2 An ARPA should present within one minute of steady state tracking the relative motion trend of a target with the following accuracy values (95 per cent probability values).

Data Scenario	Relative Course (degrees)	Relative Speed (knots)	CPA (nm)
1	11	2·8	1·3
2	7	0·6	▬
3	14	2·2	1·8
4	15	1·5	2·0

3.8.3 An ARPA should present within three minutes of steady state tracking the motion of a target with the following accuracy values (95 per cent probability values).

Data Scenario	Relative Course (degrees)	Relative Speed (knots)	C.P.A. (nm)	TCPA (mins)	True Course (degrees)	True Speed (knots)
1	3·0	0·8	0·5	1·0	7·4	1·2
2	2·3	0·3	▬	▬	2·8	0·8
3	4·4	0·9	0·7	1·0	3·3	1·0
4	4·6	0·8	0·7	1·0	2·6	1·2

3.8.4 When a tracked target, or own ship, has completed a manoeuvre, the system should present in a period of not more than one minute an indication of the target's motion trend, and display within three minutes the target's predicted motion, in accordance with paragraphs 3.4.7, 3.6, 3.8.2. and 3.8.3.

3.8.5 The ARPA should be designed in such a manner that under the most favourable conditions of own ship motion the error contribution from the ARPA should remain insignificant compared to the errors associated with the input sensors, for the scenarios of Appendix 2.

3.9 *Connections with other equipment*

3.9.1 The ARPA should not degrade the performance of any equipment providing sensor inputs. The connection of the ARPA to any other equipment should not degrade the performance of that equipment.

3.10 *Performance tests and warnings*

3.10.1 The ARPA should provide suitable warnings of ARPA malfunction to enable the observer to monitor the proper operation of the system. Additionally test programmes should be available so that the overall performance of ARPA can be assessed periodically against a known solution.

3.11 *Equipment used with ARPA*

3. 1.1 Log and speed indicators providing inputs to ARPA equipment should be capable of providing the ship's speed through the water.

ARPA Appendix 1

DEFINITIONS OF TERMS TO BE USED ONLY IN
CONNECTION WITH ARPA PERFORMANCE STANDARDS

Relative course   -   -   The direction of motion of a target related to
                                  own ship as deduced from a number of measure-
                                  ments of its range and bearing on the radar.
                                  Expressed as an angular distance from North.

Relative speed -   -   -   The speed of a target related to own ship, as
                                  deduced from a number of measurements of its
                                  range and bearing on the radar.

True course   -   -   The apparent heading of a target obtained by
                                  the vectorial combination of the target's relative
                                  motion and ship's own motion.*
                                  Expressed as an angular distance from North.

True speed   -   -   -   The speed combination of a target obtained by
                                  the vectorial combination of its relative motion
                                  and own ship's motion.*

Bearing   -   -   -   -   The direction of one terrestrial point from
                                  another. Expressed as an angular distance from
                                  North.

Relative motion display -   The position of own ship on such a display
                                  remains fixed.

True motion display -   -   The position of own ship on such a display moves
                                  in accordance with its own motion.

Azimuth stabilisation   -   Own ship's compass information is fed to the
                                  display so that echoes of targets on the display
                                  will not be caused to smear by changes of own
                                  ship's heading.

        /North-up   -   The line connecting the centre with the top of
                                  the display is North.

        /Head-up   -   The line connecting the centre with the top of
                                  the display is own ship's heading.

        /Course-up   -   An intended course can be set to the line connect-
                                  ing the centre with the top of the display.

Heading   -   -   -   -   The direction in which the bows of a vessel are
                                  pointing. Expressed as an angular distance from
                                  North.

Target's predicted motion -   The indication on the display of a linear extra-
                                  polation into the future of a target's motion,
                                  based on measurements of the target's range and
                                  bearing on the radar in the recent past.

Target's motion trend   -   An early indication of the target's predicted
                                  motion.

Radar Plotting -   -   -   The whole process of target detection, tracking,
                                  calculation of parameters and display of
                                  information.

Detection -   -   -   -   The recognition of the presence of a target.

Acquisition   -   -   -   The selection of those targets requiring a track-
                                  ing procedure and the initiation of their tracking.

Tracking	-	-	-	-	The process of observing the sequential changes in the position of a target, to establish its motion.

Tracking - - - - The process of observing the sequential changes in the position of a target, to establish its motion.

Display - - - - The plan position presentation of ARPA data with radar data.

Manual - - - - An activity which a radar observer performs, possibly with assistance from a machine.

Automatic - - - An activity which is performed wholly by a machine.

* For the purpose of these definitions there is no need to distinguish between sea or ground stabilisation.

<h2 style="text-align:center">ARPA Appendix 2</h2>

OPERATIONAL SCENARIOS

For each of the following scenarios predictions are made at the target position defined after previously tracking for the appropriate time of one or three minutes :—

Scenario 1

Own ship course	000°
Own ship speed	10 kt
Target range	8 nm
Bearing of target	000°
Relative course of target	180°
Relative speed of target	20 kt

Scenario 2

Own ship course	000°
Own ship speed	10 kt
Target range	1 nm
Bearing of target	000°
Relative course of target	090°
Relative speed of target	10 kt

Scenario 3

Own ship course	000°
Own ship speed	5 kt
Target range	8 nm
Bearing of target	045°
Relative course of target	225°
Relative speed of target	20 kt

Scenario 4

Own ship course	000°
Own ship speed	25 kt
Target range	8 nm
Bearing of target	045°
Relative course of target	225°
Relative speed of target	20 kt

ARPA Appendix 3

SENSOR ERRORS

The accuracy figures quoted in paragraph 3.8 are based upon the following sensor errors and are appropriate to equipment complying with performance standards for shipborne navigational equipment.*
Note : σ means "standard deviation".

Radar

Target glint (Scintillation) (for 200 metres length target)
    Along length of target = 30 m   (normal distribution)
    Across beam of target   = 1 m   (normal distribution)

Roll-pitch bearing. The bearing error will peak in each of the four quadrants around own ship for targets on relative bearings of 045°, 135°, 225° and 315° and will be zero at relative bearings of 0°, 90°, 180° and 270°. This error has a sinusoidal variation at twice the roll frequency. For a 10° roll the mean error is
    0·22° with a 0·22° peak sine wave superimposed.

Beam shape   —assumed normal distribution giving bearing error with
                    σ = 0·05°.
Pulse shape   —assumed normal distribution giving range error with
                    σ = 20 metres.
Antenna backlash—assumed rectangular distribution giving bearing error
                    ± 0·5° maximum.

* In calculations leading to the accuracy figures quoted in paragraph 3.8, these sensor error sources and magnitudes were used. They were arrived at during discussions with national government agencies and equipment manufacturers and are appropriate to equipments complying with the Organisation's draft performance standards for radar equipment.

Independent studies carried out by national government agencies and equipment manufacturers have resulted in similar accuracies, where comparisons were made.

Quantizations

    Bearing—rectangular distribution ± 0·01° maximum.
    Range —rectangular distribution ± 0·01 n.m. maximum.
    Bearing encoder assumed to be running from a remote synchro giving bearing errors with a normal distribution σ = 0·03°.

Gyro compass

    Calibration error 0·5°.
    Normal distribution about this with σ = 0·12°.

Log

    Calibration error 0·5 kt.
    Normal distribution about this, 3 σ = 0·2 kt.

ARPA Appendix II

MINIMUM REQUIREMENTS FOR TRAINING IN THE USE OF AUTOMATIC RADAR PLOTTING AIDS (ARPA)

1. Every master, chief mate and officer in charge of a navigational watch on a ship fitted with an automatic radar plotting aid shall have completed an approved course of training in the use of automatic radar plotting aids.

2. The course shall include the material set out below.

MINIMUM TRAINING REQUIREMENT IN THE OPERATIONAL USE OF AUTOMATIC RADAR PLOTTING AIDS (ARPA)

1. In addition to the minimum knowledge of radar equipment, the masters, chief mates and officers in charge of a navigational watch on ships carrying ARPA shall be trained in the fundamentals and operation of ARPA equipment and the interpretation and analysis of information obtained from this equipment.

2. The training shall ensure that the master, chief mate and officers in charge of a navigational watch has :—

(a) Knowledge of :

(i) the possible risks of exclusive reliance on ARPA ;

(ii) the principal types of ARPA systems and their display characteristics ;

(iii) the IMO performance standards for ARPA ;

(iv) factors affecting system performance and accuracy ;

(v) tracking capabilities and limitations of ARPA ;

(vi) processing delays.

(b) Knowledge and ability to demonstrate in conjunction with the use of an ARPA simulator or other effective means approved by the administration :

(i) setting up and maintaining ARPA displays ;

(ii) when and how to use the operational warnings, their benefits and limitations ;

(iii) the system operational tests ;

(iv) when and how to obtain information in both relative and true motion modes of display, including :
—identification of critical echoes ;
—speed and direction of targets relative movement ;
—time to and predicted range at targets closest point of approach ;
—course and speed of targets ;
—detecting course and speed changes of targets and the limitations of such information ;
—effect of changes in own ship's course or speed or both ;
—operation of the trial manoeuvre ;

(v) manual and automatic acquisition of targets and their respective limitations ;

(vi) when and how to use true and relative vectors and typical graphic representation of target information and danger areas ;

(vii) when and how to use information on past positions of targets being tracked ;

(viii) application of the International Regulations for Preventing Collisions at Sea.

RECOMMENDED TRAINING PROGRAMME IN THE OPERATIONAL USE OF AUTOMATIC RADAR PLOTTING AIDS (ARPA)

1. *General*

(a) In addition to the minimum knowledge of radar equipment, masters, chief mates and officers in charge of a navigational watch on ships carrying ARPA should be capable of demonstrating a knowledge of the fundamentals and operation of ARPA equipment and the interpretation and analysis of information obtained from this equipment.

(b) Training facilities should include the use of simulators or other effective means capable of demonstrating the capabilities, limitations and possible errors of ARPA.

(c) The simulation facilities mentioned above should provide a capability such that trainees undergo a series of real-time exercises where the displayed radar information, at the choice of the trainee or as required by the instructor, is either in the ARPA format or in the basic radar format. Such flexibility of presentation will enable realistic exercises to be undertaken, providing for each group of trainees the widest range of displayed information available to the user and thus consolidating his ability to use effectively either basic radar or ARPA systems.

(d) The ARPA training programme should include all items listed in paragraphs 3 and 4 below.

2. *Training programme development*

Where ARPA training is provided as part of the general training requirements, masters, chief mates and officers in charge of a navigational watch should understand the factors involved in decision making based on the information supplied by ARPA in association with other navigational data inputs, having a similar appreciation of the operational aspects and of system errors of modern electronic navigational systems. This training should be progressive in nature commensurate with responsibilities of the individual and the certificate issued.

3. *Theory and demonstration*

3.1 *The possible risks of exclusive reliance on ARPA*

Appreciation that ARPA is only a navigational aid and that its limitations including those of its sensors make exclusive reliance on ARPA dangerous, in particular for keeping a look-out ; the need to comply at all times with the basic principles and operational guidance for officers in charge of a navigational watch.

3.2 *The principal types of ARPA systems and their display characteristics*

Knowledge of the principal types of ARPA systems in use ; their various display characteristics and an understanding of when to use ground or sea stabilised modes and north up, course up or head up presentations.

3.3 *IMO performance standards for ARPA*

An appreciation of the IMO performance standards for ARPA, in particular the standards relating to accuracy.

3.4 *Factors affecting system performance and accuracy*

(a) Knowledge of ARPA sensor input performance parameters—radar, compass and speed inputs ; effects of sensor malfunction on the accuracy of ARPA data.

(b) Effects of the limitations of radar range and bearing discrimination and accuracy ; the limitations of compass and speed input accuracies on the accuracy of ARPA data.

(c) Knowledge of factors which influence vector accuracy.

3.5 *Tracking capabilities and limitations*

(a) Knowledge of the criteria for the selection of targets by automatic acquisition.

    (*b*) Factors leading to the correct choice of targets for manual acquisition.

    (*c*) Effects on tracking of "lost" targets and target fading.

    (*d*) Circumstances causing "target swop" and its effects on displayed data.

3.6   *Processing delays*

The delays inherent in the display of processed ARPA information, particularly on acquisition and re-acquisition or when a tracked target manoeuvres.

3.7   *When and how to use the operational warnings, their benefits and limitations*

Appreciation of the uses, benefits and limitations of ARPA operational warnings ; correct setting, where applicable, to avoid spurious interference.

3.8   *System operational tests*

    (*a*) Methods for testing for malfunctions of ARPA systems, including functional self testing.

    (*b*) Precautions to be taken after a malfunction occurs.

3.9   *Manual and automatic acquisition of targets and their respective limitations*

Knowledge of the limits imposed on both types of acquisition in multi-target scenarios, effects on acquisition of target fading and target swop.

3.10   *When and how to use true and relative vectors and typical graphic representation of target information and danger areas*

    (*a*) Thorough knowledge of true and relative vectors ; derivation of targets true courses and speeds.

    (*b*) Threat assessment ; derivation of predicted closest point of approach and predicted time to closest point of approach from forward extrapolation of vectors, the use of graphic representation of danger areas.

    (*c*) Effects of alterations of courses and/or speeds of own ship and/or targets on predicted closest point of approach and predicted time to closest point of approach and danger areas.

    (*d*) Effects of incorrect vectors and danger areas.

    (*e*) Benefit of switching between true and relative vectors.

3.11   *When and how to use information on past position of targets being tracked*

Knowledge of the derivation of past positions of targets being tracked, recognition of historic data as a means of indicating recent manoeuvring of targets and as a method of checking the validity of the ARPA's tracking.

4.   *Practice*

4.1   *Setting up and maintaining displays*

    (*a*) The correct starting procedure to obtain the optimum display of ARPA information.

    (*b*) Choice of display presentation ; stabilised relative motion displays and true motion displays.

    (*c*) Correct adjustment of all variable radar display controls for optimum display of data.

    (*d*) Selection, as appropriate, of required speed input to ARPA.

(e) Selection of ARPA plotting controls, manual/automatic acquisition, vector/graphic display of data.

(f) Selection of the time scale of vectors/graphics.

(g) Use of exclusion areas when automatic acquisition is employed by ARPA.

(h) Performance checks of radar, compass, speed input sensors and ARPA.

4.2 *System operational tests*

System checks and determining data accuracy of ARPA including the trial manoeuvre facility by checking against basic radar plot.

4.3 *When and how to obtain information from ARPA display*

Demonstrate ability to obtain information in both relative and true motion modes of display, including :

—identification of critical echoes ;
—speed and direction of targets relative movement ;
—time to and predicted range at target's closest point of approach ;
—course and speed of targets ;
—detecting course and speed changes of targets and the limitations of such information ;
—effect of changes in own ship's course or speed or both ;
—operation of the trial manoeuvre.

4.4 *Application of the International Regulations for Preventing Collisions at Sea*

Analysis of potential collision situations from displayed information, determination and execution of action to avoid close quarter situations in accordance with International Regulations for Prevention of Collisions at Sea.

MILITARY FREQUENCY BAND DESIGNATIONS

The following frequency band designations have been agreed for military purposes within NATO, but are also fairly commonly known within the Electronic Warfare sphere. These designations, however, have not been adopted in the civil field as yet.

BAND	FREQUENCY MHz	CHANNEL WIDTH MHz
A	0 — 250	25
B	250 — 500	25
C	500 — 1000	50
D	1000 — 2000	100
E	2000 — 3000	100
F	3000 — 4000	100
G	4000 — 6000	200
H	6000 — 8000	200
I	8000 — 10,000	200
J	10,000 — 20,000	1000
K	20,000 — 40,000	2000
L	40,000 — 60,000	2000
M	60,000 — 100,000	4000

For example, F_1 band : 3000–3100 MHz
F_2 band : 3100–3200 MHz, etc.
I_1 band : 8000–8200 MHz
I_7 band : 9200–9400 MHz, etc.

Therefore, a vessel transmitting at 9300 MHz (3·2 cm.) is transmitting in the I_7 band. A vessel transmitting at 3150 MHz (9·5 cm.) is using the F_2 band.

APPENDIX VII

GUIDANCE ON MANOEUVRES TO AVOID COLLISION
(R.I.N. W.P. 1972)

Course Alteration Diagram

Intended primarily for use in avoiding a vessel detected by radar and out of sight.

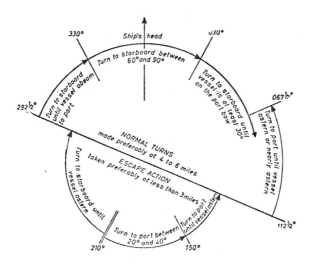

Resumption of Course

After turning to starboard for a vessel on the starboard side, keep the vessel to port when resuming course.

Escape Action

A vessel approaching from the port side and rear sector can normally be expected to take early avoiding action. The suggested turns are recommended for use when such a vessel fails to keep out of the way. As an alteration to put the bearing astern may not complement subsequent action by the other vessel, it is recommended that further turns be made to keep the vessel astern until she is well clear.

371

SPEED CHANGES IN RESTRICTED VISIBILITY

Reductions of Speed

A vessel can reduce speed or stop at any time, and such action is recommended when the *compass* bearing of a vessel on the port bow is gradually changing in a clockwise direction (increasing). A reduction of speed should be made as an alternative to, and not in conjunction with, the suggested turn to starboard for avoiding a vessel either on the port bow or ahead. Normal speed should be resumed if it becomes apparent that a vessel on the port side has either subsequently turned to starboard in order to pass astern or stopped.

Increases of Speed

It will sometimes be advantageous to increase speed if this is possible within the limitations of the requirement to proceed at a safe speed. An increase of speed may be appropriate when the vessel to be avoided is astern or on the port quarter, or near the port beam, either initially or after taking the alter course action indicated in the diagram.

Limitations

The presence of other vessels and/or lack of sea room may impose limitations on the manoeuvres which can be made, but it should be kept in mind that small changes of course and/or speed are unlikely to be detected by radar

Caution

IT IS ESSENTIAL TO ENSURE THAT ANY ACTION TAKEN IS HAVING THE DESIRED EFFECT. If not, the recommended turns can normally be applied successively for newly-developed situations with the same vessel.

SPEED AND DISTANCE TABLE

Min.\Knots	1	2	3	4	5	6	7	8	9	10	11	12	13	14	15
1	—	—	0·1	0·1	0·1	0·1	0·1	0·1	0·2	0·2	0·2	0·2	0·2	0·2	0·3
2	—	0·1	0·1	0·1	0·2	0·2	0·2	0·3	0·3	0·3	0·4	0·4	0·4	0·5	0·5
3	0·1	0·1	0·2	0·2	0·3	0·3	0·4	0·4	0·5	0·5	0·6	0·6	0·7	0·7	0·8
4	0·1	0·1	0·2	0·3	0·3	0·4	0·5	0·5	0·6	0·7	0·7	0·8	0·9	0·9	1·0
5	0·1	0·2	0·3	0·3	0·4	0·5	0·6	0·7	0·8	0·8	0·9	1·0	1·1	1·2	1·3
6	0·1	0·2	0·3	0·4	0·5	0·6	0·7	0·8	0·9	1·0	1·1	1·2	1·3	1·4	1·5
7	0·1	0·2	0·4	0·5	0·6	0·7	0·8	0·9	1·1	1·2	1·3	1·4	1·5	1·6	1·8
8	0·1	0·3	0·4	0·5	0·7	0·8	0·9	1·1	1·2	1·3	1·5	1·6	1·7	1·9	2·0
9	0·2	0·3	0·5	0·6	0·8	0·9	1·1	1·2	1·4	1·5	1·7	1·8	2·0	2·1	2·3
10	0·2	0·3	0·5	0·7	0·8	1·0	1·2	1·3	1·5	1·7	1·8	2·0	2·2	2·3	2·5
11	0·2	0·4	0·6	0·7	0·9	1·1	1·3	1·5	1·7	1·8	2·0	2·2	2·4	2·6	2·8
12	0·2	0·4	0·6	0·8	1·0	1·2	1·4	1·6	1·8	2·0	2·2	2·4	2·6	2·8	3·0
13	0·2	0·4	0·7	0·9	1·1	1·3	1·5	1·7	2·0	2·2	2·4	2·6	2·8	3·0	3·3
14	0·2	0·5	0·7	0·9	1·2	1·4	1·6	1·9	2·1	2·3	2·6	2·8	3·0	3·3	3·5
15	0·3	0·5	0·8	1·0	1·3	1·5	1·8	2·0	2·3	2·5	2·8	3·0	3·3	3·5	3·8
16	0·3	0·5	0·8	1·1	1·3	1·6	1·9	2·1	2·4	2·7	2·9	3·2	3·5	3·7	4·0
17	0·3	0·6	0·9	1·1	1·4	1·7	2·0	2·3	2·6	2·8	3·1	3·4	3·7	4·0	4·3
18	0·3	0·6	0·9	1·2	1·5	1·8	2·1	2·4	2·7	3·0	3·3	3·6	3·9	4·2	4·5
19	0·3	0·6	1·0	1·3	1·6	1·9	2·2	2·5	2·9	3·2	3·5	3·8	4·1	4·4	4·8
20	0·3	0·7	1·0	1·3	1·7	2·0	2·3	2·7	3·0	3·3	3·7	4·0	4·3	4·7	5·0

TABLE GIVING NEAREST APPROACH FROM TWO OBSERVATIONS

(assuming target maintains her course and speed).

Difference between 1st and 2nd bearing	Range Factor												
	0·95	0·90	0·85	0·80	0·75	0·70	0·65	0·60	0·55	0·50	0·45	0·40	0·35
2°	0·55	0·30	0·20	0·15	0·10	0·08	0·06	0·05	0·04	0·04	0·03	0·03	0·02
4°	0·80	0·67	0·39	0·27	0·20	0·16	0·13	0·10	0·08	0·06	0·05	0·04	0·04
6°	0·87	0·75	0·50	0·38	0·28	0·24	0·19	0·16	0·13	0·10	0·08	0·07	0·06
8°	0·91	0·80	0·60	0·47	0·36	0·31	0·25	0·21	0·17	0·14	0·11	0·09	0·07
10°	0·92	0·84	0·67	0·56	0·43	0·37	0·30	0·25	0·20	0·17	0·14	0·11	0·09
12°	0·93	0·85	0·73	0·59	0·50	0·42	0·35	0·30	0·24	0·20	0·16	0·13	0·10
14°	0·93	0·86	0·76	0·64	0·55	0·47	0·39	0·34	0·29	0·24	0·19	0·15	0·12
16°	0·94	0·87	0·79	0·68	0·60	0·51	0·43	0·38	0·33	0·26	0·21	0·17	0·14
18°	0·95	0·87	0·81	0·72	0 63	0·54	0·46	0·40	0·35	0·29	0·23	0·19	0·16
20°	0·96	0·88	0·81	0·74	0·66	0·57	0·50	0·43	0·37	0·31	0·26	0·21	0·17

Range Factor: Ratio of two successive ranges expressed as a fraction less than 1.
Nearest Approach: Number found in Table × Longest Range.

Example: Difference between two successive bearings of a target at 10 and 8 M respectively is 4 degrees.

Range Factor is 8/10 or 0·8
Nearest Approach = ·27 × 10 = 2·7 M.

RADAR PLOTTING PROBLEMS

Targets are assumed not to be seen visually.

An allowance of 5–10° in course angle, 1–2 knots in speed and up to ½ mile in nearest approach calculations is allowed in the answers. Relative or True Motion plots may be used.

Note that All Bearings in these Problems are Converted Already to True Bearings

If relative bearings are taken from a Head-up display, always change them into true bearings by adding the three-figure relative bearing to the true course and subtract 360° if this becomes necessary. For example:—
Relative bearing 030°, course 140° T., true bearing 030+140°=170°.
Relative bearing 330°, course 140° T., true bearing 330°+140°—360°=110°

Always make sure in examination problems whether the three-figure bearings given are relative or true. In the former case, follow the procedure shown above.

Report fully about the following echoes, recorded in Problems 1 up to and including 7.

1.   1st bearing and range   068°   9·2 M.
     2nd bearing and range   066°   7·4 M.
     Time interval 6 minutes.
     Own course and speed 040° T   12 knots.

2.   1st bearing and range   357°   9·5 M.
     2nd bearing and range   355°   7·6 M.
     Time interval 10 minutes.
     Own course and speed 040° T   6 knots.

3.   1st bearing and range   235°   8·0 M.
     2nd bearing and range   230°   5·7 M.
     Time interval 15 minutes.
     Own course and speed 140° T   4 knots.

4.   1st bearing and range   147°   8·4 M.
     2nd bearing and range   147°   5·8 M.
     Time interval 7½ minutes.
     Own course and speed 220° T   8 knots.

5.   1st bearing and range   300°   9·1 M.
     2nd bearing and range   298½°   8·1 M.
     3rd bearing and range   297°   7·1 M.
     Time interval between successive observations 7½ minutes.
     Own course and speed 280° T   8 knots.

6.   1st bearing and range   285°   8·8 M.
     2nd bearing and range   284°   8·3 M.
     3rd bearing and range   283¼°   7·8 M.
     Time interval between successive observations 6 minutes.
     Own course and speed 350° T   10 knots.

7.   1st bearing   176°   range 9 M.
     2nd bearing   176½°   range 8·1 M.
     3rd bearing   177½°   range 7·1 M.
     4th bearing   179°   range 6·0 M.
     Time interval between successive observations 4 minutes.
     Own course and speed 165° T   10 knots.

8.   The echo of a lightvessel is observed as follows:—
     1st bearing   215°   range 7·5 M.
     2nd bearing   207°   range 6·3 M.
     3rd bearing   198°   range 5·2 M.
     Time interval between successive observations 7½ minutes.
     Own course and speed 255° T   12 knots.
     Find the set and rate of the current.

9. The following observations of an echo of a target were recorded:—

   0400 : Bearing   348°   range 9·3 M.
   0405 : Bearing   347°   range 7·9 M.
   0410 : Bearing   346°   range 6·5 M.
   Own course and speed 020° T   14 knots.
   Own ship is stopped at 0410.

   Find the course and speed of the target and the distance and time of its nearest approach.

10. The following observations of an echo of a target were recorded:—

    1000 : Bearing   166°   range 8·8 M.
    1006 : Bearing   165°   range 7·0 M.
    1012 : Bearing   162°   range 5·2 M.
    Own course and speed 110° T   12 knots
    Compile a report for 1012.
    Speed was reduced at 1012 from 12 to 3 knots.

    Find the new nearest approach and the time it occurs.

11. The following observations of an echo of a target were recorded:—

    0800 : Bearing   122°   range 8·2 M.
    0805 : Bearing   122°   range 7·1 M.
    0810 : Bearing   122°   range 6·0 M.
    Own course and speed 190° T   6 knots.
    At 0810 speed is reduced to 3 knots.
    At 0820 speed is reduced to 1½ knots.
    At 0830 ship is stopped.

    Find the nearest approach and the time it occurs.

12. The following observations of an echo of a target were recorded:—

    0600 : Bearing   290°   range 8·7 M.
    0605 : Bearing   289°   range 7·6 M.
    0610 : Bearing   288°   range 6·5 M.
    Own course and speed 260° T   12 knots.
    At 0610 order was given to reduce speed and at 0620 the vessel had attained a steady speed of 6 knots.
    At 0625 target bore 276° range 4 M.
    At 0630 order was given to reduce speed again and at 0640 the vessel had attained a steady speed of 3 knots.

    How far does the target cross ahead?

    What is her nearest approach and at what time will it occur?

13. The following observations of an echo of a target were recorded:—

    1000 : Bearing   008°   range 12·3 M.
    1006 : Bearing   008°   range 10·3 M.
    1012 : Bearing   008°   range 8·3 M.
    Speed of own ship is 12 knots. Course is 000° T.

    Find the distance and the time of the nearest approach.

    What is the speed and aspect of the target at 1012?

    At approximately 1018 own vessel reduced speed to 3 knots and at 1030 the engines were put on full astern in order to take the way off the vessel.

    What time approximately can one expect to hear the fog signal of the other vessel, assuming that the maximum range at which a fog signal can be heard, is two miles?

14. Own vessel's course and speed are respectively 240° T, 8 knots. The following bearings and ranges of a target were taken at 1000, 1007½ and 1015 hrs. respectively:—

220°, 7 M; 221°, 6½ M. and 222°, 6 M.

Find the distance and the time of the nearest approach; also the target's aspect at 1015.

At 1040 the master of the observing vessel gave order to reduce speed to 4 knots and at 1050 the vessel had attained a steady speed of 4 knots.

Estimate the *least* time interval after 1050 hrs. before the original speed can be resumed, assuming that the plot shows that the target maintains her original course and speed.

It is the intention to keep 3 M clear of the target.

15. The course of own ship is 260° T and speed 10 knots. The following bearings and ranges were taken at 0400, 0406 and 0412 hrs. respectively:—

233°, 10·3 M; 234°, 9·1 M and 235°, 8·0 M.

Find the nearest approach, the time of the nearest approach; also the target's aspect and speed at 0412 hrs.

At 0418 hrs. the master of the observing vessel gives orders to reduce speed and at 0430 hrs. a steady speed of 3 knots is attained.

Estimate the *least* time interval allowed after 0430 hrs. before the original speed can be resumed. It is the intention that both ships will be at least 4 M clear of each other when the closest distance of approach takes place. Close observation of the plot shows that the target has maintained her course and speed throughout.

16. Own ship's speed is 6 knots. Course 180° T.

The following bearings and ranges of a target were taken from the display.

0800 : 217°   8·4 M.
0805 : 217°   8·0 M.
0810 : 217°   7·7 M.
0815 : 217°   7·4 M.
0820 : 217°   7·1 M.

Find the distance of the nearest approach and the time when this will take place; determine also aspect and speed of target at 0820 hrs.

At 0835 order was given to reduce speed and at 0845 own vessel has attained a steady speed of 3 knots.

At what time approximately do you expect that the original speed can be resumed, assuming that the plot shows that the target maintains her course and speed throughout? The nearest approach should not become less than 3½ M.

17. A ship steering 308° at 13 knots, recorded the following radar observations of another vessel on her radar screen:—

1000 : 273°, 10·0 M.   (Heading and bearings are true.)
1005 : 273½°, 8·9 M.
1010 : 274½°, 7·8 M.
1015 : 275°, 6·8 M.

The headreach for a trailing stop is 2 miles in 20 minutes.

Find the time to stop engines so that the other vessel will pass 3 miles ahead,

     (*a*) using a relative motion plot only;
     (*b*) using a true motion plot only.

Comment on the two methods.

18. Compile a report about the target from the three following observations of its echo. After the course alteration find the new nearest approach and the time interval between the nearest approach and the last observation.

1st bearing 272° range 9·7 M.
2nd bearing 271° range 8·9 M.
3rd bearing 270° range 8·1 M.

Time interval between successive observations 6 minutes.
Own course and speed 230° T 9 knots.
At the last observation course is altered 60 degrees to starboard.

19. The following observations of an echo of a target were recorded:—

0800 : Bearing 188° range 9·2 M.
0805 : Bearing 187° range 7·9 M.
0810 : Bearing 186° range 6·6 M.

Own course and speed 160° T 9 knots.
Compile a report at 0810.
At 0813 course was altered 55 degrees to starboard and at 0817 the ship was on her steady course.
At 0820 bearing and range of target were 173° 4 M.
Find the new nearest approach and the time it occurs.

20. The following bearings and ranges of a target were observed on a radar display.

1200 : Bearing 194° range 8·9 M.
1206 : Bearing 193½° range 8·4 M.
1212 : Bearing 193° range 7·9 M.
1218 : Bearing 192½° range 7·4 M.

Own course and speed 242° T 6 knots.
Compile a report at 1218.
At 1234 course was altered to bring target on the port beam and at 1238 the vessel was steady.
At 1254 echo bore 183° at 5·1 M range.
Find the new nearest approach and the time it occurs.

21. The following bearings and ranges of a target were observed on a radar display.

0600 : Bearing 344° range 9 M.
0607½ : Bearing 343½° range 8 M.
0615 : Bearing 343° range 7 M.

Own course and speed 330° T 8 knots.
Compile a report for 0615.
Later course was altered 60 degrees to starboard.
At 0635 target bore 334° range 4·4 M.
At 0640 target bore 325° range 4·1 M.
Find the new nearest approach and time it occurs. What is the relative bearing when the nearest approach takes place?

22. The following bearings and ranges of a target were observed on a radar display.

0400 : Bearing 270° range 8·2 M.
0404 : Bearing 269° range 6·8 M.
0408 : Bearing 267° range 5·4 M.

Own course and speed 330° T 15 knots.
At 0412 course was altered 40 degrees to starboard.
When the range to the target was 3 miles, own vessel was stopped.
Find the nearest approach and the time it occurs.

378   RADAR OBSERVER'S HANDBOOK

23. Compile a report from the following observations of an echo:—

    1st bearing 302°   range 9·5 M.
    2nd bearing 301°   range 8·6 M.
    3rd bearing 300½°   range 7·7 M.
    4th bearing 300°   range 6·8 M.

    Time interval between successive observations 4 minutes.

    Own course and speed 030 degrees T. 10 knots.

    At the last observation course is altered to starboard to bring other vessel 30 degrees on port quarter. The intention is to make the closest distance of approach not less than 3 miles.

    What, approximately, is the earliest time that own ship can go back to her original course, assuming that the other vessel maintains its course and speed?

    Explain why the avoiding action lasts so long.

24. Compile a report from the following observations of an echo's movement over a radar display.

    1st bearing   093°   range 8·3 M.
    2nd bearing   093½°   range 7·6 M.
    3rd bearing   093°   range 6·9 M.

    Time interval between successive observations 4 minutes.

    Own course and speed 048° T 15 knots.

    At the last observation course is altered so as to bring the target 30 degrees on port bow. Ten minutes later the original course is resumed.

    Find the closest distance of approach and the time interval between the time when this nearest approach takes place and the time of the original course resumption (the target maintains her course and speed throughout).

25. Own speed is 6 knots, course 315° T.

    Echo is observed as follows:—

    1000 : 311°   distance 9 M.
    1005 : 310°   distance 8 M.
    1010 : 309°   distance 7 M.

    At 1015 hrs. course is altered 60 degrees to starboard. The intention is to keep 3 M away from the other ship. What time, approximately, can the original course be resumed, assuming the target maintains course and speed? What other considerations besides the distance of the nearest approach, should be made before resuming course after making a bold alteration?

26. Own speed is 6 knots, course 270° T.

    Echo is observed as follows:—

    0600 : 276½°   distance 9·3 M.
    0606 : 278°   distance 7·8 M.
    0612 : 280°   distance 6·3 M.

    At 0618 hrs. course is altered 90 degrees to port. The intention is to have a nearest approach of 3 M. At what time approximately can the original course be resumed, assuming target maintains course and speed?

27. Own speed 12 knots, course 090° T.
Echo is observed as follows:—

0800 : 085° distance 7·0 M.
0806 : 084° distance 6·5 M.
0812 : 083½° distance 5·9 M.

At 0818 hrs. course is altered 50 degrees to starboard. The intention is to keep at least 3 M away from the target. What approximately, is the earliest time when the original course can be resumed, assuming that the target maintains her course and speed?

28. The following radar bearings and ranges of an echo of a target were recorded on a ship, steering 245°, speed 11 knots.

1000 : 020° 3·5 M. (Heading and bearings are true.)
1006 : 022° 2·7 M.
1012 : 024° 2·0 M.

At 1012 the other vessel is brought right astern and this action is repeated at 1018 and at 1024.

At 1030 the original course is resumed.

(a) Explain the reasons for this type of action.
(b) Calculate the speed of the other vessel.
(c) How far will the other vessel pass ahead, provided she maintains her course and speed?

29. Speed of own ship is 11 knots, course 225° T. Speed and course were maintained.
Bearings and ranges of an echo were taken every 2½ minutes. These were as follows:—

259° 9 M.
259° 8·3 M.
259° 7·7 M.
259° 7·1 M.
259° 6·4 M.
262° 5·9 M.
264° 5·5 M.
266° 5·2 M.
269½° 4·9 M.

State, first, without plotting, what information can be deduced from the recordings. Next make a complete plot and explain the relative movement on the screen.

30. Speed of own ship is 11 knots, course 135° T. Speed and course were maintained.

Bearings and ranges of an echo were taken every 3 minutes. These were as follows:—

103° 10·1 M.
103½° 9·4 M.
104° 8·8 M.
104½° 8·2 M.
105° 7·6 M.
103° 6·7 M.
100° 5·9 M.
095° 5·0 M.
090° 4·1 M.

State, first without plotting, what information can be deduced from the recordings. Next make a complete plot and explain the relative motion on the screen.

31.   Speed of own ship is 11 knots, course 045° T.   Speed and course were maintained.

Bearings and ranges of an echo were taken every 4 minutes.

These were as follows:—

077$\frac{1}{2}$°	10·7 M.
077°	9·8 M.
076$\frac{1}{2}$°	9·0 M.
076°	8·2 M.
075$\frac{1}{2}$°	7·3 M.
075°	6·9 M.
074$\frac{1}{2}$°	6·5 M.
074°	6·1 M.
073$\frac{1}{2}$°	5·7 M.

State, first without plotting, what information can be deduced from the recordings.   Next make a complete plot and explain the relative motion on the screen.

32.(i) Own ship's speed is 6 knots, course 000° T.

1000 :	Bearing of target	043°	distance 8 M.
1006 :	Bearing of target	043°	distance 7 M.
1012 :	Bearing of target	043°	distance 6 M.

Own ship alters course 50 degrees to starboard at 1016 and at 1020 she is on a steady course.

Estimate nearest approach and time of nearest approach.

What is speed of other ship?

(ii) At 1024 : Target bearing   043°   distance 3·8 M.
At 1030 : Target bearing   043°   distance 2·6 M.

Own ship stops at 1030.

Explain what has happened.   What do you estimate the nearest approach will be now and at what time will it occur?

33.(i) Own ship's speed is 8 knots, course 060° T.

The following bearings and ranges of a target were observed:—

1000 : 064°	10·3 M.
1003 : 063$\frac{1}{2}$°	9$\frac{1}{4}$ M.
1006 : 063$\frac{1}{2}$°	8·3 M.
1009 : 063°	7$\frac{1}{4}$ M.

Compile a report for 1009 hrs.

(ii) At 1010 order is given to alter course 90° to starboard and at 1014 own vessel is steady on her new course.

The following bearings and ranges of the target were then obtained:—

1015 : 058°	5·6 M.
1018 : 052$\frac{1}{2}$°	5·1 M.
1021 : 044°	4·6 M.

Has the target maintained her course and speed or has she taken avoiding action?

34.(i) Own course and speed is 140° T, 5 knots.

A target was observed as follows:—

1600 : 140°	10 M.
1603 : 140°	9$\frac{1}{4}$ M.
1609 : 140°	7$\frac{3}{4}$ M.

Find course and speed of target.

At 1612 hrs.   course was altered 90° to starboard.

(ii) Plotting continued:—

    1618 : $128\frac{1}{2}°$ 6·7 M.

    1624 : 116°   6·6 M.

At 1630 hrs. own course was resumed (140° T).

Determine action taken by target, its CPA and TCPA after 1630 hrs.

What will be its relative bearing when it is at its CPA?

35.   Speed own ship 10 knots course 230° T.

    The following radar bearings and ranges were taken:—

Time	Ship A		Ship B		Ship C		Ship D	
1000	$190\frac{1}{2}°$	5·6 M.	273°	6·9 M.	003°	3·0 M.	140°	3·3 M.
1006	190°	5·0 M.	288°	5·5 M.	003°	3·0 M.	$133\frac{1}{2}°$	3·5 M.
1012	189°	4·4 M.	309°	4·8 M.	003°	3·0 M.	$128\frac{1}{2}°$	3·7 M.

Find for each target:—

    (a)   Nearest approach and time it will occur;

    (b)   Speed and course;

    (c)   Aspect at 1012.

36.(i) An observing vessel is steaming 000° T with a speed of 12 knots.

    The following radar bearings (true) and ranges were taken.

Time	Target A		Target B		Target C	
0000	090°	6 M.	270°	5 M.	000°	6 M.
0006	092°	6 M.	270°	4·7 M.	000°	5·5 M.
0012	094°	6 M.	270°	4·4 M.	000°	5·0 M.

Determine the courses and speeds of targets A, B and C.

(ii) Assuming that the visibility is half a mile and that avoiding action is going to be taken at 0018, select for the observing vessel what you think is the most suitable form of action from the following alternatives:—

    (a)   Course alteration of 60° to starboard.

    (b)   Course alteration of 60° to port.

    (c)   Course alteration of 30° to starboard.

    (d)   Course alteration of 30° to port.

    (e)   Reduce speed to 6 knots.

(iii) After you have decided upon your action determine the predicted nearest approach between the observing vessel and each of the targets A, B and C, assuming the observing vessel is at her new speed or on her new course.

37 (i) Speed own ship 8 knots, course 135° T.

    The following radar ranges and bearings were taken.

Time	Ship A		Ship B		Lightvessel C		Ship D	
0600	130°	7·6 M.	140°	8·0 M.	248°	3.3 M.	355°	4·6 M.
$0607\frac{1}{2}$	130°	5·8 M.	141°	7·8 M.	264°	4·0 M.	350°	3·9 M.
0615	130°	4·0 M.	142°	7·6 M.	273°	4·9 M.	342°	3·2 M.

Find for targets A, B and D:—

    (a)   Nearest approach and time it will occur;

    (b)   Speed and course;

    (c)   Aspect at 0615.

What is the set and rate of the current?

37 (ii) Study plot 37 (i).

    What action would you take and why?

    (a)   Stand on;

    (b)   Alter boldly to starboard;

    (c)   Reduce speed;

    (d)   Stop the ship.

38.(i) The course of the observing ship is 090° T, the speed is 10 knots.

    Two targets were observed as follows:—

Time	Lightvessel		Ship	
0000	157°	9·0 M.	073°	11·5 M.
0003	160°	8·8 M.	074°	10·7 M.
0006	163°	8·7 M.	075°	9·9 M.
0009	166°	8·6 M.	077°	9·1 M.
0012	169·5°	8·5 M.	078½°	8·2 M.

Determine the predicted nearest approach of the ship, her course and her speed.

  (ii) At 0012 course was altered to starboard by the observing vessel in order to pass the lightvessel two miles on the starboard side.

    Find the distance between the two vessels when the observing vessel has the lightvessel abeam and also the time when this takes place.

 (iii) How far will the other vessel pass from the lightvessel, assuming she is maintaining her course?

39.   Speed own ship 10 knots, course 120° T.

    The following radar bearings and ranges were taken:—

Time	Target A		Target B		Target C	
0000	084°	4·8 M.	133°	7·0 M.	171°	6·4 M.
0006	076°	4·0 M.	134°	5·8 M.	169°	5·7 M.
0012	064°	3·4 M.	134½°	4·7 M.	166°	5·1 M.

Find for each target:—

    (a)   Nearest approach and time it will occur.

    (b)   Aspect and speed at 0012 hrs.

What is the set and rate of the current?

Which target(s) should particular attention be paid to?

40. (i) A ship steering 127° T, speed 13 knots, had the following radar observations recorded:—

	Ship		*Lightvessel*	
1000	092°	11·1 M.	150°	5·5 M.
1005	093°	9·4 M.		
1010	094°	7·7 M.		
1015	095½°	6·1 M.	180°	3·5 M.

Compile a report about the other vessel and deduce the set and rate of the tidal current.

(ii) The observing vessel alters course at 1017½ (mean of execution time and termination time of manoeuvre) in order to pass the lightvessel at 1·5 miles to starboard.
The visibility, meanwhile, has improved to 3 miles.
From the plot, deduce the following:—

    (a) The alteration of course made by the observing vessel.
    (b) The closest distance of approach between the two vessels and the time of closest point of approach.
    (c) The time the observing vessel will pass the lightvessel.
    (d) The time the other vessel will pass the lightvessel and her passing distance.
    (It was observed that the other vessel maintained its course and speed.)

41. On an observing ship, steaming 330° T at 10 knots, the bearings and ranges of an echo were as follows:—

1000 : 357°   11·2 M.
1006 : 359°   10·3 M.
1012 : 002°   9·4 M.
1018 : 006°   8·5 M.
1024 : 006°   7·4 M.
1030 : 006°   6·1 M.

At 1030 the observing vessel altered course to 030° T and at 1042 resumed her original course. Just after the resumption of course, the echo was on the heading marker, 3·8 M away.

    (a) Do you agree with the alteration to starboard by the observing vessel?
    (b) What other type of avoiding action could have been taken by the observing vessel?
    (c) Why did the observing vessel wait 30 minutes before taking action?
    (d) Was the observing vessel justified in resuming her original course at 1042?
    (e) What will be the nearest approach according to your prediction?
    (f) Give a possible explanation for the behaviour of the other vessel.

It is assumed that the action is taking place in fog.

42. On a ship, steaming in thick fog, 040° T at 10 knots, the following bearings and ranges were taken of two ships X and Y at 0000, 0006, 0012 0018 and 0024 hrs:—

Ship X:	022°	8·8 M.	*Ship Y:*	110°	7·9 M.
	022½°	8·1 M.		110°	7·0 M.
	023½°	7·4 M.		111°	6·1 M.
	025°	6·3 M.		118°	5·7 M.
	029°	5·2 M.		124°	5·3 M.

(All bearings are true.)

(a) State, without plotting, what information can be deduced from the recordings.

(b) Plot the apparent motion line. then give your opinion if you think that avoiding action should be planned.

(c) Next complete the triangles and find out if the other vessels took action.

(d) If action was taken by one or both vessels, might this have been for the observing vessel (Own Ship) or for any other vessel (detected or undetected by Own Ship)?

43.(i) On an observing ship, steering 070° T at a speed of 12 knots, the following radar bearings and ranges were taken:—

Time	Target A		Target B		Target C	
0000	077·5°	9·7 M.	041°	10·5 M.	000°	11·1 M.
0006	079°	8·2 M.	042·5°	8·8 M.	000°	9·5 M.
0012	081°	6·8 M.	044·5°	7·3 M.	000°	7·9 M.

Meanwhile visibility reduced rapidly.

At 0012 a racon signal identified target A as a light vessel. Report on targets B and C and find the direction and rate of the current.

(ii) A speed reduction order is issued at 0012 and the vessel has attained a steady speed of 6 knots at 0024.

Predict the CPA and TCPA of all targets.

(iii) The following observations were subsequently made:—

Time	Target A		Target B		Target C	
0024	087°	4·7 M.	056°	4·8 M.	007°	4·8 M.
0030			067°	4·0 M.	341°	3·6 M.
0036	097°	3·2 M.	083°	3·3 M.	306°	3·6 M.
0042			105°	3·0 M.	277°	4·7 M.
0048	120°	1·7 M.	127°	3·2 M.	269°	3·6 M.
0054			145°	3·8 M.	257°	2·7 M.

Show that the echo of target B follows the predicted motion, and explain fully the behaviour of ship C.

44.(i) During reduced visibility, the following radar observations (bearing and range) were taken on a vessel steering 120° T at a speed of 10 knots.

Time	Target A		Target B		Target C	
1015	154°	11·6 M.	129·5°	9·8 M.	113°	11·0 M.
1021	154°	10·2 M.	130·5°	8·1 M.	114·5°	8·7 M.
1027	154°	8·7 M.	132°	6·4 M.	116°	6·4 M.

Report on each target.

(ii) Find the minimum instantaneous alteration to starboard at 1033 so that each target has a predicted CPA of 2 miles or more.

(iii) At approximately 1034 an alteration to starboard is ordered and the ship is steadied on 210° T.

Further observations were as follows :—

Time	Target A		Target B		Target C	
1036	153°	6·7 M.	134°	4·1 M.	099°	3·8 M.
1042	138°	5·7 M.	117°	3·2 M.	072°	4·6 M.
1048	121°	5·3 M.	092°	2·8 M.	055°	6·3 M.

Check if the predicted motion of the echoes of the targets is followed. If not, estimate the alteration of course/speed, stating the reasons you think are responsible for the action.

45.(i) In a ship steering 240° T at a speed of 12 knots, the following radar bearings and ranges were made of a fixed navigational mark.

Time	Target A	
1230	230°	9·3 M.
1236	229·5°	7·8 M.
1242	229°	6·5 M.

Find CPA and TCPA of this target, and deduce direction and rate of current.

(ii) Find the alteration of course to starboard at 1248 to pass 2 miles of target A, assuming the alteration to be instantaneously effective. Further observation were as follows :—

Time	Target A		Target B	
1254	237°	3·6 M.	97°	8·2 M.
1300	256°	2·6 M.	94°	7·2 M.
1306	286°	2·2 M.	90°	6·1 M.

Find the CPA, TCPA, speed and aspect of the new target at 1306, assuming the current is uniform over the area covered by the PPI.

At 1306 the original course is resumed by the observing vessel.

(iii) Observations on the display are continued up to 1318 while the echoes of the targets follow their predicted movements.

After 1318, recorded observations were :—

Time	Target A		Target B	
1318	348°	2·2 M.	90°	5·0 M.
1324	004°	2·7 M.	93°	4·6 M.
1330	013·5°	3·4 M.	97°	4·3 M.

Find the course and speed of target B during the last recorded time interval and estimate its CPA.

Give an estimate of any change in the direction and the rate of the current from 1230 until 1330.

46.(i) In poor visibility, the following radar observations (bearings and ranges) were made in a ship steering 320° T at a speed of 12 knots.

Time	Target A		Target B		Target C	
1705	320°	9·0 M.	327°	7·8 M.	050°	5·0 M.
1710	320°	8·0 M.	327·5°	7·3 M.	050°	5·0 M.
1715	320°	7·0 M.	328°	6·8 M.	050°	5·0 M.

One of the targets ahead is expected to be a light vessel. Report on the two moving targets.

(ii) At 1720 course is altered to 000° T (assume an instantaneous alteration). It is the intention to go back to the original course when the distance to the ship on the beam has been reduced to 2 miles.

What will be the expected CPA and TCPA of targets A and B after the original course has been resumed.

(iii) Later the following observations were made :—

Time	Target A		Target B		Target C	
1730	303·5°	4·6 M.	317°	5·5 M.	043·5°	3·7 M.
1735	293°	4·2 M.	306°	5·0 M.	037·5°	3·1 M.
1740	279·5°	3·9 M.	287·5°	4·7 M.	030°	2·6 M.
1745	266°	3·9 M.	272°	4·8 M.	018°	2·0 M.

Ascertain whether any action has been taken by the other vessels. If so, find the new CPA of that vessel.

47.(i) Own ship is steering 055° T at a speed of 12 knots. The visibility is poor and engines are on 'stand-by'. The tide is setting in an easterly direction.

The echoes of three vessels appear ahead on the PPI, one it is thought from V.H.F.R/T broadcasts is a ship at anchor with engine break-down.

The following radar observations (bearings and ranges) were made:—

Time	Target A		Target B		Target C	
1000	065°	10·7 M.	053·5°	10·9 M.	044°	11·4 M.
1005	065·5°	9·4 M.	054°	9·0 M.	045°	10·2 M.
1010	065°	8·2 M.	054·5°	7·3 M.	046°	8·9 M.
1015	065°	7·0 M.	054°	5·4 M.	047°	7·5 M.

Report on the targets which are underway.

(ii) At 1020 an alteration of 90° is made to starboard (assume the effect of the alteration is instantaneous).

Predict the CPA and TCP of the moving targets while on the new heading.

Predict the time that the original course can be resumed on the understanding that the CPAs when back on the original course will not be less than those reported above (i.e. after taking avoiding action, but before resumption of course).

(iii) The following radar observations were subsequently made :—

Time	Target A		Target B		Target C	
1030	039°	5·3 M.	011·5°	2·8 M.	035·5°	6·0 M.
1035	026°	5·5 M.	347°	3·1 M.	029·5°	6·1 M.
1040	014°	6·1 M.	330°	3·7 M.	023·5°	6·2 M.
1045	004·5°	6·7 M.	318·5°	4·8 M.	017°	6·3 M.

Report if any action is taken by one of the targets.

Find the CPA and TCPA of the vessel at anchor when back again on course 055° T.

48.(i) Own ship is steering a course of 180° T and has a speed of 18 knots. A large vessel which has been overtaken is on port quarter at a distance of approximately 4 miles. Time : 2000. Ten minutes later, the vessel runs into a fog bank and engines are rung to 'Stand-by'. Twenty minutes later, telegraph is put on 'Slow Speed' and at 2030 the ship has settled down to a speed of 10 knots.

The following radar observations (bearings and ranges) were recorded :—

Time	Target A		Target B		Target C	
2030	040·5°	4·0 M.	250°	9·0 M.	175°	11·3 M.
2033	042°	3·8 M.	250°	8·0 M.	176°	10·2 M.
2036	046°	3·6 M.	250°	7·1 M.	177°	9·3 M.
2039	050°	3·3 M.	250°	6·1 M.	178°	8·2 M.

Give a report for each target valid for 2039.

(ii) Engines are ordered to emergency stop at 2045. At 2100 own ship is stopped in the water and her heading has slewed round to 210°. Radar observations gave the following information :—

Time	Target A		Target B		Target C	
2045	059·5°	2·9 M.	250°	4·2 M.	183°	6·3 M.
2100	115·5°	3·1 M.	211°	5·0 M.	205·5°	3·0 M.

Estimate CPA and TCPA of most dangerous target.

(iii) Further radar information gave the following :—

Time	Target A		Target B		Target C	
2103	126·5°	3·5 M.	205·5°	5·0 M.	213·5°	2·6 M.
2106	135°	4·1 M.	200°	4·9 M.	226°	2·4 M.
2109	141·5°	4·6 M.	195·5°	4·9 M.	238·5°	2·3 M.
2112	147°	5·3 M.	190°	4·9 M.	251°	2·3 M.
2115	151°	6·0 M.	186°	4·9 M.	263·5°	2·5 M.

Estimate any changes in course and/or speed executed by the targets between 2045 and 2115. By studying the overall situation try to find an explanation for their behaviour.

O

49.(i) The following radar bearings and ranges were taken in an observing ship steering 250° T and doing a speed of 12 knots :—

Time	Target A		Target B	
1500	290°	11·1 M.	217°	11·0 M.
1506	289·5°	9·8 M.	219·5°	9·5 M.
1512	290°	8·5 M.	223°	8·0 M.

Write a report for each of the targets at 1512.

(ii) At 1518 the order is given to alter course to 310° T and to reduce speed to 6 knots. This manoeuvre is completed at 1524, the ship then having a steady speed.

Subsequent radar observations are as follows :—

Time	Target A		Target B	
1524	286°	5·8 M.	235°	5·4 M.
1530	275°	5·0 M.	235°	5·0 M.
1536	262°	4·5 M.	229°	4·7 M.

Explain the echo's motion of ship B. Check that the echo of ship A follows the predicted motion in direction and rate of change of displacement.

(iii) Estimate the *earliest* time when own ship can resume a course of 250° T at 12 knots so that a CPA of not less than 3 miles can be maintained from target A.

(iv) At 1542 the position is as follows :—

Target A		Target B	
248°	5·3 M.	221°	4·6 M.

Explain why the echo of target A has also now deviated from its predicted motion.

50.(i) Own Ship heading 122° T, speed 14 knots. The following radar bearings and ranges were observed:—

	Lightvessel		Ship	
0406	160°	9 miles	153°	9·2 miles
0412	166½°	8·2 miles	153°	8·4 miles
0418	174°	7·5 miles	153½°	7·7 miles

Report on the other vessel. What is the set and rate of the tidal current? Find the course and speed made good by own ship.

What is the course and speed made good by the other vessel and to what purpose might this information be useful?

(ii) At 0422 own ship ordered helm to starboard and at 0426 she steadied on a course designed to pass the other vessel at a closest distance of 3 miles.

Determine the new course of own ship.

(iii) At 0428 the telegraph was run Half Ahead and at 0432 own ship had settled down to 11 knots.

At 0435 the other vessel bore 140°   distance 5·5 M.
At 0440 the other vessel bore 140°   distance 5·0 M.
At 0445 the other vessel bore 140°   distance 4·5 M.

Explain the echo movement of the other vessel.

(iv) At 0449 own ship alters course to maintain the closest distance of approach at 3 miles to port, requiring two minutes to swing round to the new course.

At 0510 own ship's radar broke down.

Estimate the position of the other vessel when the fog signal of the light-vessel is heard on the beam and course and speed of Own Ship are maintained.

What fundamental errors were made by own ship while taking avoiding action, and relate these to Rule 8.

ANSWERS TO RADAR PROBLEMS

Note : A positive CPA signifies an anti-clockwise change in bearing of the target; similarly a clockwise change will provide a negative CPA (see solutions to problems 43–49).

1. 066°, ahead. 7·4, decreasing. 1 M in 24 minutes.
    294 degrees. 11 knots. Aspect R.46°.

2. 355°, astern. 7·6, decreasing. 1·5 M in 37 minutes.
    160 degrees. 7½ knots. Aspect G.15°

3. 230°, ahead. 5·7, decreasing. 1·5 M in 34 minutes.
    085 degrees. 11 knots. Aspect R.35°.

4. 147°, steady. 5·8, decreasing. 0 in 16 minutes.
    305 degrees. 20 knots. Aspect G.22°.

5. 297°, ahead. 7·1, decreasing. 1·7 M in 50 minutes.
    201 degrees. 4·5 knots. Aspect R.85°.

6. 283¼°, astern. 7·8, decreasing. 1·4 M in 1 hour 33 minutes (approx.)
    020 degrees. 9·5 knots. Aspect G.84°.

7. 179°, astern. 6·0, decreasing. ·7 M in 24 minutes.
    003 degrees. 6 knots. Aspect R.4°.

8. 151 degrees. 2 knots.

9. 119 degrees. 7 knots.
    4·7 M at 0447 (approx.).

10. 162°, ahead. 5·2 M, decreasing. 1 M at 1029.
    032 degrees. 16½ knots. Aspect R.50°.
    3·3 M at 1028.

11. 2 M at 0838.

12. Crossing ahead 3·2 M.
    Nearest approach 3·1 M at 0638.

13. Nearest approach 0 at 1038.
    8 knots, Red 11°.
    1047.

14. 1·7 M at 1140.
    Aspect G.136°.
    Approx. 30 minutes after 1050.

15. 1·7 M at 0452.
    G.70°, 7 knots.
    33 minutes.

16. Nearest approach 0 at 1008.
    R.101°, 3·75 knots.
    1022.

17. Stop engines at 1027.
    Very simple construction when using a true motion plot.
    When using a relative motion plot, find the relative motion of the target for a speed of 6 knots (2 miles in 20 minutes) and transfer this line through the intersection of the heading marker and the 3-mile range ring. The point where the latter line meets the original *OA* line gives an indication of the time for stopping engines.

18.   270°, ahead.   8·1, decreasing.   1·5 M in 60 minutes.
      172 degrees.   7½ knots.   Aspect R.82°.
      New nearest approach 5·8 M in 25 minutes.

19.   186°, ahead.   6·6, decreasing.   ·8 M at 0834.
      047 degrees.   9 knots.   Aspect R.41°.
      3 M at 0828.

20.   192½°, astern.   7·4, decreasing.   1 M at 1345 (approx.).
      297 degrees.   4 knots.   Aspect G.75°.
      4·9 M at 1344 (approx.).

21.   343°, ahead.   7 M, decreasing.   ·9 M at 0709.
      249 degrees.   3 knots.   Aspect R.86°.
      3·9 M at 0647.   Red 82°.

22.   Nearest approach 1·1 M at 0427.

23.   300° astern.   6·8, decreasing.
      ·8 M in 30 minutes.   088°.   15½ knots.   Aspect G. 32°.
      About 60 minutes after the last observation.
      After the alteration of course, both ships were practically on parallel
      courses.

24.   093°, steady.   6·9, decreasing.   0 in 37 minutes.
      003 degrees.   10½ knots.   Aspect R.90°.
      3 M, 30 minutes after the resumption of course.

25.   About 1040.

26.   About 0637.

27.   About 0834.

28.   (a) Fast ship coming from astern on a near-collision course, making no
           attempt to take avoiding action.
      (b) 17 knots.
      (c) 3·8 miles.

29.   Target reduced speed from 9 to 2 knots (approx.).

30.   Target altered 68 degrees to starboard.

31.   Target altered 52 degrees to starboard.

32.(i) 1·9 M at 1040.
        Speed 7 knots.
   (ii) Target altered 44 degrees to port.
        3 cables at 1054.

33.(i) 063°, steady.   7¼ M, decreasing.   ·2 M at 1031.
        252 degrees.   12 knots.   Aspect R.9°.
   (ii) Target has maintained her course and speed.

34.(i) 320 degrees.   10 knots.   (ii) After 1612: 20 degrees, 10 knots.
        Nearest approach 6·7 M at 1636 on bearing R.50°.

35.   A   (a)   ·5 M at 1055;   (b) 265 degrees 6 knots;   (c) G.104°
      B   (a) 4·7 M at 1014;   (b) 050 degrees 10 knots;   (c) G.79°.
      C   (a) 3 M;   (b) 230 degrees 10 knots;   (c) R.47°.
      D   (a) Passed n.a.;   (b) 216 degrees 6½ knots;   (c) G.93°.

36.(i) A   000°, 10 knots.
       B   017°, 12 knots.
       C   000°, 7 knots.

(ii) Reduce speed to 6 knots.

(iii) 6 M for A; 3·3 M for B; 4·5 M for C at 0018.

37(i).　A　(a) 0 at 0632;　　　(b) 304 degrees 6 knots;　(c) G.6°.

　　　B　(a) 4·5 M at 0915　　(b) 145 degrees 6 knots;　(c) About
　　　　　　(approx.);　　　　　　　　　　　　　　　　　　　180.

　　　D　(a) 1·9 M at 0640;　(b) 162 degrees 12 knots;　(c) zero.

　　　Set and rate. East 1·5 knots.

37(ii).　Alter boldly to starboard.

38.(i) 2·7 M at 0046; 0040;
　　　207°, 10 knots;

　(ii) 2 M at 0105;

　(iii) 0·8 M on starboard side.

39.　A　(a) 2·8 M at 0024;　(b) Uncertain aspect and no speed.

　　　B　(a) ·4 M at 0037;　(b) R.51°, 3 knots.

　　　C　(a) 1·9 M at 0053;　(b) R.92°, 10 knots.

　　　Pay particular attention to target B.

40.(i) 095½°, passing ahead, 6·1 M, decreasing.
　　　0·8 M at 1033. 227°, 13 knots. Aspect G. 48½°.
　　　Tide 013°, 2·8 knots.

　(ii) (a) Own ship altered course to 169° T.

　　　(b) 1042, distance 1·1 miles.

　　　(c) 1034.

　　　(d) 1054 at ·9 M distance.

41.　(a) Yes, although a bolder alteration would have been better.

　　　(b) Stopping.

　　　(c) Because for the first 18 minutes the target was stationary and it
　　　　took a further 12 minutes to realize that she had started to make
　　　　way through the water and was crossing on a collision course with
　　　　own vessel.

　　　(d) Only if intermediate observations between 1030 and 1042 had
　　　　confirmed that the other vessel was maintaining her course and
　　　　speed.

　　　(e) 2·3 M at 1058.

　　　(f) Non-radar ship initially stopped, because of dense fog and then
　　　　deciding to proceed because the fog appears to be clearing in her
　　　　vicinity though, because of darkness she is not fully aware of the
　　　　extent of the visibility. Non-radar ship stopped with engine trouble
　　　　etc. Ship outside fog-bank and not using radar, initially stopped
　　　　with engine trouble etc., etc.

42.　(a) Both ships have taken manoeuvring action at about 0012. For ship
　　　　X the rate of change of range is changing after 0012; for ship Y the
　　　　bearing started to change fairly rapidly after 0012.

　　　(b) Avoiding action might become necessary to avoid a close quarters
　　　　situation with "X".

　　　(c) "X" altered about 50 degrees to starboard; "Y" reduced speed
　　　　from 10 knots to about 2 knots.

　　　(d) "Y" probably reduced speed to avoid own ship. "X" probably
　　　　altered course to avoid some other vessel not detected by own ship.

43. (i)

	B	C
CPA	—1·4	0
TCPA	00–39	00–41
Course	168°	136°
Speed	9·5	16·5

(ii)

	A	B	C
CPA	—1·4	—2·9	—2·3
TCPA	00–56	00–44	00–40

Current 070° 2·5.

(iii) C altered course 90° to starboard to avoid a close-quarters situation with B, then at about 0042 altered course, heading for the L.V.

44. (i)

	A	B	C
CPA	0	—0·8	—0·8
TCPA	1103	1048	1044
Course	015°	314°	281°
Speed	8·5	7·5	13·5

(ii) For A : 30° to starboard
For B : 60° to starboard } Approximately (Answer 120°).
For C : 120° to starboard

(iii) Ship C took avoiding action (90° to starboard) because she expected to run into difficulties with Own Ship and Ship B.

45. (i) CPA 0·5 M to port at 1310.

(ii) CPA +2·7 M, TCPA 1336, speed 20 knots, aspect G 20°.

(iii) 240° 20 knots. Current changing anti-clockwise from a southerly to an easterly direction and is increasing.

46. (i) Ships B and C same course as own ship, ship B doing 6 knots, ship C doing 12 knots.

(ii)

	A	B
CPA	+3·5 m	+2·5 m
TCPA	1757	1832

Predicted time of course resumption is 1747 (actual 1745).

(iii) B has altered 90° to starboard.   CPA +4·7 miles.

47. (i)

	B	C
CPA	—0·2	—1·2
TCPA	1030	1043
Course	230°	178°
Speed	10	5

(ii)

	B	C
CPA	+2·6	+6·0
TCPA	1029	1032

At 1048

(iii) No action was taken by targets.
Ship at anchor :   CPA   +6·5
                            TCPA   1110

48. (i)

	A	B	C
CPA	—2·5	0	—1·6
TCPA	2101	2058	2102
Course	180°	101°	336°
Speed	16	18	11

(ii) Target C

CPA	—2·3
TCPA	2110

(iii) B altered 90° to starboard at about 2045, then came back on course but had speed reduced to approximately 8·5 knots.

49. (i)

	A	B
CPA	0	—2·8
TCPA	1552	1540
Course	161°	343°
Speed	8	14·5

(ii) Ship A and B are practically on opposite courses.
Ship B made an emergency stop, slewing to starboard while doing so.

(iii) At approximately 1542.

(iv) Although ship B is stopped, she is still in the path of ship A, and ship A made a 90° alteration to starboard.

50.(i) 153½°, near-collision.   7·7 miles, decreasing.
0 in 60 minutes.   094°, 8½ knots.   Aspect R. 121°.
350° at 2½ knots.
114°, 12½ knots.
079°, 8 knots.   Useful for plotting on the chart but not for anti-collision purposes.

(ii) New course 142°.

(iii) Target has altered course to 145° and reduced speed to 6 knots.

(iv) About 14° forward of the port beam, 3·1 M at 0515.

Action by own ship contradicts Rule 8: " Any alteration of course and/or speed to avoid collision shall, if the circumstances of the case admit be large enough to be readily apparent visually or by radar."

REVISION QUESTIONS

*Answers should include supporting
sketches and diagrams where possible.*

Principles of Radar

1. Give a simple explanation of range measurement by radar.
2. Give a simple explanation of bearing measurement by radar.
3. Briefly outline how a radar set functions.
4. Define the terms beam-width and pulse-length. State the factors which determine the beam-width and pulse-length.

Characteristics of Set

1. Mention the characteristics of the set affecting the maximum range.
2. Mention the characteristics of the set affecting the minimum range.
3. On what factors does the range accuracy depend? How would you determine the range accuracy of the fixed range rings and of the variable range marker?
4. On what factors does the bearing accuracy depend? How can you check that the heading marker truly represents the fore and aft line of the ship?
5. What is meant by range discrimination and on what factors does it depend?
6. What is meant by bearing discrimination and on what factors does it depend?
7. What types of characteristics of a set are determined by pulse-length and beam-width?
8. What is meant by "Pulse Repetition Frequency" (p.r.f.)? Explain why the horizontal beam-width, the scanner rotation speed and the p.r.f. are closely related to each other.

Siting of Units

1. Part of the transmitter is sometimes fitted externally. Explain the advantages and disadvantages of this.
2. Careful thought should be given to the siting of the scanner unit. Explain this and mention the different factors to be considered.
3. What considerations should be taken into account when choosing a site for the radar display(s)?
4. Give a brief outline of the Interswitching System.

Operations and Controls

1. What is the Stand-by Switch and under what conditions is it used?
2. What is the purpose of the Centring Controls?
3. What is the purpose of the Heading Marker-Off Switch?
4. What is the function of the "Picture-Rotate (Alignment) Control"?

5. The Marine Radar Performance Specification, 1968, states:—
"Satisfactory means shall be provided to minimise the display of unwanted responses from precipitation and the sea." Explain what is meant by these "means" and how they are used.

6. Mention three reasons for clutter suppression. What do you consider more dangerous, too much or too little suppression?

7 When would you use the Variable Range Marker and what precautions would you take? What is the Range Calibration Switch?

8. What is meant by "testing the performance"? How could this be carried out and why is it important to do this when switching the set on?

9. If there are no land targets in sight and there is no Performance Indicator, how could you check on whether the set is working.

10. What is meant by "tuning the set"? State several methods how a set can be tuned by the operator.

11. What is the purpose of the Variable Range Delay switch?

12. Describe how you would take a bearing of a target.

13. On what factors does the accuracy of a bearing depend?

14. What precautions would you take to reduce centring error?

15. Compare the accuracy of bearings taken on a Relative Motion Un-stabilized Display and a True Motion Display.

16. Describe how you would take the range of a target. Compare range accuracy and bearing accuracy in this connection.

17. What are Parallel Index Lines? Deal briefly with some applications.

18. What additional controls are provided if a True Motion Unit is in-corporated in the Display Unit?

19. Explain two purposes of the Zero Speed Switch.

20. What methods can be used to check bearing and range accuracy in poor visibility?

Characteristics of the Target and Echo Strength

1. What is meant by "aspect" of a target? Show in sketches how aspect can contribute to echo strength. Give some practical examples.

2. What is meant by "specular reflection"? Sketch a diagram and give a practical example.

3. What is the difference between "shape" and "aspect" of a target? Explain how shape and surface texture influences the echo strength and detection range. Give examples.

4. Explain why the composition of a target is related to echo strength.

5. How do width, height and depth of a target affect the detection range?

6. What two target characteristics influence the amount of re-radiation from an object and what target characteristics determine the proportion of this re-radiation which is returned to the scanner?

7. Explain what you know about the radar echoing properties of different types of buoys.

8. At medium ranges, a buoy may vary in echo strength. Explain why this is so and include in your discussion buoys which are equipped with radar reflectors.

9. Discuss the echoing strength of ships. On what factors does this depend? What type of craft yield poor detection.

10. State several methods how you would distinguish between the echo of a buoy and the echo of a ship.
11. How would you detect if a target is moving?
12. Write a brief discussion on how different types of ice may show up on the display.
13. Write a short discussion about the echoing response of large and small icebergs. Compare Arctic and Antarctic icebergs in this connection.
14. The picture on the radar display often differs considerably with the view one might expect when consulting the chart. State the most obvious factors which are responsible for this discrepancy.

Atmospheric Conditions and Detection Range

1. Compare radar "vision" and ordinary vision so far as the range is concerned.
2. What is the effect of the earth's curvature on the radar range?
   What is the effect of the wavelength on the range?
3. What is meant by sub-refraction? What meteorological conditions favour sub-refraction?
4. What is meant by super-refraction? What meteorological conditions favour super-refraction?
5. Ice and low craft are sometimes not displayed as an echo. Explain what atmospheric conditions can handicap the radar performance in this way.
6. Echoes from much greater range than the maximum range of the set might be displayed when atmospheric conditions favour them. Briefly explain what conditions favour such extreme ranges and why the echoes are displayed.
   How can you recognize such echoes?
7. Show how Second-Trace Returns can be eliminated by p.r.f. 'wobbulation'.

Weather Conditions and Detection Range

1. Explain the limitations imposed by wind on a radar set on a ship at sea. What precautions should the radar observer and shipmaster take with regard to these limitations?
2. Explain how precipitation may affect the detection of objects in two ways. Explain the function of the Rain Switch (Differentiator).
3. How does fog affect detection ranges?
4. If the radar set is used as a navigational aid near land after a heavy snowfall, what limitations must be reckoned with?
5. Discuss the technique to produce a display practically clear of wind clutter echoes ("Clearscan").
6. Explain "Circular Polarization"; its use, advantages and disadvantages.
7. Compare 3 and 10 cm. radar and explain why, gradually, the addition of a 10 cm. radar set becomes more common.

Blind and Shadow Sectors

1. What are blind and shadow sectors? Give some examples. What precautions should the Master take in connection with these sectors when sailing in reduced visibility?
2. How can the angular width of a shadow sector be determined? How is the information recorded?

Unwanted Echoes

1. What is the cause of multiple echoes? How are they recognized and what remedy can be applied, if any?
2. What is the cause of indirect echoes? How are they recognized and how can their effect be reduced?
3. What is the cause of side echoes? How are they recognized and how can their effect be reduced?
4. Describe the effect of external and internal interference on the display. What can be done about it?
5. After leaving a port, land is dead astern. At the same time a spurious echo appears ahead near the heading marker on the display. State a possible explanation and predict the motion of the false echo across the screen.
6. What is meant by "reflection effects"? How are they recognized?
7. Which are more liable to produce false echoes, the funnel or the cross-trees? Explain.
8. Explain the production of a radar free interference display by means of pulse-correlation technique.

Use of Radar for Navigation

1. Mention some features, relating to radar detection, which are now incorporated in navigational charts.
2. "Radar information is not always reliable when making a landfall" Explain this statement.
3. Explain how you would obtain a fix by radar ranging. What type of target does one look for in this connection?
4. How can radar be usefully employed in narrow waters? Explain why the chart must be thoroughly studied for particular features likely to produce radar echoes.

Presentation of Display for Navigational Use

1. Compare the display presentations Unstabilized Relative Motion Display, Head-Up and Stabilized Relative Motion Display, North-Up.
2. Explain the display presentation which uses a Stabilized Azimuth (Bearing) Ring. Compare this display with a Stabilized Relative Motion Display.
3. What is an Expand-Centre Display? When is it used and on what range scale? Why is it no longer fitted nowadays?
4. When can it be advantageous to off-centre the display? What are the disadvantages?
5. Give a brief description of a True Motion Display. What are the advantages and disadvantages of such a display? "There are circumstances when it is extremely valuable to use a True Motion Display". Explain this statement.
6. What additional controls are provided to operate a True Motion Display and what are their functions?
7. Give a brief description of the True Motion Head-Up Display.
8. Own ship is steering 000° T. at a speed of 15 knots in clear weather and is yawing badly. Another vessel is sighted at four points on starboard bow, steering 270° T. at an approximate speed of 15 knots.

Describe and explain by means of constructional diagrams the echo motion of the other vessel on:—

    (*a*) An Unstabilized Relative Motion Head-Up Display.
    (*b*) A Stabilized Relative Motion North-Up Display.
    (*c*) A True Motion North-Up Display.

9.   The observer's vessel has a speed of 15 knots, course 090° T. It is clear weather and another vessel is seen at a distance of 10 miles, bearing 107° T. Her speed is 12 knots, her course 000° T.

    When this ship was observed the display presentation was " True Motion, North Up " with the start of the time-base coincident with the geometrical centre of the tube. Twelve minutes after the observation the display presentation is switched over to " Relative Motion Stabilized, North-Up" and 24 minutes after the first observation the display presentation is switched to "Relative Motion Unstabilized, Head-Up". Draw a diagram showing the motion of the echo over the tube face from 0000–0012 hours, 0012–0024 hours and from 0024 hrs. to 0036 hrs.

10.   Enumerate and discuss the pitfalls which must be guarded against when using true motion radar in particular.

Navigational Techniques

1.   Describe one method of using the Parallel Index for radar navigation using a Stabilized Relative Motion Display. Explain the advantages attached to such a technique.

2.   How can the Parallel Index Technique for navigation be used on a True Motion Display?

3.   Illustrate by means of an example how the Parallel Index Technique can be used when leaving a fairway to approach an anchor place.

Radar Aids

1.   How can the echoing properties of buoys and small boats be improved?

2.   Describe some arrangements (or grouping) or corner reflectors. Compare the different types of clusters and mention the purposes for which each type is used.

3.   Give a brief description of the Ramark and Racon beacon. Illustrate in a sketch what can be seen on the display in each case. Why is the range of the Racon signal shown on the screen a little greater than the range of the Racon station?

4.   Compare the relative merits of the Ramark and Racon beacon.

5.   Explain the differences between Racons, Echo Enhancers and Transponders.

6.   What is meant by "axis of symmetry" when discussing radar reflectors? Then discuss the best attitude of clusters for detection purposes.

7.   Discuss the various types of Racons which are available or under development. State their purposes.

Plotting

1.   How can you detect movement of targets on a Relative Motion Display?

2.   How would you report on the movement of ships? How would you deduce if there exists danger of collision?

3.   How would you find the rate and the set of the current by observing the range and the bearing of a lightvessel?

4. What is meant by the "aspect" of a ship?

5. What types of plot do you know and what is the essential difference between them?

6. Why is it so important to observe an echo closely *after* avoiding action has been taken?

7. What information can be derived from the direction on the screen of the "tadpole tails"
   (a) when shown on a Relative Motion Display?
   (b) when shown on a True Motion Display?

8. Mention and describe briefly some simple plotting devices.

9. Describe the use and the advantages of a Reflection Plotter.

10. Describe the use of a Vector Display as a Plotting Device.   State precautions when using this presentation.

11. How can the Parallel Index be used in the estimation of the nearest approach?

12. Mention four factors which can cause errors in plotting. Show by means of a diagram how a slow target ship can yield a poor plot if own ship is yawing and an Unstabilized Display is used.

13. Discuss the construction of Possible points of Collisions (PPCs) and Predicted Areas of Danger (PADs).

14. Explain the advantages of PPCs and PADs in connection with collision-avoidance tactics.

15. Explain the existence of dual PPCs. What will be the result when own ship increases or decreases her speed?

16. What is meant by Sectors of Danger (SODs) and Sectors of Preference (SOPs) in connection with collision-avoidance?

17. Discuss some useful rules to understand the relative motion of ships on the radar screen without actual plotting?

Radar Log

1. Give reasons why a radar log is useful.

2. Mention some items you would record in the radar log.

Automatic Radar Plotting Aids (ARPAs)

1. Discuss some of the methods of obtaining a "Bright Radar Display".

2. In what ways does an ARPA display picture differ from the conventional radar display?

3. What is the difference between the following three display presentations :
   (a) Head-up Unstabilized ;
   (b) Head-up Stabilized ;
   (c) Course-up ?

4. Explain the terms "Acquisition" and "Tracking" of a target.

5. Relating to "Acquisition", under what circumstances would you use the Manual Mode and when the Automatic Mode?

6. Explain some of the uses of the Joystick when operating an ARPA display.

7. Explain the following terms and their uses all related to control found on ARPA displays :—

Vector Length Control ;
History Track ;
Set CPA and Set TCPA ;
Guard Ring ;
Exclusion Zone Boundaries ;
Minimum Acquisition Range.

8. State various Alarms and their uses found on ARPA displays.

9. Explain the "Trial Manoeuvre", mandatory for ARPAs.

10. Explain the terms and their uses for Navigators related to controls found on ARPA displays :—

NAV. lines ;
Radar Map.

11. Explain the term "Quantized Display".

12. Explain the term "Smoothing Process" related to ARPA displays.

13. Explain some of the errors associated with ARPA displays.

14. Discuss the precautions an operator has to take and the limitations he has to keep in mind when using an ARPA display.

Use of Radar for Anti-Collision

1. When using radar as an anti-collision aid, what factors should you take into account when selecting range scales?

2. Under what conditions would you use the Stand-by Switch?

3. Describe your actions in fog in open waters when one echo is seen on the PPI from a vessel at a range of 8 miles forward of the beam on a collision course. Amplify your answer by referring to the various *Rules*.

4. What are the advantages of the Stabilized Relative Motion Display over the Unstabilized Relative Motion Display when using radar for anti-collision?

5. Mention some advantages and disadvantages of the True Motion Display as compared with the Relative Motion Display. What type of plot would evolve on the reflection plotter on a True Motion Display and indicate by means af a sketch how the nearest approach can be obtained?

6. Discuss the terms " safe speed " and " close quarters situation ".

7. State, in your own words, the contents of *Rule* 19, the recommendations made therein and the restrictions imposed on manoeuvring actions (*a*) for vessels forward of the beam, (*b*) for vessels abeam or abaft the beam.

8. *Rules* 11 to 18 apply only to vessels in sight of one another. Why is it that they cannot be applied in fog by a vessel using her radar?

9. What is meant by the term " Cumulative Turn " and describe the danger attached to this type of action?
Explain the following wording:—" in a cumulative turn the increments of positive action by one vessel reduces the original negative miss distance."

10. Explain the expression " all available means " in " Every vessel shall use all available means appropriate to the prevailing circumstances and conditions to determine if risk of collision exists." (*Rule* 7 (*a*)) and " Every vessel shall at all times maintain a proper look-out by sight and hearing as well as by all available means appropriate in the prevailing circumstances and conditions so as to make a full appraisal of the situation and of the risk of collision." (*Rule* 5.)

11. Explain the statement in *Rule* 8, " Any alteration of course and/or speed to avoid collision shall, if the circumstances of the case admit, be large enough to be readily apparent to another vessel observing visually, or by radar; a succession of small alterations of course and/or speed should be avoided."

12. Explain the statement in *Rule* 8, " If there is sufficient sea room, alteration of course alone may be the most effective action to avoid a close-quarters situation, provided that it is made in good time, is substantial and does not result in another close-quarters situation."

13. Explain in your own words the contents and practical application of *Rule* 19 (*e*).

14. *Rule* 19 (*e*) states, " Except where it has been determined that a risk of collision does not exist, every vessel which hears apparently forward of her beam the fog signal, etc." Explain the first part of this sentence and illustrate a case where it can be argued that risk af collision does *not* exist.

15. *Rule* 15 only applies to vessels in sight of one another. Is it implied, therefore, that a vessel can pass ahead of another vessel, when it is out of sight? Discuss.

16. Explain the procedure when using V.H.F. R/T communication in order to arrive at an agreement with another vessel about anti-collision tactics. Stress the precautions to be taken so that no errors in identification and mutual understanding may take place.

17. Assumptions shall not be made on the basis of scanty information, especially scanty radar information (*Rule* 7 (*c*)). Explain.

REVISION QUESTIONS

(Multi-choice)

Radar observer examinations in the United Kingdom and abroad include multi-choice questions.

Students undergoing this type of test are strongly advised to study the statements very carefully. Only ONE of the choices, i.e. the MOST APPROPRIATE choice should be indicated by marking the relevant box. If a student marks two, then this will automatically be marked wrong, even if both choices are appropriate and the correct answer is one of the two chosen.

Attempt 10 questions from Section I plus 30 questions from the other Sections (time allowed is one hour).

Select the option which you consider most appropriate and indicate your choice by marking the relevant box thus

SECTION I - True/False

1. Doubling the plotting interval will halve the chance of making errors in the estimation of the nearest approach.

    TRUE
    FALSE

2. The output power of a radar is a function of peak power and p.r.f.

    TRUE
    FALSE

3. The D.o.T. specification for marine radars (1968) requires that the radar beam in the horizontal plane shall not be less than 2°, measured to the half-power points.

    TRUE
    FALSE

4. The vertical lobe pattern for an "S" band radar is denser than that for an "X" band radar.

    TRUE
    FALSE

5. "X" band radar operates at a frequency of approximately 9,400 MHz and wavelength of 3·2 cm., while "S" band radar operates at a frequency of approximately 3,000 MHz and a wavelength of 10 cm.

    TRUE
    FALSE

6. An aerial having an aperture of 300 cms. in the horizontal plane will have better bearing discrimination when used with waves of 10 cm. length than with waves of 3 cm. length.

    TRUE
    FALSE

403

7. The figure shown stands for the IMO symbol for the control which takes the radar equipment from "Standby" to "Operate".

TRUE

FALSE

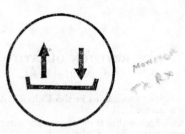

8. The "Zero Speed" or "Hold" switch is useful when a vessel does not have a log input for the radar true motion unit.

TRUE

FALSE

9. Radar interference, which is observed when only one ship's echo is on the display, indicates that the vessel concerned is using its radar.

TRUE

FALSE

10. The rain clutter on 3-cm. radar sets is less dense than on 10-cm. radar sets, given that external conditions are identical.

TRUE

FALSE

11. The height of the tide may affect the intensity of echoes on a marine radar display.

TRUE

FALSE

12. The VRM is inherently more accurate than the range rings.

TRUE

FALSE

13. A larger aerial aperture will improve bearing discrimination.

TRUE

FALSE

14. If the origin of an 'Interscan' marker is displayed as an ellipse it will lead to inaccuracies in range and bearing measurement.

TRUE

FALSE

15. Second-trace echoes are unlikely when using a high p.r.f.

TRUE

FALSE

16. Unwanted responses from a Racon at close range can be minimized by using the differentiator control.

TRUE

FALSE

17. With both vessels proceeding at the same speed the motion of own ship as seen on the target's Relative Motion Display is drawing aft.

TRUE
FALSE ✓

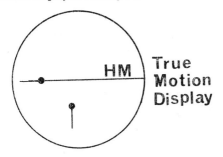

HM True Motion Display

18. One of the factors favouring sub-refraction is relative humidity increasing with height.

TRUE ✓
FALSE

19. The displayed pattern from a 'remote echo-box' is a sun.

TRUE
FALSE

20. Switching on the FTC control aids the detection of targets beyond rain.

TRUE
FALSE ✓

21. Range discrimination depends on pulse duration.

TRUE ✓
FALSE

22. A 10-cm. radar will have better bearing discrimination than a 3-cm. radar, if their aerials have the same aperture.

TRUE
FALSE ✓

23. The figure stands for the IMO symbol to indicate that the radar is on the stand-by position.

TRUE
FALSE

24. The brilliance control adjusted to its optimum setting can *usually* be left unadjusted when range scales are changed.

TRUE
FALSE ✓

25. Some marine radar sets have special circuits which eliminate radar-to-radar interference.

TRUE
FALSE

26. Using the parallel index technique on a relative motion display the observer is able to constantly monitor the effect of a cross tide by observing the motion of a fixed target.

TRUE ✓
FALSE

27. While proceeding at a moderate speed on a true course of 270°, tide setting 090° T, the echo shown in the diagram (Relative Motion Display) is of a ship target (plotted over a period of time) proceeding on the same true course and speed as the observer.

TRUE ✓
FALSE

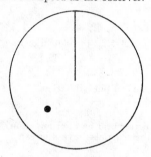

28. Under sub-refraction conditions the radar range of a target is likely to be less than its true range.

TRUE
FALSE ✓

29. A reliable method of ensuring optimum performance of a radar set is to tune for maximum response of sea clutter.

TRUE ✓
FALSE

30. A 10-cm. radar set transmits in the X-band.

TRUE
FALSE ✓

31. Targets outside the angle subtended by half-power points of the transmitted beam, may return echoes.

TRUE ✓
FALSE

32. The figure below stands for the IMO symbol used to indicate the VRM.

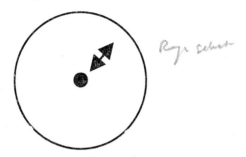

33. With the origin of the PPI trace off-centre, only two lines of bearing are correct when using the cursor.

TRUE
FALSE

34. Multiple echoes can be minimized by the fitting of Radar Absorbent Material to obstructions causing blind arcs.

TRUE
FALSE

35. Whilst proceeding at a moderate speed on a true course of 180° with the tide setting 090° true, the echo shown in the diagram below (Relative Motion Display) is of a target having a red aspect.

TRUE
FALSE

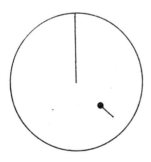

36. One of the factors favouring super-refraction conditions is cool air blowing over a warm sea.

TRUE
FALSE

37. Equal spacing of range rings confirms that they are accurate for range measurements.

TRUE
FALSE

38. The detection range of a target is improved by the use of a larger horizontal beam-width.

TRUE
FALSE

39.   The tuning control varies the transmitted frequency.

40.   The reliability of a radar set will deteriorate far less by continuous running than by frequently switching it on and off.

41.   The figure below stands for the IMO symbol used to indicate the EBI.

42.   If gain is reduced and echoes A and B fade (see diagram below) before echo C, then C must be the echo from a real target.

TRUE
FALSE ✓

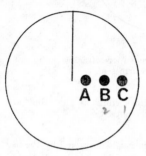

43. Whilst proceeding at a moderate speed on a course of
000° T and using a Relative Motion Display, the echo
of a lighthouse is observed as indicated in the diagram.
The current is setting westward.

TRUE
FALSE

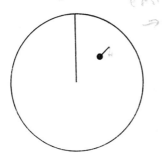

44. Whilst proceeding at a moderate speed on a true course
of 180° with tide setting 090° true, the echo shown in
the diagram (Relative Motion Display) is of a target
having zero aspect.

TRUE
FALSE

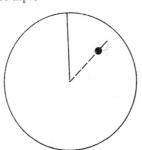

45. Range discrimination depends upon the horizontal
beam-width.

TRUE
FALSE

46. Marine radar transmissions are beamed, broad in the
horizontal plane, narrow in the vertical plane.

TRUE
FALSE

47. The figure below stands for the IMO symbol used to indicate a north-up stabilized display.

48. Snow clutter appearing near the maximum of the 6-mile range scale can be effectively reduced by careful operation of the sea clutter control.

TRUE / FALSE

49. When side echoes are observed the true echo can be determined by temporarily reducing the gain.

TRUE / FALSE

50. On a true motion display having a non-linear time-base the observed course of a target will always be in error.

TRUE / FALSE

SECTION II - Multi Choice

51. The time taken for a radar pulse to travel out to and the echo returning from a target at a range of 24 n.m. is (approximately)

(A) 100 micro seconds
(B) 200 micro-seconds
(C) 300 micro-seconds
(D) 400 micro-seconds.

Working (if required)

A B C D

52. The D.o.T. specification for marine radars (1968) requires that ". . . the echo of a GPC buoy shall be visible . . . down to a range of"

(A) 15 yards
(B) 50 yards
(C) 75 yards
(D) 125 yards.

A B C D

53.   Using Digiplot in the "forecast mode" the echo on the display moves in

    (A)   Relative motion displaying a relative vector

    (B)   Relative motion displaying a true vector

    (C)   True motion displaying a relative vector

    (D)   True motion displaying a true vector.

    A B C D

54.   On the "Predictor" display system the time-base on the 48-mile range scale is (approximately)

    (A)   4·64 micro-seconds

    (B)   9·28 micro-seconds

    (C)   593 micro-seconds

    (D)   1186 micro-seconds,

    Working (if required)

    A B C D

while on the 3/8-mile range scale it is (approximately)

    (A)   4·64 micro-seconds

    (B)   9·28 micro-seconds

    (C)   593 micro-seconds

    (D)   1186 micro-seconds.

    Working (if required)

    A B C D

55.   On the shorter range scales it is usual to increase the p.r.f.

    (A)   to avoid the possibility of second-trace returns

    (B)   to increase the transmission energy

    (C)   to improve the minimum detection range

    (D)   to increase the number of lines per scan.

    A B C D

56.   The strength of a returning echo depends upon

    (A)   Radar peak power only

    (B)   Pulse duration only

    (C)   Aerial efficiency only

    (D)   All of the above

    (E)   Setting of the gain control.

    A B C D E

57.   Which one of the following is it most appropriate to observe while tuning a radar set?

    (A)   Fineness of rings

    (B)   Output of Performance Monitor

    (C)   The speckled background of the display

    (D)   The extent of sea clutter.

    A B C D

58. The 'Overall Performance Monitor' will give warning when

   (A) there is condensation in the waveguide
   (B) detection ranges are reduced due to atmospheric attenuation
   (C) ranging errors only exist
   (D) bearing errors only exist
   (E) both ranging and bearing errors exist.

59. "Beam-width Distortion" occurs as a result of

   (A) "squinting" of modern radar aerials
   (B) diffraction of waves
   (C) refraction of waves
   (D) the radar beam having a practical finite width in the horizontal plane.

60. Within the Raleigh distance, bearing discrimination is

   (A) better than at greater distances
   (B) worse than at greater distances
   (C) the same as at greater distances
   (D) improved at the expense of range discrimination.

61. A small ship, dead ahead visually, at about 2 miles bears 357° (relative) on the radar display.

   (A) The heading marker represents 3° Red
   (B) The heading marker represents 3° Green
   (C) The gyro error is 3° high
   (D) The gyro error is 3° low
   (E) The aerial is rotating 3° per revolution faster than the trace
   (F) The aerial is rotating 3° per revolution slower than the trace.

62. A "Balanced Mixer" is used in radar to

   (A) eliminate thermal noise originating in the klystron, so that the chance of poor response targets can be improved
   (B) provide the speckled background for use in setting of the gain control
   (C) speed up the T/R switch-over time and thus improve the minimum range
   (D) allow the tuning to occur at the midpoint of travel of the tuning control.

63. The F.T.C. control can be useful for

   (A) detection of targets beyond rain

   (B) bringing receiver frequency into line with that of the transmitter

   (C) improving range discrimination

   (D) the suppression of "radar-to-radar" interference.

A B C✓ D

64. A target vessel which is on a collision course can appear to change its compass bearing if

   (A) the mechanical curzor is used when the electronic centre (start of time-base) is displaced

   (B) own ship is yawing and relative bearings are observed on an unstabilised display

   (C) a constant compass error is not allowed for

   (D) the heading marker is not correctly aligned on the bearing scale.

A✓ B C D

65. Inlets, lock entrances, etc., are frequently not detected on a radar display.
They are more likely to be observed as echoes

   (A) on a radar with a smaller aerial aperture

   (B) when closer to them

   (C) using a longer pulse duration

   (D) by putting the display "off-centre".

A B✓ C D

66. "Beam-width Distortion" can be partly compensated for by

   (A) lowering the setting of the brilliance control

   (B) finer tuning

   (C) using a shorter pulse-length

   (D) lowering the setting of the gain control.

A B C D✓

67. An echo is first detected at a range of 40 miles. If the height of the scanner is 20 metres, what—under favourable conditions—is the approximate height of the target?

   (A) 120 m.

   (B) 160 m.

   (C) 200 m.

   (D) 240 m.

Working, if required

A B C✓ D

68. A "Radaflare" produces a good radar response by ejecting a

   (A) parachute of fine metallic fabric

   (B) cloud of short dipoles which are tuned to 3-cm. waves

   (C) cloud of water and foam particles

   (D) parachute supporting a mesh Octahedral reflector.

A B✓ C D

69. On a DECCA "Clearscan" display an observer can see all ships' and land echoes

    (A) displayed at full gain and radially elongated
    (B) displayed at full gain and differentiated
    (C) logarithmically amplified and radially elongated
    (D) logarithmically amplified and differentiated.

    A B C D

70. During freak propagation conditions, false echoes have been observed on the display as a result of such conditions. These are referred to as:—

    (A) Indirect echoes
    (B) Multiple echoes
    (C) Second-trace echoes
    (D) Side echoes.

    A B C D

71. The time interval between the transmission of a radar signal and the return of the echo from a target is 0·1 micro-seconds. What is the approximate range of the target?

    (A) 10 nautical miles
    (B) 1·2 nautical miles
    (C) 300 metres
    (D) 15 metres.

Working (if required)

    A B C D

72. In dense fog an echo is displayed under the heading marker at a range of 5 miles. A break in the fog reveals the target 3° on the starboard bow. The probable cause of this error is

    (A) Heading marker not aligned to zero on bearing scale
    (B) Heading marker contacts closing late
    (C) Aerial sited to starboard of the centre-line
    (D) Heading marker contacts closing early.

    A B C D

73. The main factor governing range discrimination is:—

    (A) Horizontal beam-width
    (B) p.r.f.
    (C) Vertical beam-width
    (D) Pulse duration.

    A B C D

74. A remote echo-box performance monitor showing a plume on the PPI enables the operator to assess the performance of the

    (A) Display unit only
    (B) Receiver only
    (C) Transmitter only
    (D) Transmitter and receiver.

    A B C D

75. Approximately how many times will a radar pulse strike a
     target which is just detectable if the horizontal beam-width
     is $1\frac{1}{2}°$, the p.r.f. is 1,000 p.p.s. and the aerial rotates at 24
     r.p.m.

     (A) 6 times

     (B) 8 times

     (C) 10 times

     (D) 12 times.

Working (if required)

     A
     B
     C
     D

76. In ships having two independent radar sets not fitted with
     automatic interference suppression circuits mutual interfer-
     ence can best be minimized by :—

     (A) increasing receiver bandwidth

     (B) fitting identical magnetrons

     (C) synchronising aerial rotation rates but with direction
          180° apart

     (D) mounting one aerial higher than the other.

     A
     B
     C
     D

77. A guide to the optimum setting of the tuning control can be
     "Adjust the control and observe":–

     (A) A nearby buoy with a radar reflector

     (B) the output power monitor

     (C) the extent of sea clutter

     (D) the range rings.

     A
     B
     C
     D

78. An echo lies on the one-mile range ring when using the three-
     mile range scale. If the radar set has been type-tested, the
     maximum permissible error in range is :—

     (A) $1\frac{1}{2}\%$ of 1 mile

     (B) $\pm$ 75 yards

     (C) $\pm$ 125 yards

     (D) $1\frac{1}{2}\%$ of 3 miles.

     A
     B
     C
     D

79. Your own ship alters course to port. On a North-up stabilized
     display :

     (A) Land echoes turn clockwise on the face of the display

     (B) land echoes turn anti-clockwise on the face of the display

     (C) the heading marker turns clockwise

     (D) the heading marker turns anti-clockwise.

     A
     B
     C
     D

80. The *measured* minimum range of a radar set is 18 metres. Using the same pulse duration, the expected range discrimination of two small targets should be :—

(A) More than 18 metres

(B) 18 metres

(C) less than 18 metres

(D) 36 metres.

Ⓐ Ⓑ Ⓒ Ⓓ

81. Multiple echoes displayed on the PPI should be reduced by :-

(A) Careful adjustment of the 'tune' control

(B) careful adjustment of the 'gain' control

(C) changing the range scale

(D) changing the p.r.f.

Ⓐ Ⓑ Ⓒ Ⓓ

82. The effect of echoes caused by radiation outside the main radar beam, should be reduced by :—

(A) Careful adjustment of the 'gain' control

(B) careful adjustment of the tuning control

(C) changing the range scale

(D) selecting a different pulse duration.

Ⓐ Ⓑ Ⓒ Ⓓ

83. When choosing the ideal site for the radar aerial, which of the following considerations is the most important :—

(A) Clear of obstructions which may cause shadow sectors

(B) on or near to the ship's centre-line

(C) to provide for the shortest waveguide run

(D) to provide ease of access for maintenance purposes?

Ⓐ Ⓑ Ⓒ Ⓓ

84. An echo just discernible at 10 miles on a correctly adjusted PPI may be enhanced by :—

(A) Readjusting the brilliance

(B) increasing the gain

(C) decreasing the gain

(D) cannot be enhanced.

Ⓐ Ⓑ Ⓒ Ⓓ

85. The maximum brightness of PPI 'paint' capable of being displayed is determined by :—

(A) Video limiter

(B) maximum power of the transmitted pulse

(C) setting of the 'gain' control

(D) reflecting power of the target.

Ⓐ Ⓑ Ⓒ Ⓓ

86. A smooth textured target placed in the path of a radar beam will produce the strongest echo when its shape is :—

(A) Cylindrical
(B) plane at 45° to the beam
(C) plane at 90° to the beam
(D) spherical.

A
B
C ✓
D

87. Charted RACON beacons in U.K. waters are only displayed on marine radar sets if the transmitted energy is :—

(A) Vertically polarized X-band
(B) horizontally polarized X-band
(C) vertically polarized S-band
(D) horizontally polarized S-band.

A
B ✓
C
D

88. The presence of heavy localized snowfalls can reduce the detection of targets :—

(A) Within and beyond the area of snow
(B) within but not beyond the area of snow
(C) beyond but not within the area of snow
(D) all targets before, within and beyond the area of snow.

A ✓
B
C
D

89. An atmospheric condition favouring sub-refraction is :—

(A) A decrease of relative humidity with height
(B) an increase of temperature with height
(C) an increase of relative humidity with height
(D) no change of temperature with height.

A
B
C ✓
D

90. A correctly adjusted true motion display operating in the 'sea stabilized' mode indicates, with reference to a target ship :—

(A) The motion relative to own ship
(B) the motion relative to the land
(C) the motion relative to the sea.

A ✓
B
C

91. The duration of the time-base on the 12-mile range scale is :-

(A) 123·6 micro-seconds
(B) 148·3 micro-seconds
(C) 186·0 micro-seconds
(D) 328·6 micro-seconds.

Working (if required)

A
B ✓
C
D

92. When a magnetron is changed in a radar set having a slotted waveguide aerial, it is essential to check :—

(A) The heading marker accuracy

(B) range accuracy

(C) for changes in the pattern of side-lobe echoes

(D) for increased radar interference.

93. Range discrimination is the ability of the set to :—

(A) Show separately two targets close together on the same bearing

(B) show separately two targets close together at the same range

(C) measure the range of a target to within $1\frac{1}{2}\%$ of its true range

(D) show which way a target is heading when at close range.

94. The horizontal beam-width as displayed on the PPI can be reduced by :—

(A) Reducing the aerial size

(B) reducing the gain

(C) switching the differentiator on

(D) reducing the p.r.f.

95. An echo lies on the 1-mile range ring when using the 6-mile range scale. If the set has been type-tested, the true range of the target must lie within :—

(A) 1 mile ± 30 yards

(B) 1 mile ± 75 yards

(C) 1 mile ± 90 yards

(D) 1 mile ± 180 yards.

96. Your own ship alters course to port. On a 'Head-up' compass stabilized display :—

(A) Land echoes turn clockwise on the face of the display

(B) land echoes turn anti-clockwise on the face of the display

(C) heading marker rotates clockwise

(D) heading marker rotates anti-clockwise.

97. A dual radar installation comprises 3 cm. and 10 cm. radars. One would expect the 3 cm. radar to provide, generally:—

(A) Better bearing discrimination

(B) better target-to-sea clutter discrimination

(C) inferior range discrimination

(D) better target-to-rain clutter discrimination.

98. If an echo appears in a shadow sector on your PPI, you would check if it was a real echo by :—

    (A) Changing the range scale

    (B) altering course

    (C) by careful plotting

    (D) altering speed.

A   B   C   D

99. Interference patterns being displayed on the PPI can best be reduced by :—

    (A) Adjusting the gain control

    (B) adjusting the tune control

    (C) selecting a lower range scale

    (D) selecting a higher range scale.

A   B   C   D

100. To aid the detection of targets beyond rain :—

    (A) Use the longest pulse available

    (B) turn the differentiator on

    (C) reduce the gain

    (D) retune the receiver.

A   B   C   D

101. Second-trace echoes are most likely to occur when :—

    (A) Super-refraction conditions are present

    (B) sub-refraction conditions are present

    (C) a second radar is operating on your own ship

    (D) another radar is operating on a vessel close-by.

A   B   C   D

102. When using True Motion facilities on some radars, the origin will re-position if :—

    (A) 'zero speed' is selected

    (B) own ship alters course

    (C) tidal corrections are set in

    (D) one of the range scales is changed.

A   B   C   D

103. The theoretical detection range of an object 49 m. high, from a ship with an aerial 16 m. above sea level, is about:—

    (A) 16 n.m.

    (B) 20 n.m.

    (C) 24 n.m.

    (D) 28 n.m.

Working (if required)

A   B   C   D

P

104. 'Anti-clutter' controls the :—

(A) Local oscillator
(B) CRT
(C) amplifier
(D) limiter.

[A] [B] [C] [D]

105. Which of the following surfaces is the poorest radar reflector?

(A) Wood
(B) wood with anti-fouling coating
(C) G.R.P.
(D) G.R.P. with anti-fouling coating.

[A] [B] [C] [D]

106. Inaccuracy of bearing measurement due to heading marker misalignment will affect :—

(A) Only bearings of targets ahead of own ship
(B) bearings on either beam
(C) all bearings
(D) only bearings of echoes close to the PPI centre.

[A] [B] [C] [D]

107. Large icebergs may not produce substantial echoes on the PPI principally because :—

(A) Ice is 'transparent' to electro-magnetic energy
(B) the greater proportion of the iceberg is below water and is thus not "seen" by radar
(C) icebergs always have a smooth surface so that the radar energy is reflected away
(D) icebergs present varying aspects to the radar energy, some of which may be poor.

[A] [B] [C] [D]

108. A charted RACON beacon enables a RACON echo to be 'painted' on the PPI :—

(A) Continuously, when within range
(B) at regular intervals, when within range
(C) at irregular intervals, when within range
(D) continuously on any range.

[A] [B] [C] [D]

109. The accuracy of a position using radar ranges of two coastal features separated by 90° may be affected by :—

(A) The origin of the PPI trace being off-centred towards the echoes of one of the coastal features
(B) the heading marker being displaced away from the echoes of one of the features
(C) one of the features having a low-lying fore-shore
(D) both the heading marker and origin of the PPI trace being off-centred.

[A] [B] [C] [D]

110. A marine radar set has a pulse duration of 0·2 micro-seconds. The approximate minimum range is :—

(A) 32 metres

(B) 46 metres

(C) 69 metres

(D) 114 metres.

Working (if required)

A **B** **C** **D**

111. Radar is able to determine bearing measurement by :—
(A) Measuring the angle turned by the aerial between transmission and reception of a pulse
(B) use of the electronic bearing marker
(C) measuring the angle turned by the aerial between passing the fore and aft line and reception of the echo
(D) ensuring that the aerial rotation rate is constant at 20 r.p.m. in wind speeds up to 100 knots.

A **B** **C** **D**

112. The main factor governing bearing discrimination is :—
(A) Horizontal beam-width
(B) p.r.f.
(C) vertical beam-width
(D) pulse duration.

A **B** **C** **D**

113. An observer notes that the transmission monitor (neon tube) is correct, but that the internal 'echo-box' is below normal. The most likely cause is :—
(A) Subrefraction conditions are prevalent
(B) the transmitter performance is below standard
(C) the receiver performance is below standard
(D) condensation of water vapour in the waveguide.

A **B** **C** **D**

114. A radar set which has a logarithmic amplifier :—
(A) Also requires a non-linear time-base
(B) displays good response echoes brighter than weak ones
(C) requires more careful adjustment of the sea clutter control
(D) improves the detection of targets beyond rain.

A **B** **C** **D**

115. Application of sea clutter control has the effect of :—
(A) Reducing echo strength returned from the sea
(B) suppressing amplification of echoes at short range
(C) increasing the echo strength of ships' targets at short ranges
(D) differentiating between water and steel targets.

A **B** **C** **D**

116. When taking ranges with the variable range marker :—

    (A) Outer edge of marker should touch inner edge of echo
    (B) inner edge of marker should touch outer edge of echo
    (C) outer edge of marker should touch outer edge of echo
    (D) inner edge of marker should touch inner edge of echo.

    A B C D

117. A target observed visually right ahead, appears on the PPI 5° to port of the heading marker because :—

    (A) The heading marker has not been correctly set at 000° on the bearing scale
    (B) the heading marker contacts are closing at the wrong time
    (C) the foremast is causing the beam to be deflected 5° to port
    (D) the operator has failed to correctly centre the origin and is experiencing parallax.

    A B C D

118. The aerial should be placed so as to minimize the effect of :—

    (A) Side lobes
    (B) multiple echoes
    (C) indirect reflected echoes
    (D) radar interference from other vessels

    A B C D

119. Side lobe echoes can be reduced with the least detrimental effect on the picture by :—

    (A) Decreasing the gain control
    (B) increasing the anti-clutter control
    (C) decreasing the brilliance control
    (D) careful adjustment of the tuning control.

    A B C D

120. The most suitable means of determining the angular width of a shadow sector is to take the relative bearings of the disappearance and reappearance on the display of :—

    (A) Heavy sea clutter
    (B) the performance monitor plume
    (C) a small isolated object
    (D) a large land mass.

    A B C D

121. A radar aerial is sited to starboard of a ship's centre-line:—

    (A) This can be allowed for by off-centring the origin and heading marker on the display to starboard
    (B) this can be allowed for by off-centring the origin and heading marker on the display to port
    (C) this can be allowed for by use of the anti-parallax reflection plotter
    (D) no allowance can be made.

    A B C D

122. The weakest detectable echo on a properly functioning radar set is :—

    (A) A GPC buoy

    (B) small vessels, ice and other floating objects

    (C) one just stronger than receiver noise

    (D) one on the radar horizon.

A
B
C
D

123. A characteristic of a RACON beacon is that it :—

    (A) Receives a pulse and then transmits its own pulse

    (B) receives a pulse, and then re-transmits the same pulse

    (C) is transmitting continuously

    (D) transmits pulses at regular pre-set intervals.

A
B
C
D

124. On a VLCC with its radar aerial and transceiver on the forecastle and its display sited on the bridge, the observer measures :—

    (A) Range and bearing from the bridge

    (B) range and bearing from the forecastle

    (C) range from the forecastle but bearing from the bridge

    (D) range from the bridge but bearing from the forecastle.

A
B
C
D

125. Objects beyond a rain storm can sometimes be detected by using one of the following control adjustments :—

    (A) Temporarily reducing the gain

    (B) apply a little sea clutter control

    (C) temporarily turn up the gain

    (D) apply the differentiator.

A
B
C
D

126. Subrefraction conditions cause :—

    (A) A reduction in the detection ranges of targets

    (B) an increase in the detection ranges of targets

    (C) a reduction in the range accuracy of targets

    (D) an increase in the range accuracy of targets.

A
B
C
D

127. When using True Motion, the echo of a jetty is observed to be moving in the same direction as 'own' ship. You should:—

    (A) Decrease the speed setting of the radar

    (B) increase the speed of your vessel

    (C) reduce the speed of your vessel

    (D) increase the speed setting of the radar.

A
B
C
D

128. If it takes a total of 37·04 micro-seconds for a radar pulse to travel to, and the echo return from, the target, its range is:—

(A) 3 miles

(B) 13 miles

(C) 23 miles

(D) 33 miles.

Working (if required)

Ⓐ Ⓑ Ⓒ Ⓓ

129. An important factor which determines horizontal beam-width is :—

(A) P.R.F.

(B) pulse duration

(C) aerial height

(D) aerial aperture.

Ⓐ Ⓑ Ⓒ Ⓓ

130. A target is displayed at a range of 1 n.m. on a set using a p.r.f. of 500 p.p.s. It is known to be a second-trace return. The range of the real target is approximately :—

(A) 41 n.m.

(B) 82 n.m.

(C) 164 n.m.

(D) 328 n.m.

Working (if required)

Ⓐ Ⓑ Ⓒ Ⓓ

131. In ships having two independent radar sets, mutual interference can be minimized by :—

(A) Increasing receiver band-width

(B) fitting identical magnetrons

(C) synchronizing transmission pulses

(D) mounting one aerial higher than the other.

Ⓐ Ⓑ Ⓒ Ⓓ

132. Tuning controls the :—

(A) Magnetron

(B) Local Oscillator

(C) CRT

(D) amplifier.

Ⓐ Ⓑ Ⓒ Ⓓ

133. The presence of an indirect echo on the PPI caused by own ship's structure can be removed by :—

(A) Selecting a different range scale

(B) use of the clutter control

(C) alteration of own ship's course

(D) use of the tune control.

Ⓐ Ⓑ Ⓒ Ⓓ

134. Spiral interference patterns are caused by another ship's radar if it has :—

    (A) The same p.r.f. and same frequency

    (B) the same p.r.f. but different frequency

    (C) different p.r.f. and a different frequency

    (D) different p.r.f. but the same frequency.

    A **B** **C** **D**

135. A small boat constructed mainly of G.R.P. is usually considered to be a "poor" radar target because :—

    (A) The radar beam is almost totally scattered by the smooth surface of the boat's hull

    (B) the boat is small

    (C) G.R.P. is "transparent" to radar energy

    (D) G.P.R. absorbs radar energy.

    A **B** **C** **D**

136. Own ship alters course to starboard. On a Course-up stabilized display :—

    (A) The tube itself turns anti-clockwise

    (B) the tube is stationary, but echoes rotate anti-clockwise

    (C) the tube itself turns clockwise

    (D) the heading marker turns clockwise.

    A **B** **C** **D**

137. Comparison of radar and visual bearings indicates a constant $2°$ error. This error can be eliminated by :—

    (A) Adjustment of the heading marker contacts by $2°$

    (B) adjustment of the PPI heading marker by $2°$

    (C) adjustment of the E.B.M. by $2°$

    (D) rotation of the PPI tube in its mounting by $2°$.

    A **B** **C** **D**

138. An echo of a buoy being displayed at about 4 miles on the beam may fade or disappear as the range decreases. This phenomenon may be caused by :—

    (A) Loss of sky wave signal

    (B) interference by the sky wave signal

    (C) secondary return path of the echo via a smooth sea surface

    (D) secondary return path of the echo via a rough sea surface.

    A **B** **C** **D**

139. RACON beacons respond to the ship's radar transmission at :—

    (A) Own ship's frequency only

    (B) all S-band radar frequencies simultaneously

    (C) a frequency determined by the position in the RACON cycle

    (D) all X-band radar frequencies simultaneously.

    A **B** **C** **D**

140. The most accurate way of fixing a vessel's position using radar only is :—

   (A) A range and a bearing of the same target
   (B) two bearings of different targets
   (C) two ranges of different targets
   (D) a range of one target and a bearing of a different target.

   A **B** **C**✓ **D**

141. In the presence of other ships and entering heavy snow squalls you would first :—

   (A) Reduce speed and instigate full radar watchkeeping procedures
   (B) maintain speed and instigate full radar watchkeeping procedures
   (C) vary the gain and clutter controls to enhance detection of targets
   (D) increase speed to clear the area quickly.

   A✓ **B** **C** **D**

142. Under standard propagation conditions the distance to the radar horizon is approximately :—

   (A) The same
   (B) 0·7 times as great
   (C) 1·1 times as great
   (D) 2·2 times as great than the distance to the visible horizon (using the same height for observation point).

   A **B** **C**✓ **D**

143. Radiated pulse duration is determined by the :—

   (A) Aerial aperture and shape
   (B) correct tuning of the local oscillator
   (C) length of time that the magnetron is oscillating
   (D) 'change-over' time of the transmit/receive cell.

   A **B** **C**✓ **D**

144. Manual Tuning should be checked :—

   (A) On changing from one range scale to another
   (B) on changing pulse-length
   (C) after a period on stand-by
   (D) on changing display mode.

   A **B** **C**✓ **D**

145. A "paint" being displayed in a blind arc is most likely to be caused by :—

   (A) Side-echoes
   (B) sea clutter
   (C) multiple echoes
   (D) 2nd-trace echoes.

   A✓ **B** **C** **D**

146. Precipitation echoes are more prominent on a radar display operated from :—
   (A) An elliptically polarized radar
   (B) an S-band radar on long pulse
   (C) a circularly polarized radar
   (D) an X-band radar on long pulse.

   A **B** **C** **D**

147. The true range of a RACON beacon when compared to the RACON 'paint' is :—
   (A) Equal to the nearest edge of the 'paint'
   (B) shorter
   (C) greater
   (D) equal to the furthest edge of the 'paint'.

   A **B** **C** **D**

148. Side-lobe echoes can be reduced with the least detrimental effect on the PPI by :—
   (A) Careful adjustment of the tune control
   (B) decreasing the clutter control
   (C) increasing the gain control
   (D) increasing the clutter control.

   A **B** **C** **D**

149. Range discrimination between two echoes at about 6 miles range on the PPI can be safely improved by :—
   (A) Careful adjustment of the tune control
   (B) choosing a short pulse duration
   (C) choosing a longer pulse duration
   (D) careful adjustment of the sea clutter control.

   A **B** **C** **D**

150. The term "pentagonal cluster" means :—
   (A) A radar reflector comprised of five equal sides
   (B) a pattern of five radar reflector buoys
   (C) an assembly of corner reflectors mounted on the edges of a pentagon
   (D) a cluster of five corner reflectors mounted in a vertical column.

   A **B** **C** **D**

SECTION IIa
Multi-choice - Ordering Type

151. The following targets have the same equivalent echoing area also the same composition and surface texture.

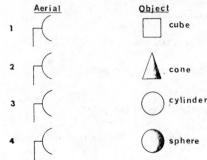

Their echo-strength in descending order is :—

(A) 1, 2, 3, 4

(B) 1, 3, 4, 2

(C) 4, 3, 2, 1

(D) 2, 3, 4, 1

$\boxed{A}$
$\boxed{B}$
$\boxed{C}$
$\boxed{D}$

152. The echo-strength and detection range of a target at medium range are related to its :—

    (1) Composition
    (2) height
    (3) aspect and shape
    (4) width.

Their degree of importance in descending order is :—

(A) 1, 2, 3, 4

(B) 1, 3, 4, 1

(C) 4, 2, 1, 3

(D) 3, 1, 2, 4

$\boxed{A}$
$\boxed{B}$
$\boxed{C}$
$\boxed{D}$ ✓

153. Bearing accuracy depends on many factors, some of which are listed below :—

    (1) Horizontal beam-width
    (2) thickness of bearing marker
    (3) aerial and PPI trace being synchronized
    (4) correct alignment of heading marker.

The three most important are :—

(A) 1, 2 and 3

(B) 1, 2 and 4

(C) 2, 3 and 4

(D) 1, 3 and 4

$\boxed{A}$
$\boxed{B}$
$\boxed{C}$
$\boxed{D}$ ✓

154. The maximum range at which a marine radar set can detect targets depends upon the following :—

    (1) Pulse repetition frequency
    (2) power output
    (3) aerial height
    (4) receiver sensitivity.

The three most significant are :—

(A) 1, 2 and 3
(B) 1, 2 and 4
(C) 1, 3 and 4
(D) 2, 3 and 4

155. The minimum range of a marine radar set is governed by the following factors :—

    (1) The position of the scanner
    (2) the vertical beam-width and wavelength used
    (3) the pulse-length
    (4) the "change-over" time of the T/R cell.

The three most significant are :—

(A) 1, 2 and 3
(B) 2, 1 and 4
(C) 2, 3 and 4
(D) 3, 1 and 2.

SECTION IIb

Multi-choice - Drawings for Completion

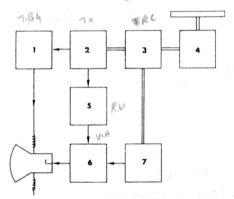

156. In the block diagram above what is the blank block numbered 5?

(A) Video Amplifier
(B) Time-base Generator
(C) I.F. Amplifier
(D) Ranging Unit.

157. In the block diagram above what is the blank block numbered 2 ?

   (A) T/R Cell
   (B) Time-base Generator
   (C) Transmitter
   (D) I.F. Amplifier.

   A
   B
   C
   D

158. In the block diagram above what is the blank block numbered 1 ?

   (A) Video Amplifier
   (B) Time-base Generator
   (C) I.F. Amplifier
   (D) Ranging Unit.

   A
   B
   C
   D

159. In the block diagram above what is the blank block numbered 6 ?

   (A) Video Amplifier
   (B) Time-base Generator
   (C) I.F. Amplifier
   (D) Ranging Unit.

   A
   B
   C
   D

160. In the block diagram above what is the blank block numbered 3 ?

   (A) T/R Cell
   (B) Time-base Generator
   (C) I.F. Amplifier
   (D) Performance Monitor.

   A
   B
   C
   D

SECTION IIc

Multi-choice - Sketch Type

161. A constant error is known to exist between the range rings and the variable range marker (V.R.M.), although neither has been independently checked.

   Cross-ranges from the V.R.M. are plotted as shown on the chart below. Would you treat the vessel as being at the intersection of

(A) a and b ?
(B) b and c ?
(C) c and a ?
(D) somewhere in the middle ?

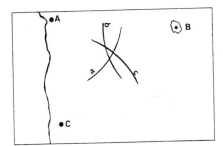

162. While plotting on a True Motion display, the plot developed is shown below.

When the echo is at A, the log input to the True Motion unit fails. Would you expect the echo to

(A) remain at A ?
(B) move toward B ?
(C) move toward C ?
(D) move toward D ?

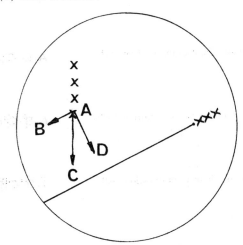

163. The radar-to-radar interference echo is

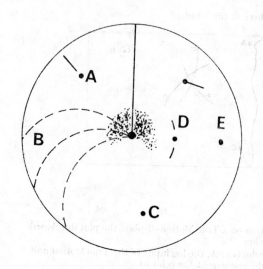

164. The indirect echo (diagram in question 163) is   A B C D E

165. The RACON echo (diagram in question 163) is   A B C D E

166. The side-lobe echo (diagram in question 163) is   A B C D E

167. The multiple echo (diagram in question 163) is   A B C D E

168. On a relative motion display the apparent motion of an echo is as shown in the diagram. Bearing errors due to using the cursor between O and A will :—

(A) Decrease
(B) increase
(C) remain constant
(D) not occur.

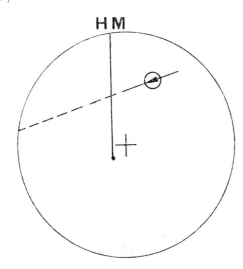

169. Own vessel is steering north at 12 knots. The target steamed east at 12 knots for 6 minutes and then stopped instantaneously. The full plotting interval is 12 minutes.
Which diagram is correct?

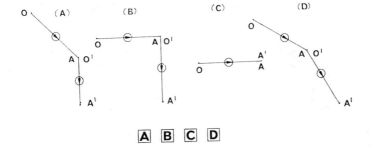

170. In the diagram below the "predicted" track of the target is the line :—

(A) OW

(B) OA

(C) O'A

(D) O'W

A
B
C
D

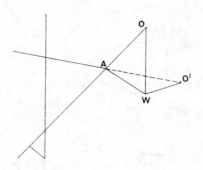

171. In the diagram below "aspect" is the angle :—

(A) α   (B) β   (C) θ   (D) ø

A
B
C
D

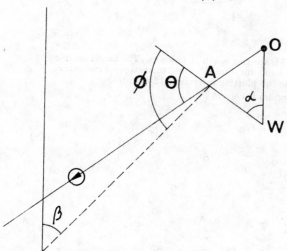

172. If own ship is going to alter course to starboard, and reduce to half speed, indicate the correct predicted plot of apparent motion.

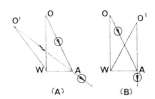

A
B
C
D

173. The correct bearing of echo *A* in the diagram is :—

(A) Green 15°
(B) 015° relative
(C) 020° relative
(D) 030° relative.

A
B
C
D

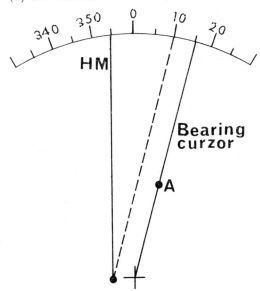

174. The figure shows the track of a ship observed visually. The observer notices on his radar screen that at *A* only a single echo is visible. The most likely cause is :—

(A) Pulse duration

(B) shadow effects

(C) vertical beam-width

(D) horizontal beam-width.

A
B
C
D

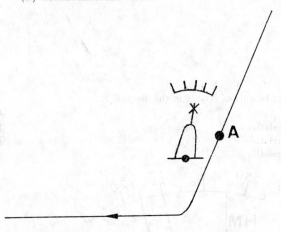

● **Observer**

175. In the PPI diagram shown, when using True Motion with log input, own ship reduces to half speed. The most likely reason for no observed change in the echo's track is :—

(A) Log has become fouled

(B) target has altered to starboard

(C) target has altered to port

(D) target has maintained its course and speed.

A

B

C

D

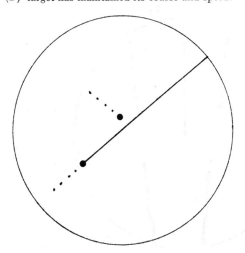

176.   While proceeding along the centre of a canal with steep high
       parallel banks, the radar picture appears as shown.
       This is probably due to :—

       (A) Shadowing
       (B) centre-burn
       (C) too much anti-clutter
       (D) expanded centre.

A
B
C
D

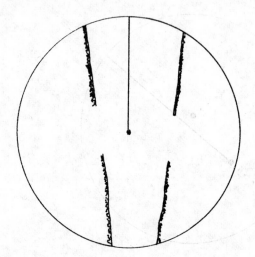

177. With which one of the PPIs is the visual scene associated?

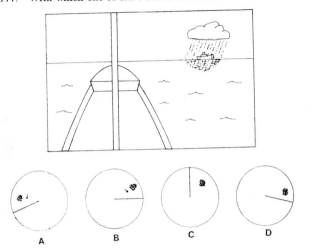

178. If own ship's speed is the same as the other ship's speed (bearing 315° rel.; aspect Green 45°), which radar plot is associated with the visual scene?

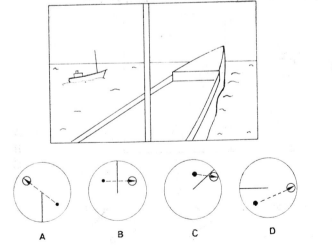

179. If own ship's speed is greater than the target's which radar
plot is associated with the visual scene?

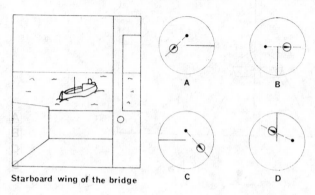

Starboard wing of the bridge

180. Which radar plot is associated with the visual scene?

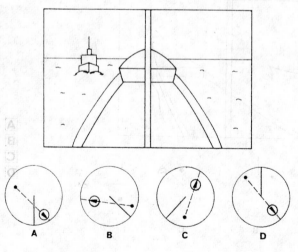

181.   If own ship's speed is slightly slower than the other ship's
        speed which radar plot is associated with the visual scene?

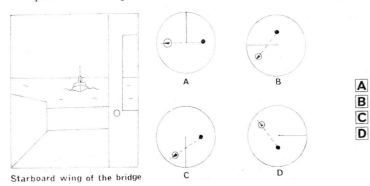

Starboard wing of the bridge

A
B
C
D

182.   If own ship's speed is greater than the target's speed, which
        radar plot is associated with the visual scene?

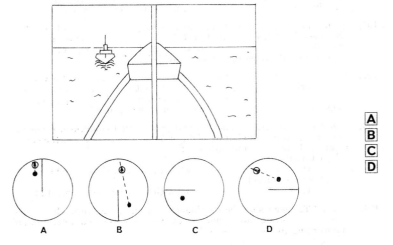

A
B
C
D

SECTION III - Completion Type

Complete the following extracts from the 1972 Collision Regulations :—

183. *Rule* 6. "Ever vessel shall at all times proceed at a ———————— speed . . ."

184. *Rule* 6 (*b*) (ii) (factor determining safe speed). "any constraints imposed by the radar ———————— ———————— in use;"

185. *Rule* 5. "Every vessel shall at all times maintain a ———————— ———————— by sight and hearing as well as by ———————— ———————— ———————— appropriate in the prevailing circumstances . . ."

186. *Rule* 19. ". . . , every vessel which hears apparently ———————— ———————— ———————— ———————— the fog signal of another vessel, . . . shall ———————— ———————— ———————— ———————— ———————— at which she can be kept on course".

187. *Rule* 19. ". . . , so far as possible the following should be avoided :—
    (i) an alteration of course to port for a vessel ———————— of the beam, other than a vessel being ————————."

188. *Rule* 7(*b*). "Proper use shall be made of radar equipment if fitted and operational, including ———————— ———————— to obtain early warning of collision and ———————— ———————— or equivalent systematic observation of detected objects."

Complete the following extracts from the Marine Radar Performance Specification 1968 :—

189. *Range Discrimination.* "On the most open scale appropriate, the echoes of two GPC buoys shall be displayed and distinct when the buoys are on the same bearing ———————— ———————— ———————— and at a ———————— ———————— ———————— ————————."

190. *Range accuracy.* "Fixed range rings shall enable the range of an object, whose echo lies on a range ring, to be measured with an error not exceeding ———————— ———————— of the maximum range of the scale in use or ———————— ————————, whichever is the greater."

191. *Minimum Range.* "When the radar is mounted at a height of 50 ft. above sea level the echo of a GPC buoy shall be visible on the radar for ———————— ———————— ———————— of the scans down to a range of ———————— ———————— on the most open scale provided, . . ."

192. *Range Accuracy.* "The accuracy of range measurement using the variable range marker, if provided, shall be such that the error shall not exceed ———————— ———————— ———————— of the maximum range of the displayed scale in use or ———————— ———————— whichever is the greater."

193. *Range Performance.* "Under normal propagation conditions when the radar aerial is mounted at a height of 50 ft. above sea level, the equipment shall give clear indication of :
    (i) Coastlines
       At ———————— ———————— ———————— when the ground rises to 200 feet" . . .
    (ii) Surface objects
       At ———————— ———————— ———————— on a General Purpose Conical buoy."

194. *Scan.* "The scan shall be continuous and automatic through ————————° ———————— ———————— at a rate of not less than ———————— ———————— in relative wind speeds of up to ———————— ————————."

ANSWERS TO MULTI-CHOICE QUESTIONS

I. True/False

1. True—see page 212
2. False—see page 2
3. False—see page 352
4. False—see page 7
5. True—see page 12
6. False—see page 142
7. False—see page 58
8. False—see page 56
9. False—see page 131
10. False—see page 143
11. True—see page 138
12. False—see page 28
13. True—see page 7
14. False—will only affect bearing accuracy
15. False—see page 111
16. True—see page 170
17. False—steady bearing
18. True—see page 107
19. False—see page 337-338
20. False—see page 47, 117
21. True—see page 33
22. False—see page 142
23. True—see page 58
24. False—see page 50, 60
25. True—see page 248
26. True—see page 156
27. True—the tide affects both target and own ship in the same way
28. False—less likelihood to detect targets at long ranges
29. True—see page 52

30. False—see page 12
31. True—see page 29
32. False—see page 58
33. True—see page 82
34. False—see page 126
35. True—see page 218
36. False—see page 107
37. False—frequency of range oscillator may be out
38. False—see page 4
39. False—see page 339
40. True—switching on and off causes current surges
41. False—see page 58
42. False—in case of false echoes, C and B should fade first; otherwise the echoes must come from different targets.
43. False—current is setting towards the east
44. False—see page 218 (tide is irrelevant)
45. False—see page 33
46. False—see page 4, 37
47. False—see page 58
48. True—snow clutter is generally less dense than wind clutter
49. True—see page 129
50. False—not always, when target is dead ahead or right astern and on a same or opposite course as own ship

II. Multi-choice

51. C—see page 11
52. B—see page 351
53. D—see page 70
54. D⎱
    D⎰—see page 232
55. C—see page 2
56. D—see page 26, 27
57. D—see page 52
58. A—see page 40
59. D—see page 29
60. B—see page 16
61. B—see page 32
62. A—see page 39
63. C—see page 47, 33

64. A—see page 83
65. B—see page 94
66. D—see page 30
67. C—see page 106
68. B—see page 175
69. A—see page 49
70. C—see page 110
71. D—see page 33
72. B—aerial rotating clockwise
73. D—see page 33
74. D—see page 40
75. C—see page 38
76. C—see page 23
77. C—see page 52

II. Multi-choice (continued)

78. D—see page 352
79. D—see page 146
80. A—see page 27
81. B—see page 126
82. A—see page 29, 30
83. A—see page 20
84. B—see page 117
85. A—see page 43
86. C—see page 87
87. B—see page 170
88. A—see page 115, 116
89. C—see page 107
90. A—always in any mode
91. B—see page 11
92. A—see page 31-32
93. A—see page 33
94. B—see page 30
95. D—see page 352
96. D—see page 148
97. A—see page 142
98. B—see page 129
99. B—see page 131
100. A—see page 2, 49
101. A—see page 108, 110
102. D—
　　　(change to long range scale)
103. C—see page 106
104. C—see page 342
105. C—see page 98
106. C—see page 32
107. D—see page 101
108. B—see page 167
109. C—see page 134, 135
110. A—see page 27, 33
111. C—see page 12
112. A—see page 35
113. C—see page 40, 41
114. B—see page 246, 247

115. B—see page 342
116. A—see page 28
117. B—see page 32, 337
118. C—see page 128
119. B—see page 129
120. C—see page 121, 122
121. D
122. C
123. B—see page 165, 166
124. B
125. C—see page 117
126. A—see page 107, 108
127. A—see page 54
128. A—see page 11
129. D—see page 15, 28, 142
130. C—see page 111
131. C—see page 23
132. B—see page 339
133. C—see page 129
134. D—see page 129, 130
135. C—see page 21, 22, 336
　　　　　　(Radomes)
136. A—see page 152
137. A—see page 32
138. C—see page 5, 97
139. C—see page 167
140. C—see page 134, 135
141. A—see page 117
142. C—see page 105
143. C—see page 330
144. C
145. A—see page 129
146. D—see page 143
147. B—see page 166
148. B—see page 129
149. B—see page 33, 34, 49
150. C—see page 162

IIa. Multi-choice—Ordering Type

151. B—see page 87, 88
152. D—(object small in relation to horizontal beam-width)

153. D—see page 28, 29
154. D—see page 26, 27
155. A—see page 27

IIb. Multi-choice. Drawings for Completion

156. D—see page 330
157. C—see page 330
158. B—see page 330

159. A—see page 330
160. A—see page 330

IIc. Multi-choice. Sketch Type

161. D—range is 'high'
162. D—see page 54, 55
163. B—see page 129-131
164. C—see page 126, 129

165. A—see page 166
166. D—see page 129
167. E—see page 125, 126
168. B

IIc. Multi-choice.   Sketch Type (continued)

169. A
170. C—see page 192-194
171. D—see page 186
172. D—see page 191-194
173. D—see page 82-83
174. A—see page 33-34
175. D
176. C

177. C
178. A—collision course.
179. C—target passing astern.
180. C—target passing astern.
     (O/S not stopped)
181. D—target passing astern.
182. D—diverging tracks.

III. Completion Type

183. safe
184. range scale
185. proper look-out
     all available means
186. forward of her beam
     reduce her speed to the
     minimum
187. forward
     overtaken
188. long-range scanning
     radar plotting
189. 50 yards apart
     range of 1 mile

190. $1\frac{1}{2}$ per cent
     75 yards
191. 50 per cent
     50 yards
192. $2\frac{1}{2}$ per cent
     125 yards
193. 20 nautical miles
     2 nautical miles
194. 360° of azimuth
     20 rpm
     100 knots

ACKNOWLEDGEMENTS

The following acknowledgements are made :—

Technical information was supplied by :
    Cossor Communications Co. Ltd.,
    Racal-Decca Marine Radar Ltd.,
    The Marconi International Marine Company Limited,
    Raytheon Marine Company Ltd.,
    Sperry Marine Systems,
    Staveley Electrotechnic Services Limited,
    The Home Office,
    S. Smith & Sons (England) Ltd., Kelvin Hughes Division,
    The Iotron Corporation,
    Norcontrol.

Figures 1.7, 1.8, 7.1 and 7.7 are based on diagrams from RADIO AIDS TO CIVIL AVIATION, Edited by R. F. Hansford, published by Heywood Books, distributed by Iliffe Books Ltd. By permission of Iliffe Books Ltd.

Figures 1.1, 3.2, 7.10, 11.1 and AIII.10 are based on diagrams from INLEIDING TOT DE RADAR TECHNIEK by J. A. Klerk, published by the Author. By permission of J. A. Klerk.

REFERENCES

AGA. "New Racon from AGA (10 cm.)", *Safety at Sea*, April 1981, 145, 31.
    "AGA SPECKTER radar reflector. Type SR 66".

J. H. Beattie. "A Marine Radar System incorporating Automatic Plotting", *J.I.N.*, 15, 429.

S. Bell and W. Dukes. "Revolutionary New Radar Reflector", *Safety at Sea*, June 1978, 111, 9.

A. G. Bole. "Radar Interference Suppression—What Else do you Lose?", *Safety at Sea*, March 1979, No. 120, 18.
    "The Elimination of Second or Multiple Trace Echoes on Marine Radar", *Safety at Sea*, June 1979, 123, 47.
    "The Simple Logic of Radar Performance Monitoring", *Safety at Sea*, No. 124 (1979), 25.
    "Interswitching—What are the Options open to you?", *Safety at Sea*, No. 126 (1979), 17.

H. G. Booker. "Elements of Radio Meteorology", *J.I.E.E.*, 93, No. 1 (1946).

R. S. H. Boulding. *Principles and Practice of Radar* (George Newnes Limited, London, 1963), pp. 385–387, 420, 421, 340, 108–112.

E. G. Bowen. *A Textbook of Radar* (Cambridge at the University Press, 1954), pp. 216–279, 405–409.
    "Radar Observations of Rain and their Relation to Mechanisms of Rain Formation", *J. Atmos. Terr. Phys.*, 1 (1951).

G. Brooke. "Decca 150—new radar of wide appeal", *Nautical Review* (The International Journal of Ship Operation and Management, 1980, Vol. 4, No. 2, 8).

W. Burger. "A Report on Marine Radar Simulator Training, 1964", *R.S.A.* (1964).

W. Burger and A. G. Corbet. "Radar Simulator Courses", *J.I.N.*, 16, 253.

E. S. Calvert. "Manoeuvres to ensure the Avoidance of Collision", *J.I.N.*, 13, 127.

M. H. Carpenter and W. M. Waldo. "Real Time Method of Radar Plotting", *Cornell Maritime Press, Inc.* (1975).

D. B. Charter. "The S- and X-Band Debate", *Seaways*, February 1981, 9.

J. I. Clucas. "A new tool for ARPA training", *Safety at Sea*, Jan. 1981, No. 142, 22.

A. N. Cockcroft and J. N. F. Lameijer. *A Guide to the Collision Avoidance Rules*, Standford Maritime, London (1976).

A. N. Cockcroft. "The S- and X-band Debate", *Seaways*, April 1981, 2.

A. G. Corbet. "Moderate Speed and Close Quarters Situation", W.C.A.T. "Collision Avoidance at Sea, with Special Reference to Radar and Radio-Telephony", *J.I.N.*, 25, 520 (1972).

A Study of the Training of Bristol Channel Pilots in the Use of Radar, M.Sc. Thesis, Department of Maritime Studies, UWIST.

"Use of Radar as an Aid for Pilotage—Parallel Index and other Techniques", *UWIST*, 1979.

Cossor Radar & Electronics Limited. "The Elimination of Rain Echoes from Radar Displays" (undated).

J. Croney. "Clutter and its Re-education on Shipborne Radars", Radar Present and Future Conference Publication No. 105, *I.E.E.* International Conference, October 1973.

DECCA RADAR LTD. *Clearscan Radar*. Pub. T427/S/1.

"New Facility for extending Displayed Range of Decca Radars". *News Release Decca Radar*, March 1973.

"The case for Decca AC Radar in 1978", MD/HLAF/LCG/14th April, 1978, *Decca House*.

"Decca Clearscan Radar. Group 12 & 16 Marine Radars". (1979).

"Interswitching Unit Type 65158".

Department of Trade. "The Use of, and Training in, Radar and Electronic Aids to Navigation". *Merchant Shipping Notice* No. M. 900/983.

Marine Division. "Azimuth Response of 'Scotchlite'," Appendix to TXR 350. (February 1975).

Radar Observer Course and Examination. Regulations and Guide (1978).

"The Use of Parallel Index Techniques as an Aid to Navigation by Radar", *Merchant Shipping Notice* No. M.860.

Merchant Shipping Radar Installations. Instructions for the Guidance of Surveyors, 1979.

"Survey of merchant shipping navigational equipment installations. Instructions for the guidance of surveyors", *H.M.S.O.* (1980).

Radar Observer Question Papers
No. 2 (10082)1
No. 111 (51838–1)1
No. 113 (46650)
No. 119 (51820)1
No. 294 (10165)1

W. S. Diehl. *Standard Atmosphere—Tables and Data* (National Advisory Committee for Aeronautics, Technical Report, No. 218, 1925).

DGON. *Radar in der Schiffahrtspraxis* (Schiffahrts-Verlag "Hansa" C. Schroedter GmbH & Co. KG, 1980).

B & W Elektronik AS. *CA-User Manual. SCANTER* 8000.

J. Evans. *Fundamental Principles of Transistors* (Heywood and Company, 1962).

F. A. Gross. "New Pathfinder Heavy Marine Radars", *Electronic Progress*, Volume XX, No. 2, Summer 1978, 2.

R. F. Hansford. *Radio Aids to Civil Aviation* (Heywood and Company, Ltd., 1960), pp. 35–49, 108–119, 426–427, 515–517.

P. L. Hartley, P. D. L. Williams. "A Review of Display Devices for Radar Applications", *Electronic Displays* 1975, Exhibition and Conference, September, October 1975.

Home Office. "Technical Parameters of Fixed Frequency Radar Beacons (Racons)", Draft Report AT/8, *Directorate of Radio Technology, Home Office*. "Frequency Requirements for Shipborne Transponders", Draft Report AU/8, *Directorate of Radio Technology, Home Office*.

H. T. Hylkema, J. A. Klerk, M. J. van Peer, P. de Hullu, J. W. Mossel, F. M. Nijhuis, J. Post, H. Unger. "ARPA Edition", De Zee, Year 11, No. 1, January 1982.

IMCO. "Report of the Ad Hoc Working Group on Electronic Navigational Aids (Radar Beacons and Transponders)", July 1975.
"Report of the Working Group on Radar Beacons and Transponders", March 1976.
"Radar Beacons and Transponders", Note by the Secretariat, November 1976.
"1974 SOLAS Convention to enter into force on 25 May 1980", *IMCO NEWS*, Number 2, 1979, 8.
"ARPA Schedule Delayed", *Safety at Sea*, May 1981, 146, 11.

IMCO NEWS. "First set of SOLAS amendments adopted", *IMCO*, Number 4, 1981, 1.

H. Jacobowitz, L. Basford. *Electronic Computers*, W. H. Allen & Company Ltd., 125–132, 187–203.

K. D. Jones and A. G. Bole. "Automatic Radar Plotting Aids Manual", *Department of Maritime Studies, Liverpool Polytechnic*, 1981.

K. D. Jones. "Implications in the Use of Marine Transponders", *Secondary Radar Seminar organised by the Nautical Institute and the Royal Institute of Navigation*, December 13, 1979.

Iotron. "Digiplot—The World's Most Advanced Computing Radar Display", *Iotron Corporation*, 1975.
"Digiplot Fully Automatic Radar Plotting" (Slide Training Program Script), *Iotron Corporation*, 1976.
"Digiplot State of the Art Report", *Iotron Corporation*, November 1979.

R. B. Keeney. "Improved Perimeter Collision Avoidance Systems", *Safety at Sea*, No. 124 (1979), 23.

Kelvin Hughes. "Anticol. The Automatic Radar Plotting Aid", *Kelvin Hughes*, 1981.

"High Performance High Definition Radar", *Marine Talk*, September 1981.

"Renaissance for Kelvin Hughes", *Safety at Sea*, Jan. 1981, No. 142, 25.

"Successful Plotting", *Marine Talk*, No. 2, 1979.

P. E. Kent. "Research and Development of Radar Transponders".

J. A. Klerk. *Inleiding tot de Radar Techniek* (J. A. Klerk, 1962), 48–53, 77–78, 120–125, 132, 137–141.

Krupp Atlas. "CAS Solution", *Safety at Sea*, Dec. 1979, No. 129, 12.

J. N. F. Lameijer. "Convention on the International Regulations for Preventing Collisions at Sea, 1972", *De Zee*, February 1973, pp. 35–39.

Liverpool Polytechnic Maritime Operations Research Unit. "A Comparison of Facilities on Computer Based Radars". Under sponsorship of *Shell International Marine Ltd.*, 1975.

Marconi International Marine Company Ltd. "Chromascan colour radar", *Mariner*, Vol. 17, No. 207, January–February 1982.

MARINER. "Marconi Marine radar now on another wavelength", *Journal of the Marconi International Marine Company Limited*, March–April 1981, 332.

K. Matsumoto. *A Study on Shipborne Automatic Radar Plotting Aids*, Dissertation submitted for the Degree of Bachelor of Science with Honours for the University of Wales.

Maritime Court Report (Dutch). "Collision of Nedlloyd Seine in Rain", *De Zee*, Vol. 10, No. 6, June 1981, 154.

I. McGeoch. "MIDAR, A Marine Identification and Recognition System", *MIDAR (Marine Systems) Ltd.*, 1978.

J. E. Meade. "Displaying the Pathfinder Marine Radar Information", *Electronic Progress*, Volume XX, No. 2, Summer 1978, 10.

Merchant Navy Training Board. "Electronic Navigation Systems Course", *MNTB*, June 1981.

"Training in the Use of Radar and other Navigation Equipment", *MNTB*, July 1981.

"ARPA Course", *MNTB*, July 1981.

"Navigation Control Course", *MNTB*, July 1981.

A. P. Milwright. "Survey of Recent Aids to Aeronautical and Marine Navigation", *Proc. I.E.E.*, Vol. 105 (1958).

W. D. Moss. *Radar Watchkeeping* (Maritime Press Limited, 1965), pp. 85–87.

Mullard Educational Service. *A simple explanation of Semi-conductor Devices* (Mullard Ltd.).

Semi-conductor Devices (mini-book).

Netherlands Maritime Institute. "Comments on currently available Computer Assisted Radar Systems", VNAV 094, *NMI*, Report No. R.54.

L. Oudet. "Radar and the Collision Regulations", *J.I.N.*, 11, 220.

W. M. Pease. "Anti-Collision Unit for the Pathfinder", *Electronic Progress*, Volume XX, No. 2, Summer 1978, 17.

J. E. Pomerleau. "Modern Marine Radar Design", *Raytheon Marine Company* (undated).

A. L. C. Quigley. "Tracking and Associated Problems", Radar Present and Future Conference Publication Number 105, *I.E.E.* International Conference October 1973.

RACAL-DECCA MARINE RADAR LTD. "Second Generation ARPA Launched", *Safety at Sea*, Sept. 1980, No. 138, 13.

"ARPA. The All-Weather Automatic Radar Plotting Aid", *Racal-Decca Marine Radar Limited*, 1980.

"ARPA Technical Manual", *Racal-Decca Marine Radar Technical Manual*.

Raytheon. "Racas Bright Display Collision Avoidance System", *Raytheon Marine Ltd.*, 1978.

"Bright Display Solid-state Modular Radar Systems", *Raytheon Marine Company*, 1978.

"Mariners Pathfinder Radar Systems, 12 and 16 Inch Displays", *Raytheon Marine Ltd.*, 1977.

"Raytheon CAS", *Raytheon Marine Company*, 1978.

"RAYCAS Collision Avoidance System Instruction Manual", *Raytheon Marine Company*, 1978.

Raytheon Marine Company. "New Collision-Avoidance System in one Cabinet", *Lloyd's List*, May 18, 1979.

E. R. Richards. "The Use of Radar Beacons and Transponders in Marine Navigation", *Secondary Radar Seminar organised by the Nautical Institute and the Royal Institute of Navigation*, December 13, 1979.

R. F. Riggs. "A modern collision avoidance display technique", *J.I.N.*, Vol. 28, No. 1, 143.

"A critique of the CAORF and Shell/Liverpool Evaluations of Marine Collision Avoidance Systems", *SPERRY Marine Systems*, JA 274–3623A, Nov. 1978.

J. B. G. Roberts and R. Eames. "Analogue Shift Register Techniques applied to Radar Signal Processing", Radar Present and Future Conference Publication Number 105, *I.E.E.* International Conference, October 1973.

E. M. Robb. "Some Practical Experience of Sea Clutter", *J.I.N.*, 7, 362. "Precise Radar Conning", *J.I.N.*, 14, 202.

A. B. Schneider and P. D. L. Williams. "Circular polarization in radars", *The Radio and Electronic Engineer*, Vol. 47, No. 1/2, 11.

Selenia. "Prora Capability increased with Autotrack", *Safety at Sea*, Dec. 1979, No. 129, 17.

"The New Generation of Processing Radars for Big Ship Marine Application", *Selenia Marine Division* and *Staveley Electrotechnic Services Limited*.

S. J. Singh. "Sectors for Collision Avoidance & Navigation by Radar", *Fourth International Symposium on Vessel Traffic Services*, Bremen, 1981.

M. I. Skolnik. *Introduction to Radar Systems* (McGraw-Hill Kogakusha, Ltd. 1975), Chapters 11 and 12.

G. J. Sonnenberg. *Radar and Electronic Navigation* (Georges Newnes Limited, London, 1978), pp. 292–299.

"Radar beacons, transponders and echo-enhancers", *De Zee*, August 1974.

Sperry. "CAS II. Collision Avoidance System", *Sperry Marine Systems*, 1979. "Features, Accomplishments and Benefits of Sperry CAS/Channel Navigation/Decca", *Sperry Marine Systems*, 1980.

Q

"Peter Griffiths talks to Advantage", *Advantage*, No. 10, Spring 1981.

Sperry Marine Systems (J. P. O'Sullivan). *Basic Instructions–CAS II.*
Sperry CAS II Presentation.
Themes for CAS. 1. Time Scale in Target Encounters. 2. Passing between PADS.
Collision Avoidance Channel Navigation and Integrated Navigation Systems.
CAS II—Developments in Radar Data Processing.
Operational Evaluation of CAS II.
The Operational Significance of PADS.
Analytical and Operational Evaluation of PAD Display.
The Speed and Heading Input Problem in Collision Avoidance Systems.

A. T. Starr. *Radio and Radar Technique* (Sir Isaac Pitman & Sons, Ltd.), pp. 80–89, 193, 228–239, 244–249, 641–644.

Sub-Committee on Safety of Navigation. "Carriage Requirements and Performance Standards for Collision Avoidance Systems and Associated Equipment. Operational Performance Standards for Automatic Radar Plotting Aids (ARPA)", *IMCO*, Jan. 1979.

R. G. Taylor. "Stationary patterns in radar sea clutter", *The Radio and Electronic Engineer*, Vol. 45, No. 3, 103.

D. H. Thomas. "Experiments in V.H.F. R/T Techniques for Collision Avoidance. An Acceptable Combination", *Safety at Sea*, March 1973.

H. Topley. "Anti-Collision Manoeuvres when using Radar", *J.I.N.*, 7, 365.
"Estimating the Nearest Approach with Radar", *J.I.N.*, 8, 50.
"The Need for Bold Alterations", *J.I.N.*, 10, 193.
"Radar Plotting Errors", *J.I.N.*, 11, 167.

J. Waterworth. "Hovercraft Transponders", *Secondary Radar Seminar organised by the Nautical Institute and the Royal Institute of Navigation*, December 13, 1979.

P. L. D. Williams. "Limitations of radar techniques for the detection of small surface targets in clutter", *The Radio and Electronic Engineer*, Vol. 45, No. 8, 379.

F. J. Wylie. *The Use of Radar at Sea* (Hollis & Carter, fourth revised edition, 1968), pp. 101–108.
The Use of Radar at Sea (Hollis & Carter, fifth revised edition, 1978), Annex, Collision Avoidance Systems, 271.
"The Operational Value of Shipborne Radar", *J.I.N.*, 2, 127.
"Radar and the Rule of the Road at Sea", *J.I.N.*, 3, 10.
"Siting the Radar Scanner", *J.I.N.*, 3, 189.
"The Use of Radar for Preventing Collisions at Sea", *J.I.N.*, 6, 271.
"A Survey of Five Years' Progress in Marine Radar", *J.I.N.*, 7, 59, 218.
"Radar and the Compass Bearing", *J.I.N.*, 7, 200.
"Radar and Sea-Sense", *J.I.N.*, 8, 45.
"The Region of Collision", *J.I.N.*, 9, 161, 453.
"The Future of Marine Electronics", *J.I.N.*, 10, 50.
"Collisions at Sea despite Radar", *J.I.N.*, 10, 320.
"Errors in Radar Appreciation", *J.I.N.*, 12, 198.
"Marine Radars for Large Ships", *J.I.N.*, 12, 198.
"Collisions at Sea in Fog : The Common Sense Approach", *J.I.N.*, 16, 253.
"Marine Radar Automatic Plotter Display Philosophy", *J.I.N.*, 27, 298.
"Escape Time : The Crucial Factor in Collision Avoidance Situations and Systems", *J.I.N.*, 31, 438.

Abbreviations

ARPA	Automatic Radar Plotting Aid.
CA	Collision Avoidance.
CAORF	Computer Aided Operations Research Facility.
CAS	Collision Avoidance System.
DGON	Deutschen Gesellschaft fur Ortung und Navigation.
H.M.S.O.	Her Majesty's Stationary Office.
I.E.E.	Institute of Electrical Engineers.
IMCO	International Maritime Consultative Organization. (changed to IMO)
J.A.T.P.	Journal of Atmospheric and Terrestrial Physics.
J.I.E.E.	Journal of the Institute of Electrical Engineers.
J.I.N.	Journal of the Royal Institute of Navigation.
MNTB	Merchant Navy Training Board.
NMI	National Maritime Institute.
R.S.A.	Royal Society of Arts.
UWIST	University of Wales Institute of Science and Technology.
W.C.A.T.	Welsh College of Advanced Technology.

INDEX